Green Manufacturing

Presenting the concept of green manufacturing in a comprehensive manner, *Green Manufacturing* covers green technologies for advanced manufacturing systems, and discusses circular economy, and waste management.

This book:

- Presents a transition from conventional manufacturing to green manufacturing and discusses innovative technical ideas for production.
- Discusses concepts, methods, and strategies for zero waste in manufacturing, implementation, and analysis of clean energy.
- Includes cutting-edge case studies adopted in the industrial sectors, quoting the need for and importance of supply chain management.
- Covers additive manufacturing for sustainable design and production and lean manufacturing.
- Explains nature-inspired biomaterials, renewable biocomposites, circular economy, and waste management.

The text is primarily written for senior undergraduates, graduate students, and academic researchers in the fields of manufacturing engineering, production engineering, industrial engineering, mechanical engineering, and aerospace engineering.

Green Manufacturing
Challenges and Applications

Edited by
Shanmugam Suresh Kumar
Sundaresan Thirumalai Kumaran

CRC Press
Taylor & Francis Group
Boca Raton London New York

CRC Press is an imprint of the
Taylor & Francis Group, an **informa** business

Front cover image: dee karen/Shutterstock

First edition published 2025
by CRC Press
2385 NW Executive Center Drive, Suite 320, Boca Raton FL 33431

and by CRC Press
4 Park Square, Milton Park, Abingdon, Oxon, OX14 4RN

CRC Press is an imprint of Taylor & Francis Group, LLC

ISBN: 978-1-032-74367-7 (hbk)
ISBN: 978-1-032-74781-1 (pbk)
ISBN: 978-1-003-47034-2 (ebk)

DOI: 10.1201/9781003470342

Typeset in Sabon
by codeMantra

Contents

Preface

The growing global emphasis on sustainability and environmental stewardship has brought green manufacturing to the forefront of industrial and academic discussions. This book, titled *Green Manufacturing: Challenges and Applications*, explores the various aspects of sustainable manufacturing practices and their role in shaping the future of production industries. It addresses key challenges, applications, and innovations that contribute to reducing environmental footprints while maintaining productivity and efficiency.

The chapters in this book cover a broad spectrum of topics, including strategies for achieving zero waste, the implementation of clean energy, and the development of renewable bio-composites. They also delve into the principles of the circular economy, the use of nature-inspired biomaterials, and innovative technologies driving sustainable practices. Furthermore, this book emphasizes the importance of international standards, the integration of Industry 4.0 to achieve Sustainable Development Goals (SDGs), and the potential of additive manufacturing for sustainable design.

Through this collection of research and case studies, we aim to provide readers with valuable insights into the current state and future directions of green manufacturing. We hope this book will serve as a comprehensive resource for scholars, practitioners, and policymakers striving for a more sustainable and eco-friendly industrial landscape.

Shanmugam Suresh Kumar

Sundaresan Thirumalai Kumaran

About the editors

Shanmugam Suresh Kumar received his M.Tech. in Manufacturing Engineering from Alagappa Chettiar Government College of Engineering and Technology, India, in 2012. He completed his Ph.D. degree in Mechanical Engineering from Kalasalingam Academy of Research and Education, India, in 2016. He has worked in various positions at Kalasalingam Academy of Research and Education since 2012. In 2018, he finished Postdoctoral Fellowship at Karadeniz Technical University, Turkey. His credentials consist of a wide range of publications in over 50 SCI/Scopus-indexed journals with over 1,500 citations. Currently, he is working as Associate Professor in the Department of Mechanical Engineering, Kalasalingam Academy of Research and Education, India.

Sundaresan Thirumalai Kumaran received his M.Tech. in Manufacturing Engineering (gold medalist) from the Government College of Technology, India, in 2008. He completed his Ph.D. degree in Mechanical Engineering from Kalasalingam Academy of Research and Education, India, in 2015. He has worked in various positions at Kalasalingam Academy of Research and Education since 2009. After completing his Ph.D., he worked as Research Professor at the School of Mechanical Engineering, Yeungnam University, South Korea. His credentials include three projects from UGC-DAE, DAE-BRNS, and CVRDE, and he has also published over 150 SCI/Scopus-indexed journals with over 2,500 citations. Currently, he is working as Associate Professor in the Department of Mechanical Engineering, PSG Institute of Technology and Applied Research, India.

Contributors

Bhuvanesh Aanathan
Department of Electrical and Electronics Engineering
PSN College of Engineering and Technology
Tirunelveli, Tamil Nadu, India

Shanawaz AM
Department of Mechanical Engineering
Sethu Institute of Technology
Kariapatti, Tamil Nadu, India

S. Arulvel
School of Mechanical Engineering
Vellore Institute of Technology
Vellore, Tamil Nadu, India

Krishna Varun Balasubramanian
Department of Aeronautical Engineering
KIT-KalaignarKarunanidhi Institute of Technology
Coimbatore, Tamil Nadu, India

Sankaranarayanan Bathrinath
Department of Mechanical Engineering
Kalasalingam Academy of Research and Education
Srivilliputhur, Tamil Nadu, India

Angshuman Chattopadhyay
Department of Mechanical Engineering
Kalasalingam Academy of Research and Education
Srivilliputhur, Tami Nadu, India

Vivek Deivasigamani
Department of Civil Engineering
KPR Institute of Engineering and Technology
Coimbatore, Tamil Nadu, India

Velmurugan Ganesan
Institute of Agricultural Engineering, Saveetha School of Engineering, Saveetha Institute of Medical and Technical Sciences (SIMATS)
Saveetha University
Chennai, Tamil Nadu, India

Shankar Ganesh
Department of Mechanical Engineering
PSN College of Engineering and Technology
Tirunelveli, Tamil Nadu, India

Ragavendran Giri
Department of Robotics and Automation Engineering
Rajalakshmi Engineering College
Chennai, Tamil Nadu, India

Arulmurugan Gnanasekaran
School of Mechanical Engineering
Vellore Institute of Technology
Vellore, Tamil Nadu, India

Prabhakaran Jaganathan
Kalasalingam Business School
Kalasalingam Academy of Research and Education
Srivilliputhur, Tamil Nadu, India

Paulchamy Jeyapandiarajan
School of Mechanical Engineering
Vellore Institute of Technology
Vellore, Tamil Nadu, India

Arunprasath Kanagaraj
Department of Mechanical Engineering
PSN College of Engineering and Technology
Tirunelveli, Tamil Nadu, India

Jayakrishna Kandasamy
School of Mechanical Engineering
Vellore Institute of Technology
Vellore, Tamil Nadu, India

Zohreh Khoshraftar
School of Chemical, Petroleum and Gas Engineering
Iran University of Science and Technology
Tehran, Iran

Lavish Kumar Singh
Department of Mechanical Engineering
Sharada University
Greater Noida, Uttar Pradesh, India

Natrayan Lakshmaiya
Department of Mechanical Engineering, Saveetha School of Engineering
Saveetha Institute of Medical and Technical Sciences, Saveetha University
Chennai, Tamil Nadu, India

Ragunath Lakshmanan
School of Mechanical Engineering
Vellore Institute of Technology
Vellore, Tamil Nadu, India

Sasi Kumar Mani
Department of Aeronautical Engineering
KIT-KalaignarKarunanidhi Institute of Technology
Coimbatore, Tamil Nadu, India

Ramaganesh Marimuthu
Department of Mechanical Engineering
Kalasalingam Academy of Research and Education
Krishnankoil, Tamil Nadu, India

Raajasekar Mathiyalagan
School of Mechanical Engineering
Vellore Institute of Technology
Vellore, Tamil Nadu, India

Kumaresan Mayakannan
Department of Mechatronics Engineering
Hindusthan College of Engineering and Technology
Coimbatore, Tamil Nadu, India

Rajesh Nasinathan
School of Mechanical Engineering
Vellore Institute of Technology
Vellore, Tamil Nadu, India

Narayanasamy Pandiarajan
Department of Mechanical Engineering
Kamaraj College of Engineering and Technology
Virudhunagar, Madurai, Tamil Nadu, India

Fathimathul Fuhaima Puthiya Purayil
Department of Civil Engineering
KPR Institute of Engineering and Technology
Coimbatore, Tamil Nadu, India

Clinton R.
Department of Mechanical Engineering
Saveetha School of Engineering, SIMATS
Chennai, Tamil Nadu, India

Satheesh Raja
Department of Marine Engineering
PSN College of Engineering and Technology
Tirunelveli, Tamil Nadu, India

Harseni Rajamohan
Department of Biotechnology
Mepco Schlenk Engineering College
Sivakasi, Tamil Nadu, India

Kamatchi Rajaram
School of Mechanical Engineering
Vellore Institute of Technology
Vellore, Tamil Nadu, India

Mala Rajendran
Department of Biotechnology
Mepco Schlenk Engineering College
Sivakasi, Tamil Nadu, India

Sundarakannan Rajendran
Department of Mechanical Engineering
Saveetha School of Engineering, SIMATS
Chennai, Tamil Nadu, India.

Vaishali Rajasekaran
Department of Aeronautical Engineering
KIT-KalaignarKarunanidhi Institute of Technology
Coimbatore, Tamil Nadu, India

Saravanakumar Ramasamy
Department of Civil Engineering
KPR Institute of Engineering and Technology
Coimbatore, Tamil Nadu, India

Priyanka Ramesh
Kalasalingam Business School
Kalasalingam Academy of Research and Education
Srivilliputhur, Tamil Nadu, India

Marimuthu Rengasamy
School of Mechanical Engineering
Vellore Institute of Technology
Vellore, Tamil Nadu, India

Gurusamy Seetharam
Department of Mechanical Engineering
Kalasalingam Academy of Research and Education
Srivilliputhur, Tamil Nadu, India

Sindhu Selvaraj
Department of Mechatronics Engineering
Hindusthan College of Engineering and Technology
Coimbatore, Tamil Nadu, India

Sambandhamoorthy Senthilrajan
Department of Robotics and Automation Engineering
Rajalakshmi Engineering College
Chennai, Tamil Nadu, India

Senthil Kumar Shanmugakani
Department of Mechanical Engineering
SRM Institute of Science and Technology
Tiruchirappalli, Tami Nadu, India

Elango Krishnan Soundararajan
Department of Civil Engineering
KPR Institute of Engineering and Technology
Coimbatore, Tamil Nadu, India

Rathinavel Subbiah
Department of Mechanical Engineering
SRM Institute of Science and Technology
Tiruchirappalli, Tami Nadu, India

Dharani Subramani
Department of Aeronautical Engineering
KIT-KalaignarKarunanidhi Institute of Technology
Coimbatore, Tamil Nadu, India

Yuvaraj Subramaniyan
Department of Civil Engineering
KPR Institute of Engineering and Technology
Coimbatore, Tamil Nadu, India

Thangaraju Tamilaban
Department of Mechanical engineering
St. Joseph College of Engineering
Chennai, Tamil Nadu, India

Manikandan Thangaraj
Department of Aeronautical Engineering
Nehru Institute of Technology
Coimbatore, Tamil Nadu, India

Senthilkumar Thangarajan Sivasankaran
Department of Mechanical Engineering
K. Ramakrishnan College of Technology
Tiruchirappalli, Tami Nadu, India

Venkata Siva Prasanth Yendeti
Texas A&M University
College Station, Texas

Eganathan Venkatesaperumal
School of Mechanical Engineering
Vellore Institute of Technology
Vellore, Tamil Nadu, India

Narayanan Venkateshwaran
Center of Excellence in Digital Manufacturing, Department of Mechanical Engineering
Rajalakshmi Engineering College
Chennai, Tamil Nadu, India

Nanssy Vijayakumar
Department of Aeronautical Engineering
KIT-KalaignarKarunanidhi Institute of Technology
Coimbatore, Tamil Nadu, India

Sinchana Shri Vinu
Department of Aeronautical Engineering
KIT-KalaignarKarunanidhi Institute of Technology
Coimbatore, Tamil Nadu, India

Siva Shankar Vinothan
Institute of Agricultural Engineering, Saveetha School of Engineering, Saveetha Institute of Medical and Technical Sciences (SIMATS)
Saveetha University
Chennai, Tamil Nadu, India

Chapter 1

Green manufacturing

Present, past, and future

Harseni Rajamohan and Mala Rajendran

1.1 INTRODUCTION

Green manufacturing refers to the holistic process of producing goods using environmentally friendly practices and materials. The objective of green manufacturing is to decrease the negative impact of manufacturing processes and conserve energy (Miragliotta and Shrouf, 2013), save natural resources (Borges et al., 2021), and promote social responsibility (Pang and Zhang, 2019).

1.2 HISTORY OF INDUSTRIALIZATION

The shift from a farming-based society to an industrialized society has come with a growing misuse of numerous resources. Industrial revolutions were defined by innovations like steam engines and electricity. There was a shift towards manufacturing by mechanized processes instead of manual processes, which was initiated around 1760 in England and the 18th century in the US. It was witnessed directly in the textile industry with the emergence of the power loom. Followed by the use of steam power, hydroelectric power was harnessed. These advancements, first developed in England, rapidly extended across Europe, followed by North America, propelling western nations into fast industrialization, enhanced efficiency, and swift economic expansion at environmental costs from air pollution and greenhouse gas emissions (Rosen, 2010).

One hundred years later, the increasing reliance on fossil fuels and electricity allowed the development of lighter, more portable technology. Later, to improve productivity, the assembly line was introduced in the car manufacturing unit. It increased the manufacturing output by 500% and reduced the price of the car ultimately (Model, 2019). The assembly line and mass production techniques, many of which are still in use today, propelled the Second Industrial Revolution. The remarkable industrial revolution from

DOI: 10.1201/9781003470342-1

the 18th to the 20th century was well known for its positive impact on mass production and mechanized agriculture in improving the standard of life at the cost of dampening the environment (Jull, 1999).

Computers, network fibre optic cables, and semiconductors first appeared in the post-war era. Robotics and factory automation witnessed their early beginnings during the Third Industrial Revolution (Mohajan, 2021). In the assembly line, robots were used. In 1961, Unimate, the first industrial robot, started working on the General Motors' assembly line. The device's inventor, George Devol, went on to start the first robot manufacturing company in history. During this period, computerized business systems designed for data management and analysis were also used for the first time. Later, lean manufacturing was developed by Toyota Motor Corporation in 1948. The lean manufacturing system identified and eliminated waste in the manufacturing unit. The principle of lean manufacturing was extended beyond the automobile industry to electronics, aerospace, healthcare, construction, and food processing units (Pearce and Pons, 2019). It emphasized the need of "producing quality products efficiently by eliminating waste, inconsistent practices, and unjustifiable demands on the production line completely."

1.2.1 Industrialization and evolution of consciousness on environment

Along with the industrial revolution, consciousness over the preservation of the environment also grew in the 20th century, an era of awakening and action. The 21st century is a pioneering century for focusing on a sustainable future powered by renewable energy. The century was defined by the massive growth in the adoption of innovative technologies, making the industrial revolution cheaper through the reduction of fossil fuels. Innovations in hydropower and geothermal energy made a tangible difference in the global energy transition. Advancements in battery technology and energy storage facilitated the exploitation of renewable energy when they were not available. The current era is marked by unimaginable green technology marvel that ensures sustainability and the application of sustainable energy. Roads are being integrated with solar cells. Air purifying buildings, ocean cleaning systems, vertical farming, wireless energy transfer, and green desalination techniques are a few more technological advancements.

1.3 GREEN MANUFACTURING PRINCIPLE

The key principles of green manufacturing emphasize the undeniable use of inherently non-hazardous raw materials and the production of environmentally benign or harmless products. In the production of non-hazardous materials, it is imperative to minimize waste generation, which guarantees the process's effectiveness and the use of energy. Green manufacturing is a holistic approach that encompasses various stages, beginning with the conception of the product idea, followed by the procurement of raw materials, design, manufacturing and finishing with the product's utilization, disposal, recycling, or remanufacturing of either the entire product or specific components (Lai et al., 2011). Advanced systems like sustainable manufacturing focus on 6Rs, including reduce, reutilize, recycle, recover, redesign, and remanufacture. The production cycle's complete system is built to maximize the efficient use of time, space, energy, and materials. Adoption of cleaner and greener technologies ensures the minimum release of greenhouse gases and hazardous chemicals into soil, water, and air.

1.3.1 Green manufacturing strategies

The major strategies associated with green manufacturing are efficient use of raw materials, energy, adopting sustainable supply chain and product design, and preventing or minimizing pollution by waste reduction, recycling, and reuse (Jayal et al., 2020). The strategy of source reduction involves the sourcing of raw materials free of hazardous substances, redesigning products to minimize raw material and energy consumption, extending product lifespan, and selecting raw materials with long-lasting properties (Haleem et al., 2023). The strategy for reusing involves retrieving products and materials to be used again after their initial life cycle, aiming to maximize the benefits from these resources. By focusing on reusing products or components after their primary function, this approach helps minimize the consumption of raw materials, leading to cost savings and reduced industrial waste. The remanufacturing strategy has become increasingly popular due to the growing environmental concerns across various industries and product life cycle stages. This strategy is essential as the resources utilized in manufacturing are substantial, leading to a significant amount of waste. Remanufacturing typically occurs towards the end of a product's life cycle. Previously, remanufacturing efforts primarily focused on extracting the economic value from used products to create items with specifications similar to new products. However, the specific procedures for remanufacturing may vary depending on the nature of the product being processed (Wang and Chan, 2013). The recycling strategy in green manufacturing refers to a set of actions aimed at gathering waste materials, sorting and processing recyclable products, and converting them into raw materials suitable for the production of new goods (Ragaert et al., 2017). Implementation of novel and advanced technologies in the product design paves the way to utilize the waste from one industry as a feedstock for another industrial process that minimizes the dumping of waste, efficient use of waste, and reduction in the operational cost. Embracing circular economy reduces waste and encourages novel product design for recyclability and durability (Barros et al., 2021).

The triple bottom line concept was introduced to address social, environmental, and economic issues in business operations (Kleindorfer et al., 2005), offering a structure to incorporate various sustainability metrics. Govindan et al. (2016) emphasized a balance among economic, environmental, and social aspects by utilizing sustainable indicators in manufacturing facilities. In fact, the three dimensions of the triple bottom line act as the foundation for maintaining equilibrium. It necessitates that various industries consider all three sustainability dimensions in a holistic manner (Ahmad and Wong, 2019). This includes numerous sub-factors under each dimension, which can serve as indicators for measuring sustainability (Agrawal and Singh, 2019).

1.4 GREEN TECHNOLOGY IN THE PAST

Ancient civilizations used green technology to harness sunlight during winter, which ensured warmth in cold months. This passive solar energy is only re-embraced now. Biogas was used in ancient China around 200BC (He et al., 2010). Profound advancements in renewable energy emerged in the medieval era where water wheels and wind mills were engineered to harness the potency of water and wind for the benefit of mankind. These renewable sources of energy were used to power grindstones, textile mills, and other machinery, thus minimizing the use of animals and reducing the human labour. It laid the foundation for today's advanced hydroelectric power. Wind energy was also used in milling grains and pumping of water from low-lying lands, which is the rudimentary engineering of today's towering wind turbines and windmills. The utilization of renewable resources increased economic growth by the increased production of grains and efficient water management.

1.4.1 Drivers of green technology

The use of coal as fuel gave rise to the Green Movement in the middle of the 20th century and the Sierra Club, the first environmental group. After the Second World War, environmental issues became increasingly important. To counteract the "Greenhouse Effect," the US government passed the Clean Energy Act in 2009 and reduced carbon dioxide emissions. The movement also addressed pollution brought on by industrialization and the preservation of rainforests.

After Rachel Carson's book "Silent Spring" revealed the detrimental effects of the insecticide dichlorodiphenyltrichloroethane (DDT) on human health, the chemical was outlawed in 1972. The book "Unsafe at Any

Speed" by Ralph Nader in 1965 focused on pollution caused by the automobile industry on air.

In the 1980s, ecopreneurs began to emerge. *The Green Consumer Guide* by John Elkington and Julia Hailes focused on green consumer choices, and it detailed the chemical constituents present in many products encompassed in different categories such as food, cleaning, and gardening. An explosive emergence of many greener products was witnessed in the late 1980s. A huge volume of waste generation was witnessed in the 1990s due to the industrial revolution and consumerism. Notably, a 10% increase in population culminated in a 40% increase in the waste generated. In the later part of the 1990s, approximately 33% of waste was recycled compared to the recycling of just 8% at the beginning of the 1980s. But ironically, there was no demand for the recycled goods.

The negative impact of industrialization was realized in the form of greater emission of greenhouse gases that spurred the scientific community and political parties on the issue of climate change. The materials sector accounts for 27% of worldwide CO_2 emissions, including energy-related emissions and process sectors. Cement (7%), iron and steel (6%), chemicals (2.6%), and oil refining (2%) are some of the industries that make up a large portion of world CO_2 emissions (Olabi et al., 2022). It becomes essential to restrict anthropogenic CO_2 emissions to levels that can be absorbed by natural sinks (Marzi et al., 2018).

During the 20th century, an exponential rise in the concern over sustainability was recorded around the globe. Many corporates started adopting sustainability strategies, and they positioned their brand accordingly. Followed by the greater awareness on sustainability, many brands reframed themselves as a greener image. High energy costs and increased expenses for food and health made consumers aware of greener alternatives. Consumers assessed the greenness of a company based on multiple factors like energy efficiency, eco-friendly packaging, energy efficiency, and the use of non-toxic chemicals at the manufacturing unit. This shift has positioned green manufacturing at the forefront of sustainable development efforts, garnering growing interest in both academic and practical spheres.

1.5 GREEN MANUFACTURING AT PRESENT

A production system evolves from a traditional manufacturing system into green manufacturing through lean manufacturing. It is a set of techniques that focus on rapidly meeting customer expectations at the lowest possible time and cost by eliminating non-value-adding activities and hidden wastes from a process. With a clear focus on environmental objectives and implementation opportunities, lean enhances process flow, reduces the risk of regulatory non-compliance, and boosts staff morale. In contrast to the traditional manufacturing system that aims at waste disposal, lean manufacturing focuses on waste reduction and better utilization of resources. Lean manufacturing techniques developed by Toyota between 1950 and 1960 (Toyota Production System) are founded on the idea of "doing more with less." Eight categories of losses (inventory, over-production, defect, waiting, motion, transportation, over-processing, and human resources) were previously believed to be detected and eliminated by lean manufacturing in order to maximize the added value (Wu and Wee, 2009). In contrast, lean manufacturing aims to increase efficiency and decrease non-value-added activities (Womack and Jones, 1996).

1.5.1 Green manufacturing and sustainability

The sustainable manufacturing philosophy seeks to tackle current requirements without endangering the resources of future generations (United Nations, 1987). The green idea provides the industry with the means to create environmentally friendly production processes that take into account social and ecological factors in addition to financial gain (Haleem et al., 2023). Therefore, in order to achieve economically, ecologically, and socially sustainable production processes, the two principles must be integrated. Enhancing productivity yields greater economic benefits, and sustainability improves as the triple bottom line: the earth, its people, and profit are prioritized (Kumar et al., 2021). By integrating green manufacturing principles, organizations can achieve cost savings, improve product quality, and enhance their reputation while opening up new markets (Xie et al., 2019).

Green manufacturing has become a crucial aspect of sustainable development, aligning profits with social and environmental well-being (Viviana D'Angelo and Francesco Cappa, 2023). Irrespective of the type of manufacturing unit, the implementation of sustainable strategy increases profitability, fosters employee engagement and loyalty, and strengthens bonds with all stakeholders. As a result, sustainability plays a significant role in modern business strategy. Companies are more likely to enhance their economic and environmental performance when they use sustainable supply chain techniques such as eco-design and green procurement (Ramanathan et al., 2014; Handayani et al., 2021). Implementing resource-saving sustainable methods can, however, provide businesses with additional difficulties, such as the requirement for a sizeable upfront investment and modifications to current procedures and practices. As a result, businesses should carefully weigh the advantages and disadvantages of adopting sustainable practices and create a strategic plan that complements their objectives. Businesses that use circular economy strategies, such as product reuse and recycling, have a higher chance of reaping financial and environmental rewards (Yin et al., 2021).

1.5.2 Design strategies for waste management

Improved waste management strategies encourage material reclamation and recycling. Design for disassembly, recyclability, remanufacturing, composting, and incineration are some of the strategies (Cayzer et al., 2017). Although waste management techniques facilitate disposal, they do not lessen pollution. Designs are intended to make something simple to reuse, recycle, compost, remanufacture, and incinerate (Pongrácz et al., 2004). Design strategies for source reduction consist of decreasing weight, changing out materials, and extending the life of the product. Reduced component thickness and the use of high strength materials can help cut down on waste production. CO_2 emissions are lowered by 9g per km for every 100kg that a vehicle can shed through material substitution or topological optimization. The mass of the vehicle and the amount of energy used in production are likewise closely related (Hirsch, 2011). For things to be recyclable and remanufacturable, they must be simple to break down into their component parts. For instance, it should be possible to separate pulp and plastics from tape tabs, elastic gathers, polyethylene pulp, and other components to make disposable diapers simple to disassemble and recover the pulp and plastics. Recycling is a challenging practice to put into effect since different materials must first be separated and cleaned before they can be used. When recycling and trash elimination are not options, designs should focus on making composting or incineration easier. Final items can be made easy to dispose of at landfills by being made entirely of biodegradable components. Reduced use of heavy metals and chlorinated organic materials allows for easier incineration of the goods. Designing for disassembly therefore involves creating multipurpose items without the need for many components. Currently, durable goods like cars, refrigerators, and kitchen equipment are made to be disassembled. Snap fittings are recommended over assembly requiring screws or glues to improve disassembly. Redesigning packaging makes sense because packaging materials make up around one-third of municipal solid waste. Packaging redesign strategies include packaging that is lightweight and made of recyclable materials (Steenis et al., 2017).

1.6 GREEN MANUFACTURING IN THE CERAMIC INDUSTRY

The word "ceramics," which refers to materials used in the pottery industry, has its root from the Greek word "keramos," meaning "burned earth" (Warren et al., 2000). Non-metallic inorganic solids are referred to as ceramics (European Commission, 2007). To be more exact, ceramics are solids made from inorganic powders fired in a furnace. Ceramic goods used as structural elements or on building surfaces are referred to as building ceramics. Ceramics are made up of different materials that are processed through pulverization, mixing, shaping, glazing, and fire.

1.6.1 Green and low-carbon technology

During the production of ceramic tiles, manufacturing and process optimization strategies use advanced technologies and equipment to reduce the amount of coal and electricity used. These technologies are evolving gradually with the aim of reducing carbon emissions. Greener technologies cover a wide range of topics, such as kiln operations, drying processes, and slurry formulation (Monfort et al., 2014).

1.6.2 Design strategies in ceramic production

Reduced thickness tiles are a significant technological advancement meant to lower production costs per unit surface area as well as packing and transportation expenses. The specific weight of 3 mm porcelain stoneware tiles (7 kg/m^2) is significantly lower than that of standard 9 mm tiles (21 kg/m^2). Low-thickness big tiles have some degree of elasticity, which makes them innovative building and construction materials. Thinner ceramic tiles use less energy to burn and require less raw materials because they have a smaller mass to fire, lower production costs per unit, and minimize transportation costs as well as those associated with packaging, shipping, and disposal (Ibáñez-Forés et al., 2011). Reduction in tile thickness reduces 4% of energy use in the spray system and 9% of energy in the firing section (Türkmen et al., 2021).

In order to remove water from the ceramic body during the drying process, ceramic companies utilize a lot of heat generated by fossil fuels. Hot air recovery is used to provide heat for drying, which lowers fuel usage and spray dryer-related emissions. Natural gas is typically used to heat dryers. Approximately 65% less natural gas is utilized in this case for the drying stage.

1.6.3 Recycling of waste from other industries in ceramic production

Waste materials from different industries like stone powder and ash are used in the ceramic industry. These practices reduce carbon emissions and conserve environmental resources. Luo (2020), for instance, created a novel process for making construction ceramics made entirely of coal ash. But before putting these strategies into practice, it is important to carefully assess how they might affect the calibre of building ceramics.

1.6.4 Technology for recycling and regeneration of waste ceramic tile

Raw materials including clay, quartz, and feldspar, as well as burned and glazed ceramic products, as well as waste from operations like shaping, drying, glazing, grinding, and polishing, are main wastes from the ceramic industries. Using various technologies, the recycling and reuse of building ceramic waste has been investigated for incorporation into different product systems, including stone materials, asphalt mixtures, and concrete admixtures. Waste sludge contains kaolinite, quartz, microcline, calcite, and albeit, which are the raw materials for the manufacturing of ceramic tiles (Elmahgary et al., 2018). Raw materials used for the manufacturing of ceramic tiles can be substituted with 1–2% of waste materials generated from the industry (Ahmed et al., 2014). Tiles manufacturing companies in the EU recycle and reuse approximately 95% of wastes in the same manufacturing unit (da Silva et al., 2014).

1.6.5 Waste reuse and clean energy technology

Recycling the leftover components and sewage sludge originating from the manufacturing of ceramic tiles (Gencel et al., 2012) and undertaking life cycle assessments of ceramic tiles are some of the sustainable measures for cleaner technology (Breedveld et al., 2007; Bovea et al., 2010). Novel techniques are essential for reducing waste and finding new ways to use resources. According to research findings, the greatest carbon reduction is achieved by switching to natural gas or green hydrogen instead of coke oven gas. Hydrogen energy's general usage within the industry is limited by its economic feasibility, safety concerns, stability issues, and compatibility issues with kiln furnaces, all of which require attention (Hou et al., 2022). Electrification and clean energy provide a decarbonization path. Utilizing a microwave, as a drying method, and an electric heater helps lower carbon emissions by 6.09% and 12.23%, respectively (Ding et al., 2023).

1.7 CHEMICAL INDUSTRIES

The chemical and petrochemical sectors rank among the biggest energy consumers. These industries' energy consumption increased at a rate of 2% yearly between 2000 and 2016, while CO_2 emissions increased at an annual rate of 2.5% over the same period (Vooradi et al., 2018). A total of 925 Mt of direct CO_2 emissions were produced by primary chemical production in 2021 (International Energy Agency, 2024). More than 7% of CO_2 emissions globally come from the manufacturing of cement, a substantial contribution that can be mitigated in a number of ways. The major energy user and CO_2 emitter is the synthetic ammonia industry. Synthesis of hydrogen and ammonia is an essential part of maintaining sustainability (Bowker et al., 2022).

1.7.1 Advanced carbon capture technologies

By the year 2050, the chemical sector wants to be net-zero emissions. As per the Net-Zero Emissions by 2050 report, till 2030, CO_2 emissions will be reduced by 17%. The three primary strategies for reducing greenhouse gas emissions are carbon collection, storage and utilization, and shifting to renewable energy and alternate feedstocks from fossil fuels.

A comprehensive preventative strategy is applied to manufacturing processes and activities by limiting wastes at the point of generation, lowering the toxicity of emissions, eliminating toxic materials, and conserving energy and raw materials. CO_2 emissions from industrial activities are captured for other purposes or stored underground by injecting them into appropriate geological formations (Kelemen et al., 2019). Technologies that transform captured CO_2 into fuels and chemicals, like electrochemical conversion (Sullivan et al., 2021), catalytic hydrogenation (Ra et al., 2020), and photocatalytic CO_2 conversion (Fu et al., 2019), possess the ability to bring CO_2 back. In addition, converting CO_2 to value-added products has gained remarkable attention (Dongare et al., 2021). To fully evaluate environmental implications, a life cycle assessment must be carried out before creating a biorefinery and its whole supply chain (Silva et al., 2022).

1.7.2 Alternative fuels

In the chemical sector, adopting a circular economy strategy entails creating products that are recyclable, biodegradable, or reusable (Aurisano et al., 2021). The decrease in the demand for fresh feedstock and the recovery of valuable resources are made possible by the development of new chemical recycling methods.

1.7.2.1 Hydrogen as a feedstock

Decarbonization must be ensured by the utilization of renewable sources rather than conventional fossil fuels. The processes can be made more energy-efficient by electrifying them (Eryazici et al., 2021) and by integrating renewable energy sources like biomass, solar, and wind (Fakayode et al., 2023). Hydrogen, when produced from renewable sources, offers a cleaner alternative as a feedstock, often referred to as green hydrogen. It can be produced via thermochemical processes (Mishra et al., 2023).

1.7.2.2 *Biological feedstock*

The industry's dependency on fossil fuels can be decreased by using feedstocks made from biomass. Through a variety of processes, including fermentation, enzymatic conversion (Gul et al., 2022), and thermochemical conversion (Rahimi et al., 2022), biomass, such as agricultural leftovers (Mahapatra et al., 2021) and food wastes (Lee et al., 2022), can be transformed into chemicals (Ewing et al., 2022). The economic and technological efficiency of renewable and alternative biomass resources is less than that of fossil fuels (Fiorentino et al., 2019). Hence, research must focus on improving efficiency to make these technologies sustainable. Research should develop more efficient process units and higher quality raw materials to enhance biotransformation and decrease feedstock consumption. Developing advanced biocatalysts (Bellabarba et al., 2023) increases the selectivity of the process and decreases energy requirement and CO_2 emission through optimized milder operating conditions (Chit et al., 2023) and heat integration.

1.7.3 Optimization of multiple objective processes

Multiple unit operations are optimized by appropriate algorithms and artificial intelligence tools (He et al., 2023). They predict and detect anomalies (Al-Anzi et al., 2022) and prevent unprecedented waste. One of the primary methodological strategies that greatly aids in the chemical industry's decarbonization is process integration. It is a sophisticated strategy that makes use of a number of techniques, including P-graphs, mathematical programming, and pinch analysis (Migo-Sumagang et al., 2023). Combining a heat pump with a combined heat and power (CHP) unit makes it possible to recycle waste heat, even at low temperatures, and boost the output of renewable energy sources for both heat and power (Drofenik et al., 2023).

1.8 CEMENT INDUSTRY

In the near future, there will be a greater need for buildings and infrastructure, making sustainable, affordable, and long-lasting concrete more and more desirable. The manufacturing of cement uses a substantial amount of natural resources. According to reports, 1.5 tonnes of raw material resources are required to make one tonne of Portland cement (Rashad, 2015). Moreover, clinker manufacture consumes a greater quantum of energy (Rashad et al., 2013). As a result, the cement industry faces numerous challenges, including the diminution of limited natural resources, growing energy expenses, the need to reduce CO_2 emissions, and the guarantee of ample supply of resources. Cleaner production is aimed at the organization of the production process with a positive impact on the environment. In order to promote environmental protection, raw materials must be used efficiently. Green manufacturing in the cement industry is achieved through resource efficiency, energy efficiency, carbon capture, product efficiency, and recycling of waste. Combined use of all these principles reduces carbon footprint and ensures sustainability.

1.8.1 Resource efficiency

Technologically, cement can be produced by substituting a major proportion of raw materials with alternative resources like fly ash (Danner and Justnes, 2020), granulated blast furnace slag (Reddy and Subramaniam, 2020), calcined clays (Kuzel and Pöllmann, 1991), red mud (Vladić Kancir and Serdar, 2022), silica fumes, and waste materials from other industries that consume less energy and emit lesser amount of carbon and other gases (Yang et al., 2015). Substitution of 10% of alternative sources will reduce 25% of gaseous emissions (Damtoft et al., 2008).

1.8.2 Energy efficiency

A cement plant's total thermal energy consumption can be reduced by using slag. Recovering waste heat for the purpose of generating electricity would not increase the thermal efficiency of clinker manufacturing, which means it would not lower direct CO_2 emissions from the process but will generate electricity (Harder et al., 2014). In cement plants, waste heat usually has a low temperature, and only 15%–25% of it can be effectively converted into electrical energy. A turbine system and heat recovery boiler are needed for the installation.

1.8.3 Fuel efficiency

Fuel oil or natural gas is used for clinker production that accounts for the emission of greenhouse gases. Alternative fuels are based on utilizing different ratios of conventional fuels and alternative fuels. It may be 40% of fuel based on biomass, 30% coal, and 10% of petcoke.

Compared to coal, biomass reduces 25% of greenhouse gas emission (Habert et al., 2010). Industrial wastes like tyres, sludge, animal residue waste, paper, and plastic residues can be reused in the cement industry as an alternative fuel. Substitution of 60% of traditional fuels with alternative fuels by 2050 is expected to reduce emissions by 35% (World Business Council for Sustainable Development, 2009). Calcined clays with limestone is a very promising supplementary cementitious material than fly ash

(Scrivener et al., 2018). Ali Naqi and Jeong Gook Jang (2019) reviewed the possibilities and advanced technological strategies for exploiting low-carbon-emitting raw materials in the cement industry. Energy consumption related to product transportation may be greatly decreased with effective logistics and transportation planning (Dai et al., 2023). Utilizing technologies can lead to the optimization of carbon-intensive tasks, including inventory management, route planning, and transportation modalities (Das, 2023).

1.8.4 Low-carbon smart manufacturing

Because of the constant demand to maintain a balance between environmental concerns, sustainable development, and competitiveness, low-carbon smart manufacturing is increasingly a prerequisite in the modern industrial environment (Vormedal et al., 2023). Low-carbon smart manufacturing is expected to advance quickly (Terry et al., 2020). The main objective of smart manufacturing is the integration of renewable energy sources to reduce the carbon footprint on the environment (Zhou et al., 2023). Conventional manufacturing methods rely heavily on conventional resources, which significantly increase greenhouse gas emissions (Jahanger et al., 2023). However, by implementing healthier production methods through the use of sustainable energy sources, conventional manufacturing methods can significantly mitigate their carbon footprints (Liang and Wang, 2023).

The goal of low-carbon supply chain manufacturing is to increase efficiency and lower emissions by implementing modern manufacturing methods and technologies (Wimbadi and Djalante, 2020). It combines low-carbon manufacturing's emphasis on eliminating greenhouse gas emissions with the features of intelligent manufacturing, which uses digital technologies and data analytics to optimize production processes (Al Shahrani et al., 2022; Zhao et al., 2023). The IE 4.0 framework comprises prominent enablers that make it easier to develop automated and digital production environments, which improve production processes (Wang et al., 2022). Intelligent sensors, live monitoring, and data analysis are used for precise regulation and maximizing production without depleting the resources (Soori et al., 2023).

Manufacturers are able to lessen the negative effects by putting waste reduction and recycling techniques into practice (Sharma et al., 2023). Adoption of key principles of circular economy, material recycling, and waste reduction yields increased profitability with eco-friendly manufacturing (Hojnik et al., 2023). Furthermore, low-carbon smart manufacturing (SM) depends on supply chain optimization (Tian et al., 2023).

1.8.5 Recycling and recarbonation of concrete

Recycling is crucial to the construction chain's overall CO_2 balance. Through recarbonation, the fines from crushed concrete—which are essentially calcium silicate hydrates—can absorb a substantial amount of CO_2 from the environment. Additionally, in specific situations, recycled concrete fines may be used as the main source material for manufacturing clinker. Since the particles are made up of calcium and silica compounds, they can be used instead of natural raw materials. Crushing the leftover material to create granules with a specific particle size is the fundamental step in concrete recycling. For this reason, crushed concrete fines are typically only utilized in the construction of roads or even deposited. Nonetheless, Portland cement clinker manufacture can make use of crushed concrete particles. Concrete can be recarbonated by reacting hardened cement paste or mortar with ambient CO_2, which causes reversible calcination reactions and CO_2 uptake. It is recognized that active carbonation allows for rapid, full carbonation of the concrete with significant technical effort (Xi et al., 2016). Additionally, CO_2 can be used as an additive in concrete manufacturing.

1.8.6 Carbon capture

Utilization of CO_2 as feedstock is a major strategy to attain net-zero greenhouse gas emissions (Huo et al., 2023; Yu et al., 2023). Creating a stream of CO_2 at a high pressure for transportation or storage is the aim of CO_2 capture. Cement kilns have special process characteristics that make it possible to both capture fuel and process CO_2, meaning that they can only be fitted with post-combustion and oxyfuel technologies. CO_2 is extracted from the waste gas after burning, allowing for 95% capture rates. Membranes with a capture yield of up to 80% can also be used for post-combustion separation of CO_2 (Lindqvist et al., 2014).

An additional technique for capturing CO_2 and replenishing the sorbent is calcium looping. In the calcium looping process, CaO and CO_2 from the flue gas are brought into contact in a carbonator at 600°C–700°C. To replenish CaO as the sorbent, calcium carbonate is split into CaO and CO_2 at temperatures greater than 900°C. CaO cycles (loops) between the carbonator and calciner can be used once again in the carbonator. After that, the CO_2-rich stream can undergo additional processing for transportation. Alkali-activated materials, which are conventional binders, can reduce carbon emissions in ordinary Portland cement (Pedraza et al., 2021). Biochar, which is produced by pyrolyzing rice husks at 550°C, can be utilized as an alternative supplementary cementitious material (Aneja et al., 2022).

1.8.7 Decarbonization and advanced technologies

Roughly 90% of the emissions from the manufacturing of cement are mostly driven by clinker (Andrew, 2018). Industry participants are actively researching alternative fuels as well as developing cutting-edge technologies like kiln electrification (Cemnet, 2024), oxy-combustion (Nhuchhen et al., 2022), and heat generation via plasma technology (Vattenfall and Cementa, 2019), in an attempt to lower emissions caused by energy use. Pure oxygen is used as the oxidizer in the kiln by oxyfuel technology. Because of this, the kiln's exit gas contains a lot of CO_2—ideally, simply CO_2 and water vapour. To maintain an optimum flame temperature, a portion of the flue gas rich in CO_2 is circulated. There are several ways to reintegrate the CO_2 emissions that are captured from production operations back into the value chain (Ferrario et al., 2023; Tanzer et al., 2023).

1.9 AUTOMOBILE INDUSTRY

Globally, one-seventh of the automobile industry pollution is produced during the car manufacturing process (Zailani et al., 2015). Pre-production and post-consumer waste creation surged in the sector with the advent of mass production in the early 20th century. Automotive industry garbage poses technical problems for disposal because of its dynamic composition such as plastics, metals, and fluids. The automotive sector has been forced by the global community and consumers to minimize pollution and adopt better environmental management practices. As a result, zero-waste programs are becoming more popular.

1.9.1 Zero-waste initiatives

Increasing concern on sustainability has compelled automobile industries to develop vehicles that are less polluting and therefore "green" to minimize waste creation in the entire vehicle production operations (Kushwaha and Sharma, 2016). These include the development of innovative technologies for trash recycling, pollution control, energy conservation management, and designing green products (Lin et al., 2014). Green innovation is thus divided into three areas such as green management innovation, green process innovation, and green product innovation. According to their definition, green product innovation is the process of creating new or substantially enhanced products in response to environmental issues. The aim of green process innovation is to produce environmentally friendly products by changing the unit operation processes (Harrington et al., 2014). The idea of a circular economy emphasizes material reduction, reuse, and recycling.

1.9.2 Circular economy

Rather than using the existing cradle-to-grave concept, the new cradle-to-cradle design is the solution. Reusable, repurposed, and refurbished products remove the need for virgin or recycled raw materials in the production process. Circular economy is beyond simple recycling as it determines the best way to manufacture goods with minimum resources and repurposing of wastes generated till the end of the life of the product (Valladares Montemayor, 2022). A four-level paradigm integrates the life of a product with a collection of practices at each level. Level 1 comprises reusing (using a product that has been rejected by another customer while it is still functioning), repairing (maintaining a product up to date with the latest level), and refurbishing (upgrading an outdated product) (Morseletto, 2020). Level 2 includes remanufacturing (using discarded parts for a new product that serves the same purpose) and repurposing (using discarded parts for a new product) that serves a different purpose (Morseletto, 2020). Recycling, composting, and anaerobic digestion are included in Level 3. Level 3 processing includes anaerobic digestion, composting, and recycling of waste to produce materials of the same or lower quality. Level 4 encompasses recovery of waste and its management. This comprises gathering, storing, and removing waste materials as well as burning them to recover energy (Ihsanullah et al., 2022; Morseletto, 2020). It aims to enhance digital manufacturing by expanding digitization and business models (European Commission, 2017).

By streamlining production procedures and minimizing the use of needless materials, lean manufacturing approaches seek to eradicate waste. Toyota, for instance, incorporates design for disassembly principles, recycling initiatives, and closed-loop technologies into its manufacturing operations. Toyota has reduced waste and increased resource efficiency significantly by reusing materials and improving component design. BMW has adopted cutting-edge waste reduction techniques like remanufacturing and put in place a thorough waste management system.

1.9.3 Design of light weight components and alternative raw materials

Sustainability measures drive the automotive sector to engineer vehicles with lightweight materials wherever possible without compromising the performance. The innovative design with lightweight materials offers enhanced performance and is cost-effective with less emission of greenhouse gases. Composite materials, which are strong and lightweight, have replaced conventional materials like steel and aluminium in the automotive sector. As a part of implementing green technology, the body panels, chassis, and interiors of vehicles are increasingly being constructed using carbon fibre-reinforced composites. These materials improve overall performance and fuel efficiency owing to their exceptional strength-to-weight ratio. Flax fibres are an effective technique for lowering CO_2 emissions. Because flax absorbs CO_2 from the environment, it has no carbon footprint when raw.

Structural engineering incorporates designs and fabrication techniques for lightweight materials, sandwich structures, and laminates that draw inspiration from natural models (Charles et al., 2021). Recently, engineered plastics have been substituted for petroleum-based plastics. They can be recycled and reused. The combination of lightweight materials saves energy and decreases greenhouse gas emissions. Natural fibres reinforced with engineered plastics are used in the production unit, ensuring recyclability and reuse (Kuppusamy et al., 2021).

When looking for environmentally responsible material substitutes, automakers may choose thermoplastics derived from more sustainable sources. They are made of recycled plastics and polymers generated from soybeans and maize starch. Though their potential applications may be limited, they represent a competitive alternative to traditional petroleum-based products. Now that additive manufacturing technology has advanced, manufacturers may create parts internally. This significantly lowers carbon emissions from transportation. Automobile manufacturers are saving money and lessening their carbon footprint by decreasing transportation.

1.9.4 Fuel efficiency

Engine performance is enhanced by the use of thermal barrier coatings, which have a beneficial effect on fuel efficiency. The piston and valve head were coated with Mo and Al_2O_3-TiO_2. These coating materials reduced fuel consumption, especially for the brakes, and greatly increased brake thermal efficiency. Al_2O_3-TiO_2 coating treatment remarkably decreased greenhouse gas emissions. Thermal barrier coatings for temperature swings can enhance performance and efficiency by effectively regulating the thermal conditions within the combustion chamber.

1.9.5 Additive manufacturing and waste reduction

Components manufactured by additive manufacturing improve material efficiency and enable the production of complex geometrical designs. Adoption of these technologies can also result in broader systemic advantages, such as more effective logistics and supply chains. Three-dimensional printing with polymer that offers less restrictions on resources (safety, training, and facilities) increases usability and accessibility to a wide range of material qualities, accompanying cost savings for all applications. Product development involves designing, building models, prototypes, mock-ups, and validating the designed concepts and products. Additive manufacturing speeds up the design and the ideation process by producing parts without additional tools. Additively manufactured materials can replicate or simulate the look and feel of other materials, such as polymers with a wide colour variety and texture capabilities. Because of this, they are the best option for concepts and prototypes that must accurately depict their final design; this is especially useful for exterior moulding, lighting, and car interiors. Easy printability, a smooth surface, and the capacity to mimic other polymers like acrylic are further considerations.

1.9.6 Recyclable materials in the automotive industry

One efficient way to reduce manufacturing-related emissions, including CO_2, CO, and NO_2, is to recycle waste materials that demand less energy. Reusing industrial aluminium waste materials has also become a vital way for producers to reduce their costs of production. Due to its advantageous combination of qualities, such as its strength, remarkable resilience, and remarkable resistance to corrosion, aluminium alloy has been widely utilized in a number of industries.

Automobile makers are always looking for innovative ways to produce ecological solutions, such as using repurposed wood and “vegan” leather. The focus of their concern has now shifted to ocean plastic garbage. These materials are being made into dashboard trims, flooring, door panels, and seat cushions, among other components. In order to address the increasing demand for environmentally friendly transportation, manufacturers are progressing by identifying new applications for these materials. Leading this initiative is the German automaker Audi, which has committed to becoming net CO_2 neutral by 2050. Their use of recycled materials, including polyethylene terephthalate (PET) bottles, in the production of their automobiles, is an essential part of this mission.

The history of plastic substitutes in auto manufacture is extensive; one of the first examples is the introduction of Henry Ford’s soybean automobile in 1941. The whole body of the vehicle was constructed from bioplastics, a material that is far safer and lighter than conventional steel. We can move towards a successful future where contemporary life coexists peacefully with the environment by embracing these materials. Automobile manufacturers have employed rattan as a substitute for plastic, such as Jaguar and Range Rover for various vehicle components. Headliners, carpets, seat cushion, and door panels are important components (Source: The New York Times).

1.9.7 Design for disassembly and recycling

Creating automobiles that are simple to disassemble and recyclable is a critical component of automotive zero-waste programs. It entails applying labelling systems, standardized connections, and modular design concepts to enable the effective disassembly and recycling of vehicle components.

Benefits of designing for disassembly include higher recycling rates, lower waste production, and enhanced resource efficiency. A number of automakers have adopted the idea of "design for disassembly," paving the way for the incorporation of environmentally friendly techniques into vehicle design.

1.10 E-WASTE

The quality of our lives has been significantly enhanced by the invention of several electronic devices, which have been made possible by rapid economic development, urbanization, and technological innovation. The European Council and Parliament's Directive defines electrical and electronic equipment as any household or industrial device that uses electric current and or electromagnetic field, including phones, computers, lighting fixtures, medical equipment, and information technology equipment. One of the main effects of relying too much on them is the generation of a massive amount of e-waste. Undesirable, outdated, non-operational, and expired appliances and equipment are referred to as "e-waste" (Baldé et al., 2017). White goods and large appliances like refrigerators and washing machines and brown goods and smaller personal electronics like laptops and cell phones are all included in waste electrical and electronic equipment (WEEE) (Needhidasan et al., 2014)

1.10.1 Categories of e-waste

The term "electronic waste" has a broad meaning that includes six categories such as 1) equipment for regulating temperature, like air conditioning, refrigeration, freezing, and heat pump systems, 2) monitor of laptops and televisions, 3) high-intensity discharge, LED, and fluorescent bulbs, 4) big appliances like photovoltaic panels, huge printing machines, dishwashers, clothes dryers and photocopiers, 5) small-sized home and medical appliances, and 6) tiny monitoring and control instruments.

1.10.2 Composition of e-waste

Hazardous materials such as mercury, brominated flame retardants, hydrochlorofluorocarbons, chlorofluorocarbons, and toxic additives are present in e-waste. Numerous useful metals as well as other priceless elements like iron, copper, and gold can be found in e-waste. Primary sources of these elements are used in the new production of electronic products when materials like these cannot be salvaged from the e-waste stream. The seven waste regions were first named using the Japanese term "mura." Waste is transportation waste, inventory waste, waste in motion, over processed waiting, over-production waste, and defect wastes (Setiyawan et al., 2013).

1.10.3 Design strategies for reducing waste in electronics manufacturing

Design for manufacturability entails taking an integrated team approach to the electrical design and ensuring that the resultant design is easily built without affecting the electrical performance. Design for testability aims to stop defective items from leaving the manufacturing line as a result of insufficient testing by identifying testing issues as early as possible. Design for recycling is the process of creating products with less material consumption, greater safety, dependability, and ease of non-destructive disassembly by using small and compact board designs, which reduces the quantity of parts utilized. Reducing the quantity of fasteners and connectors avoids the usage of hazardous materials and substances, which do not comply with restriction of hazardous substances (ROHS) regulations.. E-waste recovery process such as pyromettalurgical and hydrometallurgical methods recycle and reuse the electronic wastes.

1.10.3.1 Eco-friendly designing of products

In the design phase of products, the choice of materials to be utilized and the manufacturing procedure are included (Sarjaš, 2018). Future developments focus on using less power and less raw and renewable materials. Eco-design aims to reduce waste and pollution while using materials with less of an adverse effect on the environment, less resources, and fewer materials used in manufacturing (Immonen et al., 2022). Because they leave less environmental imprint, paper-based substrates made from renewable resources offer greener options (Jansson et al., 2022).

1.10.3.2 Reducing e-waste through innovation

The principle of circular tech economy idea entails changing the conception, manufacturing, and assembly processes of electronic products in order to reduce waste over the whole life cycle. It encourages the cautious selection of raw materials and the production of long-lasting, robust electronic products. It involves designing products that can be readily replaced and fixed.

It is imperative that the packaging of electronic equipment is recyclable and uses minimal plastic in compliance with environmental norms and standards.

1.10.3.3 Design for disassembly and repair

One important element in extending the life of electronic devices is repairability. Nowadays, producers are adopting the idea of creating items that are simple to dismantle and fix. In addition to encouraging users to fix their devices, this move towards more accessible and modular designs also lowers the quantity of electronic waste produced. Manufacturers are putting consumers in control of prolonging the life of their electronics by streamlining maintenance procedures.

1.10.3.4 Device reconditioning and refurbishment

The foremost objective of device refurbishing is to enhance the functional life of devices by inspecting, repairing and updating them to meet contemporary performance standards. Owing to their extensive use and significant repurposing possibilities, tiny consumer electronics, including desktop PCs, laptops, tablets, and cell phones, are the main targets of refurbishment. Redistribution, meticulous reconditioning, and collection are the stages of the multi-phase reconditioning process. The device needs to be inspected and assessed; any broken components need to be replaced; functionality needs to be verified by testing, cleaning, and maybe software updates; and after that, it needs to be listed for sale. The reconditioning phase is primarily manual and uses little energy or materials.

Conversely, refurbishing focuses more on fixing the device's current components, taking care of any aesthetic or technical problems without replacing many parts. It's a less complicated and expensive procedure than reconditioning. Carbon credits can be applied to both refurbishing and reconditioning, but it's crucial to distinguish between the two in this process. This distinction makes it possible to comprehend the project's methodology and the amount of effort required to get the electrical gadgets back to working order. By prolonging the life cycle of electronic goods, the generation of electronic waste is reduced and sustainability is promoted. People can also profit from technology when it is affordable. Refurbished equipment also lowers production-related energy use, gas emissions, and raw material requirements. This procedure eases the burden on resource-intensive industrial processes while permitting favourable environmental effects.

1.10.3.5 Recycling

The circular tech economy places a high value on recycling of precious elements like gold, silver, and copper found in electronic gadgets. Recycling will lessen the requirement for producing raw materials that are extremely resource-intensive and harmful to the environment. Recycling also uses less energy than creating new materials from scratch, which lowers greenhouse gas emissions and fights climate change.

Recycling electronic waste keeps hazardous materials out of landfills and lessens pollution, and the recovered materials can be used again to create a circular supply chain in production.

1.10.3.6 Additive manufacturing and waste reduction

The desire to adopt additive manufacturing and other techniques that help optimize and bring out new possibilities in electronics was sparked by the need for high performance, compact size, and optimum utilization of resources and minimizing waste generation. Additive manufacturing is used in electronics because it is a cost-effective manufacturing process that requires less labour and produces products faster and with less waste than conventional manufacturing processes. As a result, additive manufacturing plays a significant role in modern electronics. Because additive manufacturing processes can adjust to the intricate geometrical and unique design of the part that needs to be made, they offer substantial competitive benefits. Additionally, the downsizing of electronics and circuit systems has been aided by additive manufacturing. Rapid processing techniques, like additive manufacturing, significantly reduce waste.

1.10.3.7 Digitization for smart manufacturing

There are numerous chances to increase productivity, cut waste, and boost cost-effectiveness in the production of electronics using sustainable practices. The Internet of Things and artificial intelligence can help with sustainable manufacturing. Businesses can reduce waste and unnecessary expenses by automating operations with smart digital manufacturing techniques and utilizing sensor technology to identify leaks and inappropriate material usage. Sustainable Semiconductor Technologies and Systems (SSTS) is an initiative developed to mitigate the harmful impact of semiconductor production. The goal of Imec's SSTS initiative is to increase sustainability by using digital technologies to pinpoint areas for improvement in three crucial areas: water usage, greenhouse gas emissions, and energy consumption. Microsoft, Amazon, ASML, and a host of other well-known businesses are rumoured to be using Imedec's digital analysis.

1.10.3.8 Reusing electronic waste

Remanufacturing allows the replacement of worn-out parts with fresh components and increases the overall efficiency of the machine. Recycling restores their original quality. It entails removing secondary materials from any object or material that has been abandoned. These removed secondary materials can be either downcycled (as is) or upcycled (high quality or equivalent functionality). E-waste management typically consists of five main steps such as collection, detoxification, preprocessing, end processing, and disposal. The disassembly and separation phases of the process, in particular, are critical features of e-waste recycling (Lucier and Gareau, 2019). For example, hydrometallurgy, being a heat process, necessitates significant energy and financial expenditure.

1.10.3.9 Recovery of resources by urban mining

Urban mining significantly increases recycling rates and recovery of metals from e-waste that facilitates recycling in the supply chain (Debnath et al., 2019). It is a closed-loop strategy for e-waste management. It also reduces the consumption of primary minerals, which supports circularity economy (Xavier et al., 2021). In developing nations, it generates job opportunities and economic prospects at different stages of e-waste management like sorting, dismantling, and recycling.

1.10.3.10 Recovery of metals by microbial leaching

Economic expansion and the desire for quicker, more efficient consumer devices have increased the amount of garbage produced by electronics. End-of-life electronics components contain glass and metals that can be recycled or reused. However, traditional extraction techniques rely on energy-intensive processes that are ineffective for recovering e-waste that contains minute amounts of metals combined with other elements. It is necessary to develop recovery techniques that reduce the cost of metal reclamation and minimize environmental impact in order to make e-waste recycling competitive and economically viable in terms of procuring raw materials. Compared to more conventional methods of metal extraction, microbial surface adsorption can help with metal recovery at a reduced cost and energy consumption.

Concerns about how traditional mining methods affect the environment and people are becoming more widespread. Different classes of microbes with leaching properties are exploited in biomining (Trivedi et al., 2022). Hazardous metals such as lead, chromium, uranium, and cadmium can be adsorbed or accumulated by microbial biomass. Furthermore, they can be transformed by redox reactions (Narayanasamy et al., 2021).

1.10.3.11 Clean hydrogen fuel extraction by gasification

Hydrogen possesses the ability to both carry and store energy, making it a viable medium for use in various industrial processes and power generation. In an effort to create a sustainable world, hydrogen has gained popularity as a possible fuel for residential heating needs, cars, and aircraft. It can be burned directly or through a process known as "cold combustion" in fuel cells (Rauch et al., 2024). The gasification process can transform the polymeric components present in the e-waste into gaseous products (CO and H_2), which may be employed to recuperate energy or perform chemical synthesis (Gurgul et al., 2018). Utilizing steam, air, or oxygen as a gasifying agent or medium, this process is conducted under carefully regulated gasifying conditions. The optimal H_2-to-CO ratio can only be achieved by cleaning the synthetic gas after post-gasification (Shafiq et al., 2021). Syngas is utilized in gas turbines to produce electricity or utilized as a fuel (Uddin et al., 2021).

1.11 GREEN TRENDS IN THE FOOD INDUSTRY

The amount of food waste generated annually globally is more than 1.3 billion tonnes (Galanakis, 2020). Wastes are produced at several points like home consumption, retailing, manufacturing, and the beginning of the agricultural process (Garcia-Garcia et al., 2018; Kannah et al., 2020; Bhattacharya et al., 2021; Cui et al., 2022).

1.11.1 Reducing food waste

Innovative methods for reducing food waste and responsible consumption enhance food security. Sustainable packaging employing recyclable and compostable material ensures sustainability. The use of plant-based foods instead of animal-based foods will reduce the carbon footprint. Food processors are adopting the circular economy model, which involves recycling or repurposing waste streams into useful by-products. Food waste, for example, can be transformed into organic fertilizers, animal feed, or biofuels.

1.11.2 Enzymes to reduce environmental footprint

The majority of grains are processed by dehulling, washing, sorting, and grading. These foods are further processed using secondary and tertiary processing methods to create additional food items with added value. All these processes are energy-intensive and use a variety of chemicals and solvents. Extensive use of solvents and the release of used solvents and effluents pollute the environment.

Enzymatic procedures are more environmentally friendly since they operate under milder reaction conditions. Pancreatic lipase, phospholipases, pepsin, trypsin, bromelain, papain, glucose isomerase transglutaminases, pectinases, cellulases, galactosidases, and chitinases and enzymes such as glucose oxidase and polyphenol oxidases can be employed in the food processing industry. They are non-toxic biocatalysts with high specificity; they enhance the reaction rates and save time, energy, and resources in the processing of food. The use of enzymes improves the shelf life and organoleptic properties of food.

1.11.3 Green technology for sterilization

Thermal treatments improve the shelf life of foods. Since operations use a lot of energy, alternative methods that use less energy and have a smaller environmental impact should be taken into consideration. Sterilization of foods using physical principles with the use of ionizing radiation, UV, microwave and radiofrequency radiation, ohmic heating, and chemically using ozone are a few examples of novel innovations.

Ozone has been used in the processing of fruits and vegetables to get rid of harmful bacteria, mycotoxins, pesticide residue, and chemical traces (De Souza et al., 2018). To satisfy adequate drinking criteria, water quality in the food processing business is influenced by a number of elements, including financial, ecological, and technical considerations. Ohmic heat treatment is suitable for meat processing in a very short time compared to conventional smoke cooking by reducing 95% of the time (Brunton et al., 2005). Reverse osmosis and microfiltration are the two methods of water treatment that rely on plasma bioreactors.

1.11.4 Sustainable drying

A crucial unit function that needs to be taken into account when aiming for a greener process is drying, which for most businesses is a big source of energy consumption and a large financial outlay. It lowers transportation costs, improves shelf life, and adds texture. Heat recovery from exhaust fumes, drying, retrieving the dried product, and wet feed pre-treatment are the usual components of a drying operation. Osmotic dehydration can be considered as a preheat treatment or a final dehydration step. To save the environment from harmful effects, heat can be recovered and recirculated during drying. After heat recovery, greenhouse gases can be eliminated by further cleaning the exhaust gases.

1.11.5 Sustainable packaging

A distinguishing feature of green processing is the use of sustainable packaging. As consumers' concerns about plastic pollution and packaging waste grow, food producers are looking into alternatives like reusable containers, compostable materials, and biodegradable packaging. These programs improve customer loyalty and brand reputation in addition to lessening their negative effects on the environment. Fundamentally, green processing deviates from traditional approaches by utilizing cutting-edge technology and environmentally friendly procedures to solve urgent environmental problems. A lot of food processing facilities use solar energy to power their operations. To further reduce trash, consumers can now dispose of the packaging in compost bins, where it decomposes into organic matter.

1.11.6 Reuse and recycle of packaging materials

Reusability is the ability to reuse materials in their original form repeatedly as opposed to discarding them (Njomo, 2019; Obebe and Adam, 2020). The recycled use of glass bottles has shown to be a viable way to lessen the influence on the environment, according to Stefanini et al. (2021). In the event that the bottle is unharmed, it permits the full healing. The two stages of reuse affecting the process are sterilization and drying. Reuse of glass bottles reduces heat and electricity use and decreases the environmental impact by 40% (Amienyo and Azapagic, 2016; Landi et al., 2019). Plastic containers used in the food industry and processing units can be recycled. Plastics are difficult to degrade and can be used for a variety of purposes (Obebe and Adam, 2020). Although the amount of plastic garbage transported to recycling has increased since 2006, in 2018, 25% of plastic waste from consumer goods was still disposed of in landfills.

1.11.7 Waste bioconversion and energy conservation

Hybrid technology can be utilized as an alternative to fossil fuels in order to lower pollution. These include, for instance, combining solar and wind energy with electricity or using biogas as a highly effective replacement in the dairy industry (Lian et al., 2019). Solar energy is used for power generation, cooking, water heating, refrigeration, and building heating and cooling (Li and Sumathy, 2000). Systems for solar energy, both passive and active, are used. Passive solar energy systems rely on direct sun radiation heating the desired region. However, in order to collect, store, and transform solar energy into a form that is more useful, like electricity or hot water, active solar energy systems rely on outside energy sources like hot water pumps. In the non-availability of these natural renewable sources, biomass can be converted into an energy source.

Food businesses can reduce water consumption and wastewater discharge by employing water recycling systems and creative cleaning techniques, which will ease the burden on nearby water sources and ecosystems. The cleaned water is repurposed for other uses. This lowers the facility's environmental impact in addition to conserving water.

1.12 FUTURE PROSPECTS: THE GREEN HORIZON

Green technology in future will harness the extraordinary potential of quantum computing, designing a novel material for solar panels, possibility of harvesting water from the atmosphere directly, advanced technologies to transform atmospheric carbon dioxide into assets, hyper-efficient transportation like hyper-loop system, superconducting maglev trains powered by renewable energy, implementation of AI tools in all manufacturing and conservation processes and the adoption of circular economy in all dimensions of life. Even though green technology is observed in many facets of life, it is rudimentary in certain areas like energy storage applications and meeting the global demand for technologies like carbon capture. In addition to the technological barrier, the progress is also impeded by the huge capital requirement, biodiversity conservation, ethical consideration, policy approaches, and social norms. It decreases the dependency on virgin materials and cuts down the production cost of raw materials. Deep tech technologies are used directly or indirectly to optimize energy and reduce greenhouse gas emissions.

REFERENCES

Agrawal, Saurabh, and Rajesh Kr Singh. "Analyzing disposition decisions for sustainable reverse logistics: Triple Bottom Line approach." *Resources, Conservation and Recycling* 150 (2019): 104448.

Ahmad, Shamraiz, and Kuan Yew Wong. "Development of weighted triple-bottom line sustainability indicators for the Malaysian food manufacturing industry using the Delphi method." *Journal of Cleaner Production* 229 (2019): 1167–1182.

Ahmed, M. M., M. H. Roushdy, ShK Amin, and Magdi F. Abadir. "Reuse of the product obtained on grinding kiln rollers in the manufacture of ceramic wall tiles." *Ceramics Technical* 38 (2014): 60–66.

Al Shahrani, Ali M., Madani Abdu Alomar, Khaled N. Alqahtani, Mohammed Salem Basingab, Bhisham Sharma, and Ali Rizwan. "Machine learning-enabled smart industrial automation systems using internet of things." *Sensors* 23, no. 1 (2022): 324.

Al-Anzi, Fawaz S., Haitham MS Lababidi, Ghanima Al-Sharrah, Suad A. Al-Radwan, and Ho Joon Seo. "Plant health index as an anomaly detection tool for oil refinery processes." *Scientific Reports* 12, no. 1 (2022): 14477.

Amienyo, David, and Adisa Azapagic. "Life cycle environmental impacts and costs of beer production and consumption in the UK." *The International Journal of Life Cycle Assessment* 21 (2016): 492–509.

Andrew, Robbie M. "Global CO_2 emissions from cement production." *Earth System Science Data* 10, no. 1 (2018): 195–217.

Aneja, Akash, R. L. Sharma, and Harpal Singh. "Mechanical and durability properties of biochar concrete." *Materials Today: Proceedings* 65 (2022): 3724–3730.

Aurisano, Nicolò, Roland Weber, and Peter Fantke. "Enabling a circular economy for chemicals in plastics." *Current Opinion in Green and Sustainable Chemistry* 31 (2021): 100513.

Baldé, C. P., V. Forti, V. Gray, R. Kuehr, and P. Stegmann. "*The Global E-waste Monitor–2017, United Nations University (UNU)*." Bonn/Geneva/Vienna: International Telecommunication Union (ITU) & International Solid Waste Association (ISWA), 2017.

Barros, Murillo Vetroni, Rodrigo Salvador, Guilherme Francisco do Prado, Antonio Carlos de Francisco, and Cassiano Moro Piekarski. "Circular economy as a driver to sustainable businesses." *Cleaner Environmental Systems* 2 (2021): 100006.

Bellabarba, Ronan, Peter Johnston, Steven Moss, Carsten Sievers, Bala Subramaniam, Cathy Tway, Ziyuan Wang, and Hongda Zhu. "Net zero transition: possible implications for catalysis." *ACS Catalysis* 13, no. 12 (2023): 7917–7928.

Bhattacharya, Ananya, Alka Nand and Daniel Prajogo. "Taxonomy of antecedents of food waste – a literature review." *Journal of Cleaner Production* 291 (2021): 125910.

Borges, Thyago de Melo Duarte, Gilberto Miller Devos Ganga, Moacir Godinho Filho, Ivete Delai, and Luis Antonio Santa-Eulalia. "Development and validation of a cleaner production measurement scale." *Journal of Cleaner Production* 299 (2021): 126907.

Bovea, María D., Elena Díaz-Albo, Antonio Gallardo, Francisco J. Colomer, and Julio Serrano. "Environmental performance of ceramic tiles: improvement proposals." *Materials & Design* 31, no. 1 (2010): 35–41.

Bowker, Michael, Serena DeBeer, Nicholas F. Dummer, Graham J. Hutchings, Matthias Scheffler, Ferdi Schüth, Stuart H. Taylor, and Harun Tüysüz. "Advancing critical chemical processes for a sustainable future: challenges for industry and the Max Planck–Cardiff centre on the fundamentals of heterogeneous catalysis (FUNCAT)." *Angewandte Chemie* 134, no. 50 (2022): e202209016.

Breedveld, Leo, Giorgio Timellini, Giorgio Casoni, Alberto Fregni, and Graziano Busani. "Eco-efficiency of fabric filters in the Italian ceramic tile industry." *Journal of Cleaner Production* 15, no. 1 (2007): 86–93.

Brunton, Nigel P., James G. Lyng, Wenqu Li, Denis A. Cronin, Desmond Morgan, and Brian McKenna. "Effect of radio frequency (RF) heating on the texture, colour and sensory properties of a comminuted pork meat product." *Food Research International* 38, no. 3 (2005): 337–344.

Cayzer, Steve, Percy Griffiths, and Valentina Beghetto. "Design of indicators for measuring product performance in the circular economy." *International Journal of Sustainable Engineering* 10, no. 4–5 (2017): 289–298.

Cemnet. The Electrified Commercial Cement Kiln. Available online: https://www.cemnet.com/News/story/174030/the-electrified-commercial-cement-kiln.html (accessed on 9 June 2024).

Charles, Amal, Aileen Hofer, Ahmed Elkaseer, and Steffen G. Scholz. "Additive manufacturing in the automotive industry and the potential for driving the green and electric transition." In *Proceedings of the International Conference on Sustainable Design and Manufacturing*, pp. 339–346. Singapore: Springer Singapore, 2021.

Chit, Viesuieda, Lian See Tan, Peck Loo Kiew, Tomoya Tsuji, Toshitaka Funazukuri, and Serene Sow Mun Lock. "Advancing process intensification with high-frequency ultrasound: a mini-review of applications in biofuel production and beyond." *Processes* 11, no. 4 (2023): 1236.

Cui, Yue, Yunmeng Cao, Yijun Ji, I-Shin Chang, and Jing Wu. "Determinant factors and business strategy in a sustainable business model: an explorative analysis for the promotion of solid waste recycling technologies." *Business Strategy and the Environment* 31, no. 5 (2022): 2533–2545.

D'Angelo, Viviana, Francesco Cappa, and Enzo Peruffo. "Green manufacturing for sustainable development: The positive effects of green activities, green investments, and non-green products on economic performance." *Business Strategy and the Environment* 32, no. 4 (2023): 1900–1913.

da Silva, Andre L., Jucilene Feltrin, Marcelo Dal Bó, Adriano M. Bernardin, and Dachamir Hotza. "Effect of reduction of thickness on microstructure and properties of porcelain stoneware tiles." *Ceramics International* 40, no. 9 (2014): 14693–14699.

Dai, Jiapeng, Rafael Alvarado, Sajid Ali, Zahoor Ahmed, and Muhammad Saeed Meo. "Transport infrastructure, economic growth, and transport CO_2 emissions nexus: does green energy consumption in the transport sector matter?." *Environmental Science and Pollution Research* 30, no. 14 (2023): 40094–40106.

Damtoft, Jesper Sand Jacques Lukasik, Duncan Herfort, Danielle Sorrentino, and Ellis Martin Gartner. "Sustainable development and climate change initiatives." *Cement and Concrete Research* 38, no. 2 (2008): 115–127.

Danner, Tobias, and Harald Justnes. "Bauxite residue as supplementary cementitious material–efforts to reduce the amount of soluble sodium." *Nordic Concrete Research* 62, no. 1 (2020): 1–20.

Das, Kaushik Pratim. "A survey on artificial intelligence for reducing the climate footprint in healthcare," *Energy Nexus*, 9 (2023): 100167.

Debnath, Biswajit, Ranjana Chowdhury, and Sadhan Kumar Ghosh. "Urban mining and the metal recovery from E-waste (MREW) supply chain." In *Waste Valorisation and Recycling*, Springer: Berlin/Heidelberg, Germany, 2019; pp. 341–347.

Ding, Kai, Anjie Li, Jingxiang Lv, and Fu Gu. "Decarbonizing ceramic industry: technological routes and cost assessment." *Journal of Cleaner Production* 419 (2023): 138278.

Dongare, Saudagar, Neetu Singh, Haripada Bhunia, and Pramod K. Bajpai. "Electrochemical reduction of CO_2 using oxide based Cu and Zn bimetallic catalyst." *Electrochimica Acta* 392 (2021): 138988.

Drofenik, Jan, Danijela Urbancl, Darko Goričanec, Zdravko Kravanja, and Zorka Novak Pintarič. "Food Waste to Energy through Innovative Coupling of CHP and Heat Pump." *Energies* 16, no. 8 (2023): 3344.

Elmahgary, Maryam G., Shereen K. Amin, Mohamed A. Sadek, and Magdi F. Abadir. "Waste Minimisation of Ceramic Wall Tiles." *Chemical Engineering Transactions* 70 (2018): 421–426.

Eryazici, Ibrahim, Narayan Ramesh, and Carlos Villa. "Electrification of the chemical industry—materials innovations for a lower carbon future." *MRS Bulletin* 46, no. 12 (2021): 1197–1204.

European Commission. *Reference Document on Best Available Techniques in the Ceramic Manufacturing Industry.* Brussels. 2007.

Ewing, Tom A., Niels Nouse, Matthijs van Lint, Jacco van Haveren, Jeroen Hugenholtz, and Daan S. van Es. "Fermentation for the production of biobased chemicals in a circular economy: a perspective for the period 2022–2050." *Green Chemistry* 24, no. 17 (2022): 6373–6405.

Fakayode, Olugbenga Abiola, Hafida Wahia, Lei Zhang, Cunshan Zhou, and Haile Ma. "State-of-the-art co-pyrolysis of lignocellulosic and macroalgae biomass feedstocks for improved bio-oil production – a review." *Fuel* 332 (2023): 126071.

Ferrario, Daniele, Stefano Stendardo, Vittorio Verda, and Andrea Lanzini. "Solar-driven calcium looping system for carbon capture in cement plants: Process modelling and energy analysis." *Journal of Cleaner Production* 394 (2023): 136367.

Fiorentino, Gabriella, Amalia Zucaro, and Sergio Ulgiati. "Towards an energy efficient chemistry. Switching from fossil to bio-based products in a life cycle perspective." *Energy* 170 (2019): 720–729.

Fu, Zhiyan, Qi Yang, Zhan Liu, Fei Chen, Fubing Yao, Ting Xie, Yu Zhong et al. "Photocatalytic conversion of carbon dioxide: From products to design the catalysts." *Journal of CO_2 Utilization* 34 (2019): 63–73.

Galanakis, Charis. "Food waste valorization opportunities for different food industries." In *The Interaction of Food Industry and Environment*, pp. 341–422. Academic Press, 2020.

Garcia-Garcia, Guillermo, Jamie Stone, and Shahin Rahimifard. "Opportunities for waste valorisation in the food industry–a case study with four UK food manufacturers." *Journal of Cleaner Production* 211 (2018): 1339–1356.

Gencel, Osman, Cengiz Ozel, Fuat Koksal, Ertugrul Erdogmus, Gonzalo Martínez-Barrera, and Witold Brostow. "Properties of concrete paving blocks made with waste marble." *Journal of Cleaner Production* 21, no. 1 (2012): 62–70.

Govindan, Kannan, Kiran Garg, Seema Gupta, and P. C. Jha. "Effect of product recovery and sustainability enhancing indicators on the location selection of manufacturing facility." *Ecological Indicators* 67 (2016): 517–532.

Gul, Banat, Shamim Khan, and Iftikhar Ahmad. "Extraction of phytochemicals from date palm (Phoenix dactylifera L.) seeds by enzymatic hydrolysis." *Journal of Food Processing and Preservation* 46, no. 11 (2022): e17007.

Gurgul, Agnieszka, Włodzimierz Szczepaniak, and Monika Zabłocka-Malicka. "Incineration and pyrolysis vs. steam gasification of electronic waste." *Science of the Total Environment* 624 (2018): 1119–1124.

Habert, Guillaume, Catherine Billard, Pierre Rossi, Cong Chen, and Nicolas Roussel. "Cement production technology improvement compared to factor 4 objectives." *Cement and Concrete Research* 40, no. 5 (2010): 820–826.

Haleem, Abid, Mohd Javaid, Ravi Pratap Singh, Rajiv Suman, and Mohd Asim Qadri. "A pervasive study on Green Manufacturing towards attaining sustainability." *Green Technologies and Sustainability* 1, no. 2 (2023): 100018.

Handayani, Dwi Iryaning, Ilyas Masudin, Ahmad Rusdiansyah, and Judi Suharsono. "Production-distribution model considering traceability and carbon emission: a case study of the Indonesian canned fish food industry." *Logistics* 5, no. 3 (2021): 59.

Harder, J., and Onestone-consulting. "Developments in the global cement market." *Cement International* 12, no. 1 (2014): 34–41

Harrington, Donna Ramirez, George Deltas, and Madhu Khanna. "Does pollution prevention reduce toxic releases? A dynamic panel data model." *Land Economics* 90, no. 2 (2014): 199–221.

He, Chasheng, Chengwei Zhang, Tengfei Bian, Kaixuan Jiao, Weike Su, Ke-Jun Wu, and An Su. "A review on artificial intelligence enabled design, synthesis, and process optimization of chemical products for industry 4.0." *Processes* 11, no. 2 (2023): 330.

He, Pin Jing. "Anaerobic digestion: an intriguing long history in China." *Waste Management* 30, no. 4 (2010): 549–550.

Hirsch, Jürgen. "Aluminium in innovative light-weight car design." *Materials Transactions* 52, no. 5 (2011): 818–824.

Hojnik, Jana, Mitja Ruzzier, Maja Konečnik Ruzzier, Boris Sučić, and Brandon Soltwisch. "Challenges of demographic changes and digitalization on eco-innovation and the circular economy: qualitative insights from companies." *Journal of Cleaner Production* 396 (2023): 136439.

Hou, J. G., H. C. Yao, X. L. Wang, P. F. Song, Y. Q. Zhang, Y. Y. Sui and X. Y. Mu. "Construction of "natural gas+hydrogen" dual clean and low-carbon energy system and choice of technical path." *Low-Carbon Chemistry and Chemical Engineering* 47, no. 06 (2022): 1–5.

Huo, Jing, Zhanyun Wang, Christopher Oberschelp, Gonzalo Guillén-Gosálbez, and Stefanie Hellweg. "Net-zero transition of the global chemical industry with CO_2-feedstock by 2050: feasible yet challenging." *Green Chemistry* 25, no. 1 (2023): 415–430.

Ibáñez-Forés, Valeria, Maria-Dolores Bovea, and Amelia Simó. "Life cycle assessment of ceramic tiles. Environmental and statistical analysis." *The International Journal of Life Cycle Assessment* 16 (2011): 916–928.

Ihsanullah, I., Gulzar Alam, Arshad Jamal, and Feroz Shaik. "Recent advances in applications of artificial intelligence in solid waste management: a review." *Chemosphere* 309 (2022): 136631.

Immonen, Kirsi, Johanna Lyytikäinen, Janne Keränen, Kim Eiroma, Mika Suhonen, Minna Vikman, Ville Leminen, Marja Välimäki, and Liisa Hakola. "Potential of commercial wood-based materials as PCB substrate." *Materials* 15, no. 7 (2022): 2679.

International Energy Agency, IEA. Chemicals—Fuels & Technologies. Available online: https://www.iea.org/fuels-and-technologies/chemicals (accessed on 29 May 2024).

Jahanger, Atif, Ilhan Ozturk, Joshua Chukwuma Onwe, Tonuchi E. Joseph, and Mohammad Razib Hossain. "Do technology and renewable energy contribute to energy efficiency and carbon neutrality? Evidence from top ten manufacturing countries." *Sustainable Energy Technologies and Assessments* 56 (2023): 103084.

Jansson, Elina, Johanna Lyytikäinen, Panu Tanninen, Kim Eiroma, Ville Leminen, Kirsi Immonen, and Liisa Hakola. "Suitability of paper-based substrates for printed electronics." *Materials* 15, no. 3 (2022): 957.

Jayal, Anshu Dhar, Badurdeen, Fazleena, Dillon Oscar Wlliam, and Ibrahim S. Jawahir. "Sustainable manufacturing: modeling and optimization challenges at the product, process and system levels." *CIRP Journal of Manufacturing Science and Technology* 2, no. 3 (2010): 144–152.

Jull, James. "The second industrial revolution. The history of a concept." *Rivista internazionale di storia della storiografia* 36 (1999): 81–90.

Kannah, R. Yukesh, J. Merrylin, T. Poornima Devi, S. Kavitha, P. Sivashanmugam, Gopalakrishnan Kumar, and J. Rajesh Banu. "Food waste valorization: biofuels and value added product recovery." *Bioresource Technology Reports* 11 (2020): 100524.

Kelemen, Peter, Sally M. Benson, Hélène Pilorgé, Peter Psarras, and Jennifer Wilcox. "An overview of the status and challenges of CO_2 storage in minerals and geological formations." *Frontiers in Climate* 1 (2019): 482595.

Kleindorfer, Paul R., Kalyan Singhal, and Luk N. Van Wassenhove. "Sustainable operations management." *Production and Operations Management* 14, no. 4 (2005): 482–492.

Kumar, Anuj, Asif Ali Syed, and Anoop Pandey. "Adoption of online resources to improve the marketing performance of SMES." *Asia Pacific Journal of Health Management* 16, no. 3 (2021): 137–144.

Kuppusamy, Murugan, Thirumalai Ramanathan, Udhayakumar Krishnavel, and Seenivasan Murugesan. "Experimental analysis of the combustion chamber coated with MO and Al_2O_3-TiO_2 in a diesel engine." *Materials and Technology* 55, no. 4 (2021): 509–515.

Kushwaha, Gyaneshwar Singh, and Nagendra Kumar Sharma. "Green initiatives: a step towards sustainable development and firm's performance in the automobile industry." *Journal of Cleaner Production* 121 (2016): 116–129.

Kuzel, Hans Jürgen, and Herbert Pöllmann. "Hydration of C3A in the presence of $Ca(OH)_2$, $CaSO_4 \cdot 2H_2O$ and $CaCO_3$." *Cement and Concrete Research* 21, no. 5 (1991): 885–895.

Lai, Kee-Hung, Venus Y.H. Lun, Christina W.Y. Wong, and Tai Chiu Edwin Cheng. "Green shipping practices in the shipping industry: conceptualization, adoption, and implications." *Resources, Conservation and Recycling* 55, no. 6 (2011): 631–638.

Landi, Daniele, Michele Germani, and Marco Marconi. "Analyzing the environmental sustainability of glass bottles reuse in an Italian wine consortium." *Procedia CIRP* 80 (2019): 399–404.

Lee, Jechan, Wei-Hsin Chen, and Young-Kwon Park. "Recent achievements in platform chemical production from food waste." *Bioresource Technology* 366 (2022): 128204.

Li, Z. F., and K. Sumathy. "Technology development in the solar absorption air-conditioning systems." *Renewable and Sustainable Energy Reviews* 4, no. 3 (2000): 267–293.

Lian, Jijian, Yusheng Zhang, Chao Ma, Yang Yang, and Evance Chaima. "A review on recent sizing methodologies of hybrid renewable energy systems." *Energy Conversion and Management* 199 (2019): 112027.

Liang, Chao, and Qi Wang. "The relationship between total factor productivity and environmental quality: a sustainable future with innovation input." *Technological Forecasting and Social Change* 191 (2023): 122521.

Lin, Ru-Jen, Rong-Huei Chen, and Fei-Hsin Huang. "Green innovation in the automobile industry." *Industrial Management & Data Systems* 114, no. 6 (2014): 886–903.

Lindqvist, Karl, Simon Roussanaly, and Rahul Anantharaman. "Multi-stage membrane processes for CO_2 capture from cement industry." *Energy Procedia* 63 (2014): 6476–6483.

Lucier, C.A. and B.J. Gareau. Electronic waste recycling and disposal: an overview. In *Assessment and Management of Radioactive and Electronic Wastes*, edited by Hosam El-Din Saleh, pp. 1–12. Norderstedt, Germany: Books on Demand, 2019.

Mahapatra, Manisha, and Arun Kumar Pradhan. "Bioethanol production from agricultural wastes with the aid of nanotechnology." In *Bio-Nano Interface: Applications in Food, Healthcare and Sustainability,* edited by M. Arakha, A.K. Pradhan, and S. Jha, pp. 329–337. Singapore: Springer Singapore, 2021.

Marzi, Thomas, Volker Knappertsbusch, Anna Greve, Görge Deerberg, Christian Doetsch, and Eckhard Weidner. "Resources of a new carbon economy." *Chemie Ingenieur Technik* 90, no. 10 (2018): 1374–1383.

Migo-Sumagang, Maria Victoria, Kathleen B. Aviso, Dominic CY Foo, Michael Short, Purusothmn Nair S. Bhasker Nair, and Raymond R. Tan. "Optimization and decision support models for deploying negative emissions technologies." *PLOS Sustainability and Transformation* 2, no. 5 (2023): e0000059.

Mihelcic, James R., John C. Crittenden, Mitchell J. Small, David R. Shonnard, David R. Hokanson, Qiong Zhang, Hui Chen et al. "Sustainability science and engineering: the emergence of a new metadiscipline." *Environmental Science & Technology* 37, no. 23 (2003): 5314–5324.

Miragliotta, Giovanni, and Fadi Shrouf. "Using Internet of Things to improve eco-efficiency in manufacturing: a review on available knowledge and a framework for IoT adoption." In *Advances in Production Management Systems. Competitive Manufacturing for Innovative Products and Services:*

IFIP WG 5.7 International Conference, APMS 2012, Rhodes, Greece, September 24–26, 2012, Revised Selected Papers, Part I, pp. 96–102. Springer: Berlin Heidelberg, 2013.

Mishra, Kirti, Samarjeet Singh Siwal, Adesh Kumar Saini, and Vijay Kumar Thakur. "Recent update on gasification and pyrolysis processes of lignocellulosic and algal biomass for hydrogen production." *Fuel* 332 (2023): 126169.

Model T. History.com. 2 May 2019. Retrieved 8 November 2020.

Mohajan, Haradhan. "Third industrial revolution brings global development." *Journal of Social Sciences and Humanities*, 7, No. 4, (2021): 239–251.

Monfort, E., A. Mezquita, E. Vaquer, I. Celades, V. Sanfelix, and A. Escrig. "8.05-ceramic manufacturing processes: energy, environmental, and occupational health issues." *Comprehensive Materials Processing*, pp. 71–102. Oxford: Elsevier, 2014.

Morseletto, Piero. "Targets for a circular economy." *Resources, Conservation and Recycling* 153 (2020): 104553.

Naqi, Ali, and Jeong Gook Jang. "Recent progress in green cement technology utilizing low-carbon emission fuels and raw materials: a review." *Sustainability* 11, no. 2 (2019): 537.93.

Narayanasamy, Marappa, Dharumadurai Dhanasekaran, and Nooruddin Thajuddin. "Bioremediation of noxious metals from e-waste printed circuit boards by Frankia." *Microbiological Research* 245 (2021): 126707.

Needhidasan, Santhanam, Melvin Samuel, and Ramalingam Chidambaram. "Electronic waste–an emerging threat to the environment of urban India." *Journal of Environmental Health Science and Engineering* 12 (2014): 1–9.

Nhuchhen, Daya R., Song P. Sit, and David B. Layzell. "Decarbonization of cement production in a hydrogen economy." *Applied Energy* 317 (2022): 119180.

Njomo, Louis Mosake. "The ban on plastic shopping bags in Cameroon an exploratory study of resilience strategies of subsistence marketplaces." *International Journal of Innovative Science and Research Technology* 4, no. 4 (2019): 475–493.

Obebe, S. B., and A. A. Adamu. "Plastic pollution: causes, effects and preventions." *International Journal of Engineering Applied Sciences and Technology* 4, no. 12 (2020): 85–95.

Olabi, A. G., Tabbi Wilberforce, Khaled Elsaid, Enas Taha Sayed, Hussein M. Maghrabie, and Mohammad Ali Abdelkareem. "Large scale application of carbon capture to process industries – a review." *Journal of Cleaner Production* 362 (2022): 132300.

Pang, Rui, and Xiaoling Zhang. "Achieving environmental sustainability in manufacture: a 28-year bibliometric cartography of green manufacturing research." *Journal of Cleaner Production* 233 (2019): 84–99.

Pearce, Antony, and Dirk Pons. "Advancing lean management: the missing quantitative approach." *Operations Research Perspectives* 6 (2019): 100114.

Pedraza, J., A. Zimmermann, J. Tobon, R. Schomäcker, and N. Rojas. "On the road to net zero-emission cement: integrated assessment of mineral carbonation of cement kiln dust." *Chemical Engineering Journal* 408 (2021): 127346.

Pongrácz, Eva, Paul S. Phillips, and Riitta L. Keiski. "From waste minimization to resources use optimization: Definitions and legislative background." In *Proceedings of the Waste Minimization and Resources Use Optimization Conference*, June, vol. 10. 2004. Oulu, Finland: Oulu University Press.

Ra, Eun Cheol, Kwang Young Kim, Eun Hyup Kim, Hojeong Lee, Kwangjin An, and Jae Sung Lee. "Recycling carbon dioxide through catalytic hydrogenation: recent key developments and perspectives." *ACS Catalysis* 10, no. 19 (2020): 11318–11345.

Ragaert, Kim, Laurens Delva, and Kevin Van Geem. "Mechanical and chemical recycling of solid plastic waste." *Waste Management* 69 (2017): 24–58.

Rahimi, Zohra, Amrit Anand and Shalini Gautam. "An overview on thermochemical conversion and potential evaluation of biofuels derived from agricultural wastes." *Energy Nexus* 7 (2022): 100125.

Ramanathan, Usha, Yongmei Bentley, and Gu Pang. "The role of collaboration in the UK green supply chains: an exploratory study of the perspectives of suppliers, logistics and retailers." *Journal of Cleaner Production* 70 (2014): 231–241.

Rashad, Alaa M., Yun Bai, P. A. Muhammed Basheer, Neil B. Milestone, and Nicholas C. Collier. "Hydration and properties of sodium sulfate activated slag." *Cement and Concrete Composites* 37 (2013): 20–29.

Rashad, Alaa M. "An exploratory study on high-volume fly ash concrete incorporating silica fume subjected to thermal loads." *Journal of Cleaner Production* 87 (2015): 735–744.

Rauch, Reinhard, Yohannes Kiros, Klas Engvall, Efthymios Kantarelis, Paulo Brito, Catarina Nobre, Santa Margarida Santos, and Philipp A. Graefe. 2024. "Hydrogen from waste gasification" *Hydrogen* 5, no. 1: 70–101.

Reddy, K. Chiranjeevi, and Kolluru VL Subramaniam. "Blast furnace slag hydration in an alkaline medium: influence of sodium content and sodium hydroxide molarity." *Journal of Materials in Civil Engineering* 32, no. 12 (2020): 04020371.

Rosen, William. *The Most Powerful Idea in the World: A Story of Steam, Industry, and Invention*. Random House, 2010.

Sarjaš, A. Ecodesign of Electronic Devices; Technical Report UNIT 7: 2018; Ecosign: Whistler, BC, Canada, 2018.

Scrivener, Karen, Fernando Martirena, Shashank Bishnoi, and Soumen Maity. "Calcined clay limestone cements (LC3)." *Cement and Concrete Research* 114 (2018): 49–56.

Setiyawan, Danang Triagus, Sudjito Soeparman, and Rudy Soenoko. "Minimasi waste untuk perbaikan proses produksi kantong kemasan dengan pendekatan Lean Manufacturing." *JEMIS (Journal of Engineering & Management in Industrial System)* 1, no. 1 (2013): 8–13.

Shafiq, Hamza, Shakir Ul Azam, and Arshad Hussain. "Steam gasification of municipal solid waste for hydrogen production using Aspen Plus® simulation." *Discover Chemical Engineering* 1 (2021): 1–16.

Sharma, Manu, Sudhanshu Joshi, Mukesh Prasad, and Shalini Bartwal. "Overcoming barriers to circular economy implementation in the oil & gas industry: environmental and social implications." *Journal of Cleaner Production* 391 (2023): 136133.

Silva, Carla, Patricia Moniz, Ana Cristina Oliveira, Samuela Vercelli, Alberto Reis, and Teresa Lopes da Silva. "Cascading crypthecodinium cohnii biorefinery: global warming potential and techno-economic assessment." *Energies* 15, no. 10 (2022): 3784.

Soori, Mohsen, Behrooz Arezoo, and Roza Dastres. "Internet of things for smart factories in industry 4.0, a review." *Internet of Things and Cyber-Physical Systems* 3 (2023): 192–204.

Steenis, Nigel D., Erica Van Herpen, Ivo A. Van Der Lans, Tom N. Ligthart, and Hans CM Van Trijp. "Consumer response to packaging design: the role of packaging materials and graphics in sustainability perceptions and product evaluations." *Journal of Cleaner Production* 162 (2017): 286–298.

Stefanini, Roberta, Giulia Borghesi, Anna Ronzano, and Giuseppe Vignali. "Plastic or glass: a new environmental assessment with a marine litter indicator for the comparison of pasteurized milk bottles." *The International Journal of Life Cycle Assessment* 26 (2021): 767–784.

Sullivan, Ian, Andrey Goryachev, Ibadillah A. Digdaya, Xueqian Li, Harry A. Atwater, David A. Vermaas, and Chengxiang Xiang. "Coupling electrochemical CO_2 conversion with CO_2 capture." *Nature Catalysis* 4, no. 11 (2021): 952–958.

Tanzer, Samantha Eleanor, Kornelis Blok, and Andrea Ramírez Ramírez. "Scoping cost and abatement metrics for biomass with carbon capture and storage—the example of bioCCS in cement." *International Journal of Greenhouse Gas Control* 125 (2023): 103864.

Terry, Shane, Hao Lu, Ismail Fidan, Yunbo Zhang, Khalid Tantawi, Terry Guo, and Bahram Asiabanpour. "The influence of smart manufacturing towards energy conservation: a review." *Technologies* 8, no. 2 (2020): 31.

Tian, Guangdong, Weidong Lu, Xuesong Zhang, Meng Zhan, Maxim A. Dulebenets, Anatoly Aleksandrov, Amir M. Fathollahi-Fard, and Mikhail Ivanov. "A survey of multi-criteria decision-making techniques for green logistics and low-carbon transportation systems." *Environmental Science and Pollution Research* 30, no. 20 (2023): 57279–57301.

Trivedi, Amber, Anusha Vishwakarma, Bhavini Saawarn, Byomkesh Mahanty, and Subrata Hait. "Fungal biotechnology for urban mining of metals from waste printed circuit boards: a review." *Journal of Environmental Management* 323 (2022): 116133.

Türkmen, Burçin Atılgan, Şeyma Karahan Özbilen, and Tuba Budak Duhbacı. "Improving the sustainability of ceramic tile production in Turkey." *Sustainable Production and Consumption* 27 (2021): 2193–2207.

Uddin, Md Nizam, Kunza Arifa, and Eylem Asmatulu. "Methodologies of e-waste recycling and its major impacts on human health and the environment." *International Journal of Environment and Waste Management* 27, no. 2 (2021): 159–182.

Valladares Montemayor, H.. Quest Lecture: The Circular Economy [Conference Pre-sentation] January 12. Whistler Public Library Lecture Series, Whistler, BC, Canada, 2022.

Vattenfall and Cementa. Take the Next Step towards a Climate Neutral Cement. Available online: https://group.vattenfall.com/press-and-media/pressreleases/2019/vattenfall-and-cementa-take-the-next-step-towards-a-climate-neutral-cement (accessed on 9 June 2023).

Vladić Kancir, Ivana, and Marijana Serdar. "Contribution to understanding of synergy between red mud and common supplementary cementitious materials." *Materials* 15, no. 5 (2022): 1968.

Vooradi, Ramsagar, Maria-Ona Bertran, Rebecca Frauzem, Sarath Babu Anne, and Rafiqul Gani. "Sustainable chemical processing and energy-carbon dioxide management: review of challenges and opportunities." *Chemical Engineering Research and Design* 131 (2018): 440–464.

Vormedal, Irja, Julie Bjander, Mari Lie Larsen, and Marie Byskov Lindberg. "Technological change and the politics of decarbonization: a re-making of vested interests?" *Environmental Innovation and Societal Transitions* 47 (2023): 100725.

Wang, Wei, Baoqiang Yuan, Qie Sun, and Ronald Wennersten. "Application of energy storage in integrated energy systems—a solution to fluctuation and uncertainty of renewable energy." *Journal of Energy Storage* 52 (2022): 104812.

Wang, Xiaojun, and Hing Kai Chan. "An integrated fuzzy approach for evaluating remanufacturing alternatives of a product design." *Journal of Remanufacturing* 3 (2013): 1–19.

Warren, Matthew P., Paul L. Forrester, John S. Hassard, and John W. Cotton. "Technological innovation antecedents in the UK ceramics industry." *International Journal of Production Economics* 65, no. 1 (2000): 85–98.

Wimbadi, Ramanditya Wimbardana, and Riyanti Djalante. "From decarbonization to low carbon development and transition: a systematic literature review of the conceptualization of moving toward net-zero carbon dioxide emission (1995–2019)." *Journal of Cleaner Production* 256 (2020): 120307.

Womack, James P., and Daniel T. Jones. "Lean thinking—banish waste and create wealth in your corporation." *Journal of the Operational Research Society* 48, no. 11 (1996): 1148–1148.

World Business Council for Sustainable Development. Cement Technology Roadmap 2009, Carbon Emissions Reductions up to 2050.

Wu, Simon, and Hui Ming Wee. "Why the Big Three Decline Despite Their Lean Management-A Study Based on the Theory of Constraints." In *Global Perspective for Competitive Enterprise, Economy and Ecology: Proceedings of the 16th ISPE International Conference on Concurrent Engineering*, pp. 691–699. Springer London, 2009.

Xavier, Lúcia Helena, Ellen Cristine Giese, Ana Cristina Ribeiro-Duthie, and Fernando Antonio Freitas Lins. "Sustainability and the circular economy: A theoretical approach focused on e-waste urban mining." *Resources Policy* 74 (2021): 101467.

Xi, Fengming, Steven J. Davis, Philippe Ciais, Douglas Crawford-Brown, Dabo Guan, Claus Pade, Tiemao Shi et al. "Substantial global carbon uptake by cement carbonation." *Nature Geoscience* 9, no. 12 (2016): 880–883.

Xie, Xuemei, Jiage Huo, and Hailiang Zou. "Green process innovation, green product innovation, and corporate financial performance: a content analysis method." *Journal of Business Research* 101 (2019): 697–706.

Yang, Keun-Hyeok, Yeon-Back Jung, Myung-Sug Cho, and Sung-Ho Tae. "Effect of supplementary cementitious materials on reduction of CO_2 emissions from concrete." *Journal of Cleaner Production* 103 (2015): 774–783.

Yin, Shi, Nan Zhang, Baizhou Li, and Hengmin Dong. "Enhancing the effectiveness of multi-agent cooperation for green manufacturing: dynamic co-evolution mechanism of a green technology innovation system based on the innovation value chain." *Environmental Impact Assessment Review* 86 (2021): 106475.

Yu, Xiang, Carmen Otilia Catanescu, Robert E. Bird, Sriram Satagopan, Zachary J. Baum, Leilani M. Lotti Diaz, and Qiongqiong Angela Zhou. "Trends in research and development for CO_2 capture and sequestration." *ACS Omega* 8, no. 13 (2023): 11643–11664.

Zailani, Suhaiza, Kannan Govindan, Mohammad Iranmanesh, Mohd Rizaimy Shaharudin, and Yia Sia Chong. "Green innovation adoption in automotive supply chain: the Malaysian case." *Journal of Cleaner Production* 108 (2015): 1115–1122.

Zhao, Hongxia, Guangming Xu, Lu Liu, Changchun Shi, and Huijuan Zhao. "Low-carbon technology innovation decision making of manufacturing companies in the industrial internet platform ecosystem." *Sustainability* 15, no. 4 (2023): 3555.

Zhou, Yuekuan, and Peter D. Lund. "Peer-to-peer energy sharing and trading of renewable energy in smart communities – trading pricing models, decision-making and agent-based collaboration." *Renewable Energy* 207 (2023): 177–193.

Chapter 2

Green technologies for modern manufacturing

Sasi Kumar Mani, Kumaresan Mayakannan, Sindhu Selvaraj, Krishna Varun Balasubramanian, and Sinchana Shri Vinu

2.1 INTRODUCTION

Green technology, or sustainable technology, is the creation and use of products, equipment, and systems that protect the natural environment and resources, minimizing the harmful effects of human activities. This field comprises a wide range of innovative and practical approaches aimed at promoting sustainability and environmental friendliness. The fundamental tenet of green technology is to fulfill the requirements of the current generation while safeguarding the capacity of future generations to fulfill their own requirements.

The increasing global awareness of environmental issues has led to a substantial surge in the demand for green technologies. The ramifications of industrialization, including pollution, loss of resources, and climate change, have emphasized the pressing necessity for sustainable practices in all industries, especially manufacturing.

These technologies utilize renewable energy sources like solar and wind power, resulting in a substantial decrease in dependence on fossil fuels and a reduction in greenhouse gas emissions. The implementation of energy-efficient technology and procedures aims to reduce energy consumption, while the use of sustainable materials, such as recyclable and biodegradable choices, helps minimize waste. In addition, sophisticated waste management techniques, such as recycling and composting, strive to attain the goal of producing no waste. Water conservation techniques, such as closed-loop systems and rainwater collection, also improve resource utilization efficiency. Through the implementation of environmentally friendly technology, manufacturers can save the environment while simultaneously reaping the advantages of reduced expenses, adherence to regulations, and a favorable market reputation. This, in turn, promotes innovation and ensures the industry's long-term viability. Modern manufacturing places great emphasis on green technologies, which prioritize sustainability and environmental stewardship, leading to a transformation of old production practices. These technologies utilize renewable energy sources such as solar

DOI: 10.1201/9781003470342-2

and wind power, resulting in a substantial decrease in dependence on fossil fuels and a reduction in greenhouse gas emissions. Energy-efficient technology and optimized production processes reduce energy usage, while sustainable materials, including recyclable and biodegradable options, reduce waste. Advanced waste management techniques, such as recycling and composting, strive to attain the goal of producing no waste. By incorporating environmentally friendly technology, manufacturers can diminish their ecological footprint, attain financial advantages, adhere to regulations, and improve their market reputation while promoting innovation and ensuring long-term viability.

Climate change has emerged as one of the most substantial challenges to humanity. Developing countries contribute a larger percentage of carbon dioxide emissions due to the increasing need for industrial globalization in emerging nations. The dilemma arising from the conflict between the need to reduce carbon emissions and the desire for rapid industrial development causes enterprises in developing nations to be hesitant in making deliberate commitments and taking action to reduce carbon dioxide emissions from their global manufacturing processes. The research on sustainability, green, and low-carbon supply chain networks is still in its early stages from a theoretical standpoint and necessitates additional research and development [1].

The fuzzy Delphi method is utilized to optimize energy utilization, promote sustainable transportation, enhance quality of life, and ensure ecological safety. The fuzzy analytical hierarchy process (FAHP) is utilized to ascertain the relative significance of primary and secondary indicators. The data suggest that energy utilization and agriculture and forestry play significant roles as indicators [2].

Sustainability is the concept of achieving development that meets the needs of the present generation without jeopardizing the availability of resources for future generations. The globe is currently grappling with sustainability challenges that encompass multiple dimensions, including environmental, social, and economic factors. For example, there are over one billion individuals worldwide who are experiencing extreme poverty [3–5].

Industry 5.0 is a manufacturing system that prioritizes human well-being, sustainability, and resilience, with a strong focus on incorporating environmentally friendly technologies. The writers have extensive experience in the development and digitalization of the local manufacturing industry in Croatia. The current focus of their investigation is the utilization and perception of sustainable and technological advances in the interior and exterior transport networks of Croatian manufacturing businesses [6].

Implementing technologically driven sustainable manufacturing is an essential approach for achieving high-quality economic growth and improving competitiveness globally. This approach is particularly important in the context of China's economic transformation and development, which is closely linked to the ongoing technological revolution and industrial reform. The industrial sector is classified into three groups according to the level of pollution: the clean sector, the medium pollution sector, and the highly polluting sector. Research indicates that the clean manufacturing industry has the highest level of green development efficiency, followed by the heavy pollution sector. The medium pollution industry, on the other hand, has the lowest efficiency [7].

Green technology innovation is crucial for achieving sustainable development (SD), especially considering the significant impact of the Chinese environment on manufacturing firms. By leveraging green technology and fostering innovation, China aims to address its environmental crisis and subsequently make significant advancements in the global economy. The field of green technology innovation capability has effectively developed the capacity for green product innovation, which is then followed by the capacity for green process innovation. The capability for governing terminal technology is the least developed [8].

2.2 NEED FOR GREEN TECHNOLOGIES

Conventional manufacturing methods frequently require a large number of resources and harm the environment, resulting in substantial ecological footprints. Green technologies tackle these problems by implementing cutting-edge solutions that minimize waste, preserve energy, and employ sustainable materials. The transition is motivated by escalating regulatory demands, customer demand for environmentally friendly products, and the necessity to alleviate climate change's negative impacts.

The necessity for green technology in contemporary manufacturing arises from the imperative need to alleviate the detrimental environmental effects of conventional industrial processes. Due to industrialization, there has been a substantial increase in pollution, loss of resources, and climate change, creating an urgent need for sustainable solutions. Green technology tackles these difficulties by decreasing energy usage, minimizing waste generation, and preserving natural resources through inventive methods and sustainable energy sources. Additionally, it aids producers in adhering to progressively more rigorous environmental requirements and fulfilling the escalating consumer desire for environmentally responsible items. Adopting green technology not only safeguards the environment but also improves operational efficiency, cuts expenses, and fosters long-term sustainability, making it a crucial element of responsible and forward-looking manufacturing.

2.3 CLASSIFICATION OF GREEN TECHNOLOGIES

Green technologies in contemporary manufacturing can be categorized into distinct domains, each emphasizing various facets of sustainability and the mitigation of environmental damage. The classification of green technology is explained in Figure 2.1. The following are the primary categorizations:

- **Energy Utilization:** The efficient and effective use of energy resources.
- **Ecological Safety:** Ensuring the protection and preservation of the environment.
- **Environmental Quality:** Maintaining and improving the overall condition and health of the environment.
- **Resource Utilization:** The efficient and sustainable use of natural resources.
- **Life Health:** Promoting and maintaining the well-being and health of living organisms.

2.3.1 Energy utilization

The utilization of sustainable hydrogen and power-to-X technologies holds significant potential to aid the global transition toward reaching net-zero emissions in the energy sector. This can be due to the underlying assumption

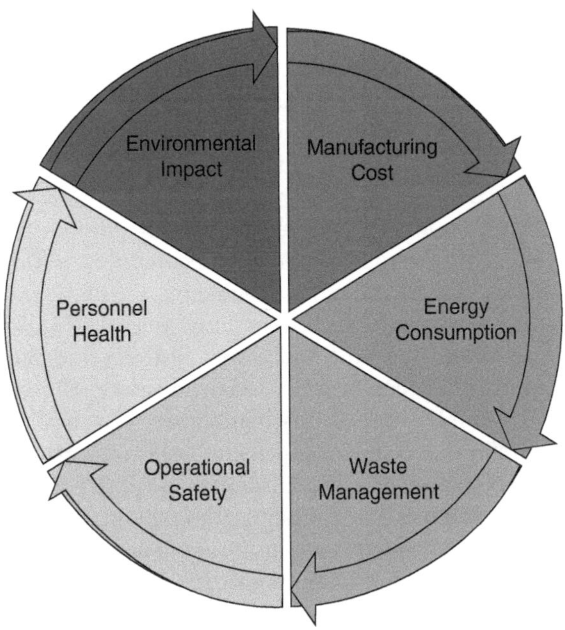

Figure 2.1 Classification of green technology.

that these technologies have the capability to reduce carbon emissions in various sectors globally. They achieve this by offering adaptable and sustainable energy sources and raw materials for industries, which can replace fossil fuels and chemicals. In previous years, an extensive comparison has been conducted on the intrinsic advantages of sustainable hydrogen and power-to-X technologies across various applications. This study investigates the potential incorporation of sustainable hydrogen and power-to-X technologies in Jordanian businesses. They have conducted a detailed analysis of the quantitative usage potential of these technologies for both current and future capacities. The findings suggest that the food processing and heavy industries in Jordan have become important sectors with tremendous potential to incorporate sustainable hydrogen and power-to-X products as alternative fuels or chemical feedstocks. This would involve substituting fossil-based fuels and chemicals. Considering this, the total potential capacity for use in these areas is around 57 thousand tons per year. The annual capacity of the cement industry, oil refinery, steel reformation, and fertilizer production is around 6.8, 11.8, 12.7, and 25.8 thousand tons, respectively. Currently, it is worth mentioning that all hydrogen utilized by Jordanian industries is obtained from fossil fuels, with an annual capacity of approximately 8.9 thousand tons [9].

Recent advancements in green and creative technology have significantly accelerated SD by introducing numerous modifications in industrial processes. The adoption rate of green innovation in manufacturing organizations remains very low, despite its multiple benefits. The Unified Theory of Acceptance and Use of Technology (UTAUT) mandates organizations to embrace cutting-edge technologies, forming the basis for the development and validation of the green innovation adoption (GIA) model. A total of 516 participants from Pakistan's industrial sector took part in a survey, and the data collected was then analyzed using artificial neural networks (ANNs) and structural equation modeling (SEM). The research model's various integrated components, which are the predictors of green behavioral intention (GBI), including performance expectancy, effort expectancy, hedonic motivation, social influence, conducive settings, and innovation cost, were identified through the use of SEM and ANN methodologies. Moreover, it was found that GBI had a strong direct and indirect impact on GIA inside integrated constructions. Furthermore, the moderation of organizational size emphasized the distinction between small, medium, and giant firms. The relevance value of green enabling circumstances is highest for GIA, while ANN assesses the durability and order of importance of all integrated components. As numerous countries strive to achieve their Sustainable Development Goals within the next decade, the conversation surrounding sustainable innovation is gaining momentum. The research has yielded distinct findings that can be considered significant contributions to the existing body of knowledge. This chapter introduces a comprehensive methodology for improving system development in contemporary business environments by applying the UTAUT model [10,11].

Here, they have explained in detail the contribution of lean manufacturing and green manufacturing to the world [12]. The industrial sector has taken proactive measures to address environmental issues by designing recyclable items and developing cleaner manufacturing practices [13]. It is committed to educating the next generation of environmentally conscious engineers to drive sustainable growth in the green manufacturing industry. Green engineering is defined as a prominent sector that aims to achieve SD in manufacturing by reducing consumption and saving resources. Green manufacturing is responsible for introducing new standards in the education of aspiring green engineers. The study delineates the core principles of instructing aspiring environmental engineers. Special attention is paid to improving the instructional system for aspiring green engineers by strengthening the teaching team, modifying the teaching methods, improving teaching practices, and offering practical training to achieve learning goals. The authors assert that it is crucial to modify the course material and clearly define the objectives for future green engineer training [14].

2.3.2 Ecological safety

Ecological safety in the realm of green technology in contemporary manufacturing pertains to the use of methods and technologies that reduce detrimental environmental effects, safeguard the well-being of ecosystems, and encourage the sustainable utilization of resources. Integrating ecological safety into green manufacturing not only safeguards the environment but also promotes innovation, effectiveness, and sustainability, guaranteeing that manufacturing methods are both economically feasible and environmentally conscious. Climate vulnerability, biodegradation, and deforestation present substantial global ecological risks and difficulties for human existence. To effectively address these global problems, policymakers, environmentalists, and governments must build a comprehensive and long-lasting ecological safety monitoring program at several spatial and temporal levels. Climate change precipitates natural calamities and poses a disconcerting threat to environmental security. An efficient emergency response strategy can mitigate the severity of a disaster and safeguard against significant human casualties. Hence, the crucial component of responding to environmental emergencies is the prompt identification and mitigation of ecological hazards. These environmental calamities necessitate technical help and specialized skills in order to safeguard human lives [2,15].

The mechanical equipment industry plays a crucial role in advancing energy efficiency and minimizing emissions in areas such as electrical power, metallurgy, and petrochemicals. Nevertheless, it encounters obstacles such as excessive energy consumption, inadequate manufacturing precision, substantial machining allowance, elevated waste rate, insufficient production efficiency, and substantial waste emissions. The majority of energy consumption and pollution in the sector is caused by thermal operations such as casting, forging, and heat treatment. These activities account for 60%–80% of the industry's emissions. The imperative of researching and expanding green manufacturing technology lies in its ability to conserve energy, reduce emissions, bolster the manufacturing industry, and foster sustainable growth. Enterprises can achieve a more sustainable future by minimizing their environmental effect and minimizing resource consumption, maintaining a balance between economic and social benefits [16].

2.3.3 Environmental quality

Understanding supplier responsibilities is crucial for enhancing performance and managing environmental impact in sustainable operations. The natural resource-based view (NRBV) framework is employed to assess the effectiveness of green operations (GO) in terms of manufacturing and supply chain governance while working with suppliers who have different levels of electromagnetic compatibility (EMC), ranging from high to low. We offer actual evidence illustrating how suppliers' EMC affects an organization's ability to successfully execute GO and attain environmental and commercial objectives. The study's findings demonstrate the influence of substantial fluctuations in supplier EMC on the performance results of product and process stewardship. We offer managerial insights into the importance of supplier environmental management system (EMS) and the optimal degree of supplier EMS that promotes effective process and product stewardship. This study establishes the groundwork for studies in environmental management. Additionally, it delves into subjects about the proficiency of providers, such as processing data, as well as supplier cooperation and dedication. These factors have an impact on the performance outcomes of GO and how manufacturing companies can utilize their suppliers' EMS to compete. This research is relevant not only in manufacturing but also in other industries such as shipping and transport logistics [17,18].

2.3.4 Resource utilization

Green recycling involves the process of recycling products or parts that have been abandoned in order to reduce environmental pollution and enhance resource utilization. If furniture products are not recycled at the end of their life cycle, it will lead to resource wastage and environmental degradation [19].

Green manufacturing research aims to achieve the practical use and adoption of environmentally friendly practices in companies, resulting in a broader shift toward sustainability in the industry. This transformation would bring about a scenario where economic, social, and environmental benefits are mutually advantageous. By using green manufacturing practices, companies can optimize the exploitation of resources and limit resource consumption. This, in turn, leads to a direct decrease in costs [20].

The utilization of resources in green technology in contemporary manufacturing emphasizes the effective and sustainable utilization of natural resources to minimize environmental harm and improve economic feasibility. This entails the implementation of sustainable energy sources, such as solar and wind power, to diminish dependence on non-renewable fuels and mitigate the release of greenhouse gases. Manufacturers prioritize the efficient utilization of materials by opting for sustainable and recyclable materials, which effectively minimize waste and decrease the need for new resources. Water conservation techniques, such as closed-loop systems and rainwater collecting, guarantee the responsible use and recycling of water within the manufacturing process. In addition, waste management strategies strive to reduce waste production by implementing recycling, reusing, and repurposing methods for resources. Green technology enhances resource efficiency, leading to the conservation of vital natural resources. Additionally, it fosters a circular economy, lowers operational expenses, and adheres to regulatory standards and customer demands for sustainable practices.

2.3.5 Life health

The relationship between the progress of green technologies and patenting has been described as follows: Improved patent enforcement facilitates the encouragement of patenting in general. However, it remains uncertain whether this results from more basic innovation or a more strategic utilization of patents. Protection of intellectual property rights (IPRs) may also lead to a shift in research focus toward areas that are eligible for patents and have practical applications. This shift could potentially hinder the progress of developing truly innovative ideas. Strong patents facilitate the transfer of technology to developing countries through the means of imports, foreign direct investment, and licensing. Enhanced patent protection has minimal impact on the transfer of technology to the most economically disadvantaged nations. The category of climate change-related technologies encompasses a diverse range of distinct technologies that address various concerns related to climate change. The utility of patents and the inclination to patent differ significantly across technological disciplines. The relationship between IPRs and the development and dissemination of green technology beyond national lines cannot be explained by a single, all-encompassing mechanism due to its highly unlikely nature. Moreover, it is highly improbable to have a singular, all-encompassing mechanism that explains the connection between IPRs and the development and dissemination of environmentally friendly technologies across different countries. This is because the suitability of various technologies varies depending on factors such as geographical location, industrial composition, and level of economic advancement [21].

2.4 SUSTAINABLE AND MODERN MANUFACTURING

Manufacturing enterprises are becoming more conscious of the significance of environmental responsibility, as public awareness of these concerns is increasingly vital to their operations. Consequently, firms are now facing heightened pressure. In order to demonstrate environmental responsibility, it is important to address public concerns, as they often lead to the creation of environmental laws. In addition, the "green" consumer movement has prompted industrial enterprises to adopt a more ecologically responsible approach. Environmentally sustainable strategies should be a fundamental component of a company's overall strategy and organizational philosophy, including all functional areas and activities, because of the interconnectedness between these areas. Given the substantial influence of environmental concerns on manufacturing activities, this is especially crucial [22].

The study and adoption of eco-friendly manufacturing processes are increasingly reflective of the trends of globalization. The correlation between green manufacturing and globalization is getting further interconnected. Manufacturing holds great relevance in the domain of huge multinational organizations, as it makes significant improvements to manufacturing scale, sales economy, and profit generation [20]. The recycling process is explained in Figure 2.2.

The primary function of Industry 4.0 is to facilitate the establishment of interconnected systems inside industrial industries. This enables the processing, analysis, and utilization of information to carry out essential tasks. The system that encompasses the industrial environment is referred to as smart manufacturing [23].

Metal matrix composites (MMCs) are utilized in high-demand industries such as automotive, aircraft, electronics, and sports equipment due to their lightweight nature and durability. They possess outstanding physical and mechanical characteristics, such as remarkable strength, resistance to corrosion, and durability against wear. Nevertheless, cutting MMCs poses challenges due to suboptimal surface smoothness, tool deterioration, and exorbitant material removal expenses. This chapter examines the sustainability of machining using various lubrication and cooling methods, as well as the economic aspects of MMC processes. The main goal is to optimize elements that affect the ease of machining, such as the quality of the surface, the production of chips, the wear of tools, and the study of sustainability. The chapter offers recommendations for choosing accurate cutting parameters for MMCs and proposes that the implementation of various

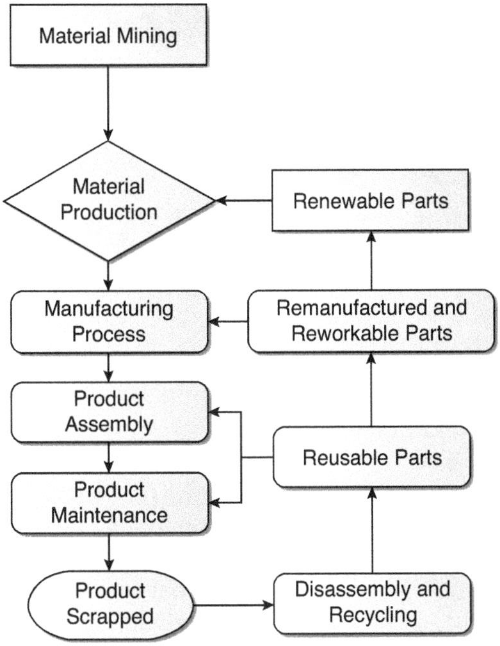

Figure 2.2 Product recycling process.

cooling/lubrication technologies can enhance sustainability, machinability properties, tool longevity, and surface quality in cutting processes.

It has been shown that the cooling/lubrication method in machining enhances the ease of machining lightweight, difficult-to-cut composite materials. Nevertheless, it is not viable in terms of both environmental and economic sustainability. Nevertheless, it demonstrates superior outcomes in machining properties such as tool longevity, surface quality, and diminished cutting force compared to the dry cutting method [24].

Disseminate knowledge about sustainable supply chain management (SSCM) and advocate for the adoption of ecologically friendly products and services. Establish a dedicated and responsible department to only handle electronic trash through efficient reverse logistics. Prolong the lifespan of the product by creating designs for disassembly or updating instead of purchasing new ones or leasing services. Eco-design development refers to the utilization of methodologies that combine technological inventiveness to consistently enhance the environmental value of the final product, with the aim of integrating environmental concerns into the process of product design and development [1].

Figure 2.3 Sustainable manufacturing process.

The implementation of environmentally friendly and sustainable manufacturing practices and methods can significantly enhance several sectors within the manufacturing industry. The primary principles that should be considered are minimizing resource consumption, utilizing environmentally friendly resources, minimizing all forms of waste, and maximizing material reuse and recycling to support the goal of our planet's self-recovery potential [25]. Figure 2.3 explains the sustainable manufacturing process in detail.

The primary objective is to assess the current and prospective states of sustainable cooling technologies, with a particular emphasis on their ability to address thermal management challenges while minimizing adverse environmental impacts [26].

2.5 DEVELOPMENT AND INNOVATIONS IN GREEN MANUFACTURING

Green innovation is synonymous with green product innovation. The data is analyzed using two assessment models and a three-step ordinary least squares (OLS) regression analysis. A particular model analyzes the effects of advanced manufacturing technologies (AMT) on the development and implementation of environmentally friendly innovations. The second model investigates the correlation between performance and both AMT and green innovations. The findings suggest that AMT has an impact on both green innovation and the overall success of the organization. We have found that technical breakthroughs play a moderating role in green developments. We provide evidence that 66% of all innovations are environmentally friendly, contradicting the prevailing belief that corporations will only engage in eco-innovation if they are given financial incentives or if strict rules are imposed. The primary obstacle to green innovation is the longevity of the firm, rather than factors such as ISO 14000 accreditation [27].

According to research, the *Journal of Cleaner Production* is the most prolific journal, as it stores publications that are connected to green manufacturing. P.R. China has the highest number of documents, with a total of 1,357, accounting for 22.66% of the total. Zhejiang University is the most productive university. The United States and People's Republic of China maintain the most robust collaborative relationships with other nations and areas, as seen by the cooperation network [28].

The implementation of environmentally friendly solutions is leading to modifications in industrial production processes. One important decision is selecting a new production solution, and implementing new technologies may lead to substantial changes in the overall configuration of the supply chain [29,30].

The advancement of environmentally friendly innovation in the industrial sector relies on the careful selection of the most suitable path for green technological innovation. The study utilizes niche theory as a framework to develop green technological innovation trend indicators for the manufacturing industry. It also applies cloud model tools to construct a multi-layer comprehensive evaluation model. An empirical examination of Jiangsu province's green technology manufacturing sector reveals that the province engaged in manufacturing activities over the period from 2012 to 2016 [31].

The Green Technology Bank (GTB) in China has implemented a three-tier categorization system for green technology (CSGT) in order to facilitate the advancement of sustainable innovation and investment. The CSGT classifies green technologies into five primary categories, 30 subcategories, and 87 sub-subcategories. The analysis examined 2,453 instances of environmentally friendly technologies, offering useful insights for policymakers and financiers. The CSGT focuses on both current environmental concerns and long-term Sustainable Development Goals by incorporating national plans and addressing local demands. It has the potential to be implemented in other countries and locations [32].

Designs and manufacturing priorities are expected to have a greater influence on the development of future ecological practices, the enhancement of ecological performance, and the advancements and progress in manufacturing. Given the importance of sustainable cultures in both the present and future, decision-makers, manufacturers, and designers who prioritize and implement sustainable practices in their businesses will undoubtedly achieve success in their respective sectors [33].

Smart manufacturing, commonly referred to as Industry 4.0, entails improving manufacturing processes by integrating systems, combining both physical and cyber capabilities, and utilizing information, especially big data [34].

2.6 CHALLENGES AND OPPORTUNITIES IN IMPLEMENTING GREEN MANUFACTURING

2.6.1 Challenges in implementing green manufacturing

Implementing green manufacturing in modern manufacturing poses various challenges, including financial limitations for establishing and developing a work methodology, research into technical issues, and other related factors. The following are the challenges:

- High initial costs
- Supply chain management complexity
- Adherence to regulations and guidelines
- Technological constraints
- Market recognition
- Financial factors
- Simple integration with existing systems.

2.6.1.1 High initial costs

2.6.1.1.1 Technological Investment

Green manufacturing calls for huge investments in cutting-edge technologies designed for minimal impact on the environment. This not only includes investment in equipment that is energy efficient but also sourcing renewable sources of energy and implementing intelligent manufacturing technologies that reduce waste and optimize the use of resources. These initiatives will involve substantial upfront costs but are relevant to long-term sustainability objectives and operational efficiency. These high upfront costs can be regained in savings on energy consumption and waste management over a period of time. Furthermore, governments also give incentives to those companies which follow eco-friendly practices. It will also provide an opportunity to leverage competitive advantages, as demand is increasing among consumers for eco-sensitive products by investing in environmentally friendly technologies.

2.6.1.1.2 Training

Providing training to staff is crucial for acquainting them with the functioning of new technology and ensuring their adherence to the new processes associated with green manufacturing. This training can be expensive and time-consuming, as it may necessitate extensive programs to ensure that all staff members are proficient in the new systems and understand the company's sustainability goals.

2.6.1.2 Supply chain management complexity

2.6.1.2.1 Responsible Procurement

Achieving the sustainable procurement of all materials and components can be a complex undertaking. Efficient tracking and certification systems are essential to guarantee that suppliers adhere to environmental standards. Overseeing a worldwide supply chain that adheres to diverse norms and regulations can provide significant challenges.

2.6.1.2.2 Supplier Collaboration

Collaborating to ensure suppliers comply with green manufacturing requirements might present difficulties. Suppliers may be located in regions with different environmental regulations or insufficient resources to implement sustainable practices, which can make it difficult to maintain a consistent green supply chain.

2.6.1.3 Adherence to regulations and guidelines

2.6.1.3.1 Diversification of Regulations

Green manufacturing must effectively navigate a diverse range of environmental requirements that vary between countries and areas. The complexity of the issue may provide challenges in guaranteeing adherence, as organizations are obligated to stay well-informed and conform to multiple sets of standards.

2.6.1.3.2 Regular and Frequent Updates

Environmental regulations are often revised to incorporate the latest scientific discoveries and community concerns. Maintaining pace with these alterations necessitates ongoing surveillance and adjustment, which can burden resources and complicate strategic planning.

2.6.1.4 Technological constraints

2.6.1.4.1 Green Technologies' Availability

Not every green technology has achieved complete development or is economically viable. Some technologies still need to be put in the experimental stages, which presents difficulties for organizations when it comes to large-scale implementation.

2.6.1.4.2 Scalability

Technologies that are effective in small-scale applications may not readily adapt to the large-scale production levels necessary in manufacturing. Addressing these scalability problems is essential for the general acceptance of green manufacturing processes.

2.6.1.5 Market recognition

2.6.1.5.1 Consumer Awareness

It can be difficult to educate consumers about the advantages of environmentally friendly items and persuade them to pay a higher price. Several consumers may lack a comprehensive understanding of the environmental consequences of their shopping choices or may be reluctant to alter their established buying patterns.

2.6.1.5.2 Competence for the Market

The market for environmentally friendly products may not be sufficiently mature, hence posing challenges in attaining economies of scale. Insufficient consumer demand may pose challenges for corporations in justifying the increased expenses linked to green manufacturing.

2.6.1.6 Financial factors

2.6.1.6.1 Return on Investment (ROI)

The future profitability of investments in green manufacturing may be uncertain. Companies must meticulously evaluate the potential long-term benefits in comparison to the present financial expenses, a challenging endeavor due to market volatility and evolving regulations.

2.6.1.6.2 Cost Competitiveness

Green products generally have higher production costs due to the use of sustainable resources and techniques. Competing with established products on price can be difficult, especially in markets that are very responsive to pricing.

2.6.1.7 Simple integration with existing systems

2.6.1.7.1 Legacy Systems

Integrating new eco-friendly technology into current production systems can be complex. Obsolete systems may not be compatible with contemporary technologies, requiring significant modifications or replacements.

2.6.1.7.2 Enhancements to the Process

Implementing green manufacturing sometimes necessitates substantial alterations to existing processes. These modifications might lead to disruptions, requiring careful planning and execution to minimize operating disruptions and ensure a smooth transition.

2.6.2 Opportunities in implementing green manufacturing

2.6.2.1 Cost reductions

Implementing environmentally friendly manufacturing techniques frequently results in substantial decreases in energy use, leading to reduced utility expenses. Adopting energy-efficient technologies and processes can yield significant cost savings in the long run. Companies can decrease disposal costs and enhance resource efficiency by minimizing trash output. Efficient waste management solutions contribute to both the preservation of the environment and the reduction of costs.

2.6.2.2 A competitive benefit

Implementing sustainable manufacturing practices can enable organizations to distinguish themselves in the market. This can appeal to environmentally sensitive consumers and bolster the brand's position as a frontrunner in sustainability. Embracing green techniques early on establishes a company as a frontrunner in the market. This can confer a competitive advantage, appealing to clients and partners who prioritize sustainability.

2.6.2.3 Innovation and growth

Green manufacturing has the potential to stimulate innovation, resulting in the development of novel and environmentally friendly products. By doing so, it has the potential to create opportunities in untapped markets and provide other sources of income, promoting the expansion of the firm. Companies can enhance their market position and tap into new customer segments by offering environmentally friendly products. Consumers are progressively looking for sustainable alternatives, which is leading to prospects for market expansion.

2.6.2.4 Hazard mitigation

By proactively embracing environmentally friendly methods, organizations can maintain a competitive edge by avoiding the risk of incurring fines and penalties associated with regulatory non-compliance. Maintaining compliance can help bolster the company's brand and foster confidence among stakeholders. Adopting sustainable sourcing and methods enhances the durability of the supply chain. This can help reduce the risks related to limited availability of resources, changes in regulations, and negative effects on the environment.

2.6.2.5 Ecological consequences

Green manufacturing contributes to the conservation of natural resources through waste reduction and the optimization of resource utilization. This fosters the enduring viability of both the corporation and the environment. Green manufacturing plays a crucial role in addressing climate change by decreasing greenhouse gas emissions. Implementing sustainable practices aids in the reduction of the company's carbon footprint and contributes to the achievement of global environmental objectives.

2.7 CONCLUSION

Green manufacturing is a significant advancement in industrial methods that focuses on sustainability, efficiency, and environmental stewardship. As stated in this review study, green manufacturing incorporates a range of technologies and practices with the goal of reducing environmental harm while yet being economically sustainable. Manufacturers may greatly diminish their dependence on fossil fuels and decrease greenhouse gas emissions by embracing renewable energy sources like solar and wind power. Energy efficiency is improved by employing high-efficiency gear, light emitting diode (LED) lighting, and optimized production processes, all of which help decrease energy use. The discussed classifications of green manufacturing technologies encompass the integration of renewable energy, enhancement of energy efficiency, utilization of sustainable materials, management and reduction of waste, conservation of water, implementation of green manufacturing processes, reduction of emissions, adoption of digitalization and smart manufacturing, and utilization of eco-friendly packaging. Every classification makes a distinct contribution toward the ultimate objective of minimizing environmental impact and advancing sustainability in the industrial sector. Green manufacturing not only tackles the environmental issues associated with conventional manufacturing methods but also provides notable advantages such as cost reduction, adherence to regulations, and improved market competitiveness. Manufacturers may attain a harmonious relationship between economic expansion and environmental accountability by embracing these technologies and practices, thus laying the groundwork for a sustainable future.

REFERENCES

1. T. D. Moshood, G. Nawanir, F. Mahmud, S. Sorooshian, and A. Q. Adeleke, "Green and low carbon matters: a systematic review of the past, today, and future on sustainability supply chain management practices among manufacturing industry," *Clean Eng Technol*, vol. 4, p. 100144, Oct. 2021, doi: 10.1016/j.clet.2021.100144
2. M. Ikram, M. Ferasso, R. Sroufe, and Q. Zhang, "Assessing green technology indicators for cleaner production and sustainable investments in a developing country context," *J Clean Prod*, vol. 322, p. 129090, Nov. 2021, doi: 10.1016/J.JCLEPRO.2021.129090
3. M. Umar, S. A. R. Khan, H. M. Zia-ul-haq, M. Y. Yusliza, and K. Farooq, "The role of emerging technologies in implementing green practices to achieve sustainable operations," *TQM J*, vol. 34, no. 2, pp. 232–249, Feb. 2022, https://doi.org/10.1108/TQM-06-2021-0172
4. J. Weng, S. Mizoguchi, S. Akasaka, and H. Onari, "Smart manufacturing operating systems considering parts utilization for engineer-to-order production with make-to-stock parts," *Int J Prod Econ*, vol. 220, p. 107459, Feb. 2020, doi: 10.1016/J.IJPE.2019.07.032
5. A. Ardanza, A. Moreno, Á. Segura, M. de la Cruz, and D. Aguinaga, "Sustainable and flexible industrial human machine interfaces to support adaptable applications in the Industry 4.0 paradigm," *Int J Prod Res*, vol. 57, no. 12, pp. 4045–4059, Jun. 2019, doi: 10.1080/00207543.2019.1572932
6. M. Mustapić, M. Trstenjak, P. Gregurić, and T. Opetuk, "Implementation and use of digital, green and sustainable technologies in internal and external transport of manufacturing companies," *Sustainability (Switzerland)*, vol. 15, no. 12, p. 9557, Jun. 2023, doi: 10.3390/SU15129557/S1
7. T. Khan, A. Khan, L. Wei, T. Khan, and S. Ayub, "Industrial innovation on the green transformation of manufacturing commerce," *J Market Strat*, vol. 4, no. 2, pp. 283–304, May 2022, doi: 10.52633/jms.v4i2.219
8. T. J. Zhang, J. Y. Zhou, M. Wang, K. Ren, M. Imran, and R. Wang, "Cultivation mechanism of green technology innovation in manufacturing enterprises under environmental regulations in China," *Front Environ Sci*, vol. 10, p. 926896, Jun. 2022, doi: 10.3389/FENVS.2022.926896/BIBTEX
9. H. Muhsen, M. Al-Mahmodi, R. Tarawneh, A. Alkhraibat, and A. Al-Halhouli, "The potential of green hydrogen and power-to-X utilization in Jordanian industries: opportunities and future prospects," *Energies (*Basel*)*, vol. 17, no. 1, Jan. 2024, doi: 10.3390/en17010213
10. M. Shahzad, Y. Qu, S. U. Rehman, and A. U. Zafar, "Adoption of green innovation technology to accelerate sustainable development among manufacturing industry," *J Innov Knowl*, vol. 7, no. 4, Oct. 2022, doi: 10.1016/j.jik.2022.100231
11. C. W. Y. Wong, K. H. Lai, K. C. Shang, C. S. Lu, and T. K. P. Leung, "Green operations and the moderating role of environmental management capability of suppliers on manufacturing firm performance," *Int J Prod Econ*, vol. 140, no. 1, pp. 283–294, Nov. 2012, doi: 10.1016/j.ijpe.2011.08.031

12. W. D. Leong, H. L. Lam, W. P. Q. Ng, C. H. Lim, C. P. Tan, and S. G. Ponnambalam, "Lean and green manufacturing – a review on its applications and impacts," *Process Integr Optim Sustain*, vol. 3, pp. 5–23, Mar. 2019, doi: 10.1007/s41660-019-00082-x
13. M. Singh, K. Singh, and A. Singh, "Green manufacturing practices in manufacturing industry: a state of art," *Asian J Eng Appl Technol*, vol. 7, no. 2, pp. 52–56, Oct. 2018, doi: 10.51983/ajeat-2018.7.2.956
14. V. A. Kirik, S. Cheng, N. I. Vyunova, O. V. Galustyan, S. S. Gamisonija, and S. D. Galustyan, "Education of future green engineers for achieving sustainable development in green manufacturing industry," *Int J Eng Pedagog*, vol. 11, no. 5, pp. 138–145, 2021, doi: 10.3991/IJEP.V11I5.22165
15. F. Islam, F. Saeed, M. Afzaal, A. Ahmad, M. Hussain, M. A. Khalid, S. A. Saewan, and A. O. Khashroum., "Applications of green technologies-based approaches for food safety enhancement: a comprehensive review," *Food Sci Nutr*, vol. 10, no. 9, pp. 2855–2867, Sep. 2022, doi: 10.1002/FSN3.2915
16. Z. Shan, S. Qin, Q. Liu, and F. Liu, "Key manufacturing technology & equipment for energy saving and emissions reduction in mechanical equipment industry," *Int J Precis Eng Manuf*, vol. 13, no. 7, pp. 1095–1100, Jul. 2012, doi: 10.1007/s12541-012-0143-y
17. C. W. Y. Wong, K. H. Lai, K. C. Shang, C. S. Lu, and T. K. P. Leung, "Green operations and the moderating role of environmental management capability of suppliers on manufacturing firm performance," *Int J Prod Econ*, vol. 140, no. 1, pp. 283–294, Nov. 2012, doi: 10.1016/j.ijpe.2011.08.031
18. S. Shrivastava and R. L. Shrivastava, "A systematic literature review on green manufacturing concepts in cement industries," *Int J Qual Reliab Manag*, vol. 34, no. 1, pp. 68–90, Jan. 2017, doi: 10.1108/IJQRM-02-2014-0028
19. X. Xiong, Q. Ma, YingyingYuan, Z. Wu, and M. Zhang, "Current situation and key manufacturing considerations of green furniture in China: a review," *J Clean Prod*, vol. 267, p. 121957, Sep. 2020, doi: 10.1016/J.JCLEPRO.2020.121957
20. S. Ning and X. Li, "A scientometric review of emerging trends in green manufacturing," in *Advances in Intelligent Systems and Computing*, Springer Verlag, 2020, pp. 234–247. doi: 10.1007/978-3-030-21248-3_17
21. B. H. Hall and C. Helmers, "The role of patent protection in (clean/green) technology transfer," National Bureau of Economic Research, Working Paper 16323, Sep. 2010.
22. J. Sarkis and A. Rasheed, "Greening the manufacturing function," *Business Horizons*, vol. 38, no. 5, pp. 17–27, Sep.–Oct. 1995, doi: 10.1016/0007-6813(95)90032-2
23. E. Shaikh, A. Bashar, and N. Mohammad, "Recent applications of computing and mobility technologies to modern manufacturing," *2020 International Conference on Communications, Signal Processing, and their Applications (ICCSPA)*, Sharjah, United Arab Emirates, pp. 1–6, 2021, doi: 10.1109/ICCSPA49915.2021.9385756
24. R. A. Laghari, N. He, M. Jamil, M. I. Hussain, M. K. Gupta, and G. M. Krolczyk, "A state-of-the-art review on recently developed sustainable and green cooling/lubrication technologies in Machining Metal Matrix Composites (MMCs),", *Korean Soc Precis Eng*, doi: 10.1007/s40684-023-00521-8

25. P. Mishra, "Moving towards sustainable and green manufacturing," *Manag Humanit*, vol. 1, no. 1, pp. 32–38, 2020.
26. P. E. Ohenhen, O. Chidolue, A. A. Umoh, B. Ngozichukwu, A. V. Fafure, V. I. Ilojianya, and K. I. Ibekwe, "Sustainable cooling solutions for electronics: a comprehensive review: Investigating the latest techniques and materials, their effectiveness in mechanical applications, and associated environmental benefits," *World J Adv Res Rev*, vol. 21, no. 1, pp. 957–972, Jan. 2024, doi: 10.30574/WJARR.2024.21.1.0111
27. I. Palčič and J. Prester, "Impact of advanced manufacturing technologies on green innovation," *Sustainability* vol. 12, no. 8, p. 3499, Apr. 2020, doi: 10.3390/SU12083499
28. Z. Pei, T. Yu, W. Yi, and Y. Li, "Twenty-year retrospection on green manufacturing: a bibliometric perspective," *IET Coll Intel Manuf*, vol. 3, no. 4, pp. 303–323, Dec. 2021, doi: 10.1049/CIM2.12038
29. M. Rinaldi, M. Caterino, M. Fera, P. Manco, and R. Macchiaroli, "Technology selection in green supply chains – the effects of additive and traditional manufacturing," *J Clean Prod*, vol. 282, p. 124554, Feb. 2021, doi: 10.1016/J.JCLEPRO.2020.124554
30. J. K. Patra and K. H. Baek, "Green nanobiotechnology: factors affecting synthesis and characterization techniques," *J Nanomater*, 2014, no. 1, 417305 (2014) , doi: 10.1155/2014/417305
31. B. Peng, C. Zheng, G. Wei, and E. Elahi, "The cultivation mechanism of green technology innovation in manufacturing industry: from the perspective of ecological niche," *J Clean Prod*, vol. 252, p. 119711, Apr. 2020, doi: 10.1016/J.JCLEPRO.2019.119711
32. R. Guo, S. Lv, T. Liao, F. Xi, J. Zhang, X. Zuo, X. Cao, Z. Feng, and Y. Zhang, "Classifying green technologies for sustainable innovation and investment," *Resour Conserv Recycl*, vol. 153, p. 104580, Feb. 2020, doi: 10.1016/J.RESCONREC.2019.104580
33. B. Elira, "Green infrastructure and manufacturing: analysis of IE and SM innovations for future generations," *J Mach Comput*, vol. 1, no. 2, pp. 97–105, Apr. 2021, doi: 10.53759/7669/jmc202101011
34. A. Ahmadi, C. Cherifi, V. Cheutet, Y. Ouzrout, and Y. Ouzrout, "Recent advancements in smart manufacturing technology for modern industrial revolution: a survey," *J Eng Inform Sci Stud*, 2020 [Online], https://hal.science/hal-03054284

Chapter 3

Concepts, methods, and strategies for zero waste in manufacturing

Arunprasath Kanagaraj, Shankar Ganesh PSP, Satheesh Raja R., Shanawaz AM, Sundarakannan Rajendran, Lavish Kumar Singh, Clinton R., and Bhuvanesh Aanathan

3.1 SOURCE REDUCTION

People create a lot of trash, which adversely affects life while utilizing air and water. Developing industrialization, populace, and trash yield without viable waste administration methods presents serious difficulties to ecological supportability. Given these issues, the ongoing review centres around the sorts and wellsprings of waste development, as well as junk decrease by empowering waste decrease, reusing, and recycling. Fast industrialization, urbanization, and the inflow of transient individuals into urban communities have all added to huge trash heaps on the planet [1]. These wastes, which contain solid and fluid contaminated with synthetics, food waste, prior waste, etc., have been thrown in open landfills consistently, causing environmental harm in metropolitan regions. This uncontrolled unloading of refuse in broad daylight spaces shows that the city's authorities' lack waste administration capacities. In this way, soil, air, and water contaminations are run-of-the-mill events in most parts of the world with bigger structures. The greatest trouble is the outrageous lack of land [2].

Centre-gathering conversations are utilized to examine the discernments, perspectives, convictions, suppositions, and thoughts of the grounds' local area individuals to identify the essential factors affecting their zero-waste supportive to an ecological way of behaving. To learn about the locals' thoughts on the topics covered by the aforementioned approach, the inside-and-out interview is used. Making adjustments and drawing on previous studies, particularly in the model, the paradigm is based on a new information disposition and supports an ecological way of functioning. Natural objectives, ecological schooling, individual waste-related insight, natural approach, ecological mindfulness, support possibilities, local area commitment, social obligations, praiseworthy authority, and virtual entertainment all affect the grounds' local area's zero waste [3].

 DOI: 10.1201/9781003470342-3

In light of its maintainable item plan, item life augmentation, and eco-accommodating methodology, the new circular economy model is impressively better for laying out maintainability in essentially all modern organizations. It fundamentally tries to reuse, fix, recuperate assets, and reuse items as much as possible. To understand the development, practices, obstructions, and advantages of round economy execution, a background marked by the circular monetary plan of action and a combination of definitions and calculated structures were offered [4,5]. The transformation and difficulties in the circular economy is not an option, but a mandatory decision. Similarly, suggested methods and criteria for the implementation of plans and assessments were cited, as were realistic, circular financial advancements.

Circular economy, CO_2 decrease, green innovation, waste minimization innovation, and wastewater treatment innovation are instances of green advancements. Focusing on a solitary idea to accomplish maintainability would not be more viable. Subsequently, existing examination misses the mark concerning coordinating and organizing all ecological assurance innovation ideas in the material business' production network. Thus, limiting the absolute expense and ecological damage is very troublesome [6]. Given the shortage of examination and economical model improvement in the quickly expanding materials and design enterprises, this study aims to fabricate a three-layer facilitated reasonable production network model in the material and clothing areas.

Current innovations have drastically changed the world by growing new materials for an extensive variety of designing applications, for example, biomedical, auto, aviation, and energy stockpiling frameworks. Generally speaking, these innovations work on everybody's personal satisfaction. An essential downside of a higher style of life is the depletion of basic regular assets and resources. Moreover, the broad use of non-biodegradable trash and different oil-based items radiates a lot of unsafe contaminations. The present worldwide wind power generation capacity is anticipated to be 743 GW, which is enough to balance 1.1 billion tonnes of CO_2 emissions. Wind energy through wind turbine cutting edges is viewed as an environmentally friendly power energy asset and is the world's quickest developing energy area. Power age is supposed to increase from 5% today to 30% by 2050. Moreover, the Worldwide Energy Office (IGA) anticipated that the wind energy would surpass 3317 terawatt hours by 2030. Europe consumes most of wind turbine blades (WTBs) because of good wind energy age conditions [7].

Improvement methodology, including the use of pineapple leaves with the advancement of excellent strands as far as primary and mechanical characteristics are concerned, has been found to yield valuable results. In the ongoing setting, distributed works for powerful advances managing the development of maintainable pineapple leaf filaments as an important waste material source are examined, with an emphasis on lessening ecological contamination through more practice at the locality, public, and global levels. The utilization of pineapple leaf fibre (PALFs) can assist with lessening the gigantic waste that should be burned or discarded. Besides, the OK assembling interaction of PALFs with a greatest sum can moderate energy and normal assets. It can likewise assist with laying out a green climate and backing monetary development in a sans waste future [8]. Pineapple leaf strands are tended to as a waste-determined biomass. They are used for the union of significant worth items, with the most useful framework in the agro-modern circle. It has now been exhibited to be an exceptionally encouraging hotspot for use in manure structure. PALFs from pineapple waste have shown potential for use in the development and auto businesses [9].

Also, all the activities have been made to examine these potential outcomes, in spite of the fact that their application truly has been restricted. This survey study explores the dissimilarity between the idea and genuine execution with various decisions in view of data from distributed chips away at wastes and their related items. Web-based reusing has been developed from online exchanges, moving from the B_2C to C_2B mode, and both relate to online-to-disconnected business techniques. In all structures, the "B" consistently has a data advantage, and they find out about the things they exchange for. In the web-based exchange of new things in the B2C mode, the merchant decides the cost of their items in light of item quality, client view of their items, etc., and the exchange starts when the customer settles on the cost. Confronted with shopper doubt, the web recycler ought to be more careful in the web-based offer of utilized items [10].

The breakdown of web recyclers' procedure in the web-based exchange of utilized items when customers have low confidence in web recyclers is discussed in this review. Reusing of fibre is useful for composite waste administration and also for the circular economy and economical innovation ideas. The expression "reusing" alludes to the reuse of handled material following mechanical, warm, or substance medicines, or any blend of these medicines. Given the pestilence, the worldwide Business 4.0 development has as of late built up momentum. A bunch of exceptionally problematic innovations changes a worldwide framework, permitting the world to turn out to be more associated while also helping economies to prosper and turn out to be more serious. While Industry 3.0 focuses on the robotization of specific machines and cycles, Industry 4.0 focuses on the start to finish digitalization of every single actual resource and their reconciliation into computerized environments in the waste area by connecting the physical and virtual universes. Artificial intelligence (AI) and machine learning (ML) are used to work with asset recuperation of non-biodegradable junk as the last option, joining computerized change and digitalization [11].

Organized research using a contextual analysis based on carbon-neutral viewpoints are outstanding, even if the organization of digital technology could promote problematic advancements in the waste sector by combining waste reduction focusses with financial goals. Regardless of how nobody has detailed the alleviation of COVID-19 ecological outcomes utilizing computerized techniques, such commitments to carbon non-partisanship merit extra request and assessment from experts in the field [12].

Makers are examining how computerized innovation applications can uphold their manageability activities by aiding the change of conceptual maintainability objectives, like net-zero outflows and circular economy, into doable and reasonable activities, achievements, and, eventually, an economical upper hand. This chapter takes an asset-based view to research the expected job of computerized innovations in the development of an assembling company's upper hand, as well as the organization of existing inward assets and centre capabilities to accomplish net-zero assembling discharges and capability enhancement. Centre skills that are normally upheld by computerized innovation inside might actually be utilized in the accomplishment of net-zero discharges [13].

To drive zero-waste esteem chains, such computerized arrangements depend significantly on quality-related data inside the production network business environment. To give significant, solid, and trustful information to zero-waste esteem chain techniques, an answer for start to finish modern information detectability, trust, and security across various cycle chains or even between hierarchical inventory chains is required. We at first proposed item, cycle, and information quality administrations to advance zero-waste esteem chain techniques in this record [14]. Following that, we present the confided in structure, a basic empowering agent for the safe and successful sharing of value-related data inside the production network business environment and hence for quality streamlining events. In green manufacturing, source reduction is the technique of eliminating waste before it is created. It is the most recommended waste management technique since it prevents garbage from being generated in the first place. This can be accomplished in a variety of ways. Source reduction is a critical component of green manufacturing. Manufacturers can decrease their environmental effect, enhance efficiency, and save money by employing source reduction methods. Figure 3.1 describes about the source reduction priority in the concept of zero waste.

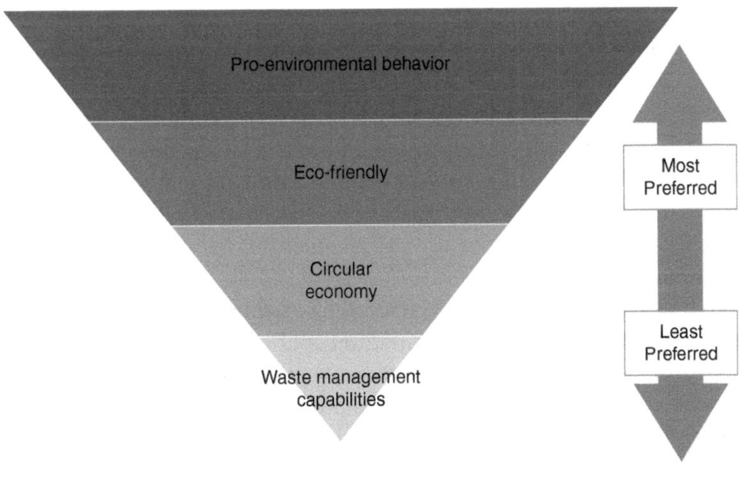

Figure 3.1 Source reduction priority in the concept of zero waste.

3.2 PROCESS OPTIMIZATION

Green manufacturing process optimization is a systematic strategy to enhance the efficiency and environmental performance of manufacturing processes. It entails finding and implementing improvements that reduce the consumption of energy, materials, and other resources, as well as waste and emissions. Process optimization in green manufacturing is a critical component of long-term manufacturing. Businesses can decrease their environmental impact, save money, enhance product quality, and become more competitive by applying process optimization approaches [15].

The characterization of obstructions to circular economy reception in the structure development industry of developing and developed countries displays the two balances and imbalances. This is basic because the uncertainty of the impacts of hindrances in existing examinations affects exhaustive and concentrated strategy plan. Accordingly, reactions from 140 specialists and 39 manufacturing and industrialized economies were investigated. The fluffy engineered assessment approach was utilized to decide the significant outcomes of the boundaries, while the Mann-Whitney U test was utilized to find critical varieties in master decisions between the two economies [16]. As per the Food Service Establishment, the most critical obstacles to overall circular economy reception are authoritative, data innovation, foundation, and coordinated operations. The Mann-Whitney U test exhibits a significant variety of master points of view on administrative, data innovation, and financial and market hindrances among emerging and established economies [17]. Thus, they are viewed as unmistakable obstacles since they influence reception in the two economies. Foundation and coordinated factors, as well as authoritative obstructions, are viewed as non-exclusive boundaries [18].

Because of the deficient space for dump yards and landfills, as well as higher junk unloading charges, modern waste administration has gotten significant consideration lately.

Despite the growing popularity of plant-based alternatives to meat, conventional slaughterhouses and the waste they generate continue to be a cause for worry. Waste valorization is a deeply grounded approach that aims to deliver a closed circle process in areas that do not create waste. In spite of being a profoundly contaminating industry, slaughterhouse waste has for quite some time been transformed into monetarily viable cowhide. In any case, the tannery area dirties as much as, while possibly not more than, slaughterhouses.

Given their poisonousness, compelling treatment of the tannery's fluid and solid wastes is basic. The damaging materials delivered infiltrate the pecking order, having long haul ramifications for the biology. A few cowhide waste change innovations are generally used in businesses, and they offer high-benefit merchandise. Notwithstanding, far reaching examination of the strategies and results of waste valorization is often disregarded as long as the changed by-product is more significant than the trash.

The most effective and naturally well-disposed waste management procedure should convert waste into a value-added use with no harmful side-effects. The zero-waste idea is an augmentation of the zero fluid release idea, where solid waste is dealt with and reused to the point that no lingering is moved to the landfill [19].

The fast development of bamboo species makes them an appealing option for eco-reclamation, and their dialect cellulosic substrate could be used to deliver high-quality green products, for example, bio-fills, synthetics, and biomaterials. This audit completely researched the thermo-synthetic and natural change of bamboo biomass to estimate added fills and synthetic compounds inside these systems. This exploration likewise gives an intensive outline of bamboo biomass lignin extraction innovations, bioengineered approaches, and their profile processing plant change systems. Besides, bamboo biomass is mostly made out of cellulose, hemicellulose, and lignin, as well as proteins, lipids, and a couple of micronutrients that are not effectively taken advantage of by current bioengineered processes. As per the discoveries, the potential for assembling high-quality products from bamboo biomass has not been entirely researched. In any case, a critical open door will make bamboo bio-processing plant advancements financially savvy and naturally supportable, as investigated exhaustively in the ebb and flow research work [20].

Fabricating lines in micro, small, and medium enterprises (MSMEs) involve various muddled cycles, and cycle varieties bring about item dismissals and money-related misfortunes and client misery. The conveyance of a great item inside the constraints of work, hardware, and other restricted assets requires the improvement of the creation line's cycle execution through quality administration. In light of this, the ongoing work presents a structure for distinguishing and focusing on deficiencies by joining multi-rules dynamic methodologies, for example, the fluffy dynamic preliminary and assessment research centre and the fluffy scientific organization cycle with quality administration practices. Preceding information assortment, the combination shifts through the most potent imperfections and focuses on them to contact fundamental assembling process issues. Moreover, it handles the board challenges like an enormous number of flaws, restricted deformity information, and dependence on indicated measures [21].

Polylactic acid (PLA)/almond shell composite is produced using expelled fibres with a breadth of 1.75 0.5 mm. This study refined the compelling system boundaries for 3D printing of PLA/almond shell composite and assessed its compressive strength. For the advancement technique, the Plan of Analysis strategy is utilized. The trial was done by changing five variables and three levels (infill design, infill thickness, printing direction, printing temperature, and printing speed). For the trial approach, a L27 symmetrical cluster is made, and the Taguchi streamlining strategy is utilized to accomplish the greatest compressive strength for the fabricated PLA/almond shell composite.

The exploratory consequences of this examination work reveal that infill thickness and printing direction have a greater effect on compressive attributes than the other printing process boundaries. The numerical models are found from the advancement consequences of the PLA/almond shell composite compressive strength research. The proposed numerical model has a blunder level of 3.70% in light of the relapse examination results and a palatable fit with the exploratory outcomes. The broken examples obviously show that the expanded infill thickness of PLA/almond shell tests does not flop rashly under compressive tension [22].

To guarantee quality in savvy manufacturing plants, there is an ongoing flood of computerized arrangements in another age in light of keen frameworks, mixture advanced twins, and simulated intelligence-driven improvement devices. To drive zero-waste esteem chains, such computerized arrangements depend considerably on quality-related data inside the production network business biological system. To give significant, solid, and trustful information to zero-waste and sustainability value chain methodologies, an answer for start to finish information recognizability, trust, and security across various cycle chains or even between hierarchical inventory chains is required. We first propose item, cycle, and information quality administrations to advance zero-waste value chain systems in this record. Following that, we present the confided in system (TF), a basic empowering influence for the safe and compelling sharing of value-related data inside the store network business environment and in this manner for quality improvement activities towards zero-deformity creation. A complete set of personality traits, the open programming interface offered to frameworks in manufacturing plants, and the interaction/item/information Product, Process, and Data (PPD) quality trademark information model are all part of the TF. Combining secure, uniform, and high-quality data vertically is the responsibility of the executive layer [23].

Customary assembling methods are tedious and wasteful in different businesses because of the lack of proper instrument to make the things. AI (ML) calculations have become more normal in assembling as of late to fabricate things and items with diminished work cost, time, and exertion. Digitalization, along with state-of-the-art fabricating advancements and immense information accessibility, has expanded the requirement for an interest in consolidating AI and streamlining ways to deal with further development of item quality [24].

Techniques for incorporating AI into creation help acknowledge new methodologies; save time, energy, and assets; and reduce waste. ML-coordinated gathering techniques add to brilliant assembling, in which innovation automatically remedies any mistakes continuously to prevent spills. However, various methodologies and apparatuses are utilized in the assembly area for personal computers (PCs); late advances, for example, AI and information mining strategies, are useful in addressing research challenges.

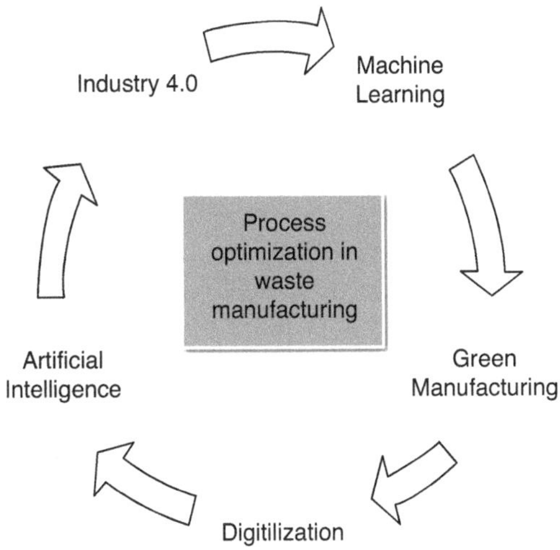

Figure 3.2 Process optimization in waste manufacturing.

Industry 4.0 is an assembly line change that considers the assortment and examination of information across machines, empowering quicker, more adaptable, and more effective cycles to fabricate better merchandise at lower costs. This new 21st century transformation will help efficiency, shift financial aspects, enhance modern development, and change the specialist profile. Digitization is carried out and empowered through an assortment of state-of-the-art innovation. Some of these innovations incorporate modern robots, modern computerization, distributed computing, the web of things and services, and AI [25]. Figure 3.2 shows the process optimization in waste manufacturing in the concept of zero waste.

3.3 RECYCLING AND COMPOSTING

The most common way of changing waste assets into new materials and things is known as reusing. An option in contrast to customary waste disposal can assist with protecting assets and reduce ozone-depleting substance emissions. Reusing may help your local area as well as the climate. Fertilizing the soil is a characteristic interaction that changes over natural waste into a supplement-rich correction. Natural materials, for example, food scraps, yard clippings, and compost, are obtained from live organic entities. Natural mixtures break down into more modest pieces, which are transformed into fertilizer. Population development has added to a huge expansion in the age of civil solid waste, which is associated with extreme natural and general medical conditions all over the planet.

This condition is worked on by purportedly overseen landfills, which were once open dumps but have been disengaged by introducing a filter, a filtrate, and a gas waste framework, and adding rubbish cover layers. From 2010 to 2015, the accompanying objections were utilized for the last removal of Master of Social Work in Brazil: clean landfill (57.6–58.7%), controlled landfill (24.3–24.1%), and open dump (18.1–17.2%). The quantity of controlled landfills in the nation has been expanding because of the conclusion of a few open dumps and the requirement for these locales to get medicinal activity. From a socio-ecological point of view, this lacks a methodology for MSW removal; yet, it is desirable over open dumps because the contamination caused is limited [26].

Imbalance in pay circulation might restrict makers' motivators to put resources into their ranches or upgrade creation processes, bringing about a less manageable production network. As of late, actions have been taken to lighten this issue and work on the force of makers, with one such arrangement being the age of money through waste reusing. Reusing waste can give entertainers an extra wellspring of money and increment their dealing influence. Pay incongruities among food inventory network entertainers can be tended to by creating pay from reusing food misfortune fibre length and fibre weight. Factors, for example, deficient unrefined components, operations network dependability, innovative abilities, coordination among entertainers, financial and natural variables, approaches, and market changes all influence store network execution and lead to misfortunes. Thus, this examination focused on the chance of delivering extra money from the reusing of new tomato waste to adjust the pay conveyance among the judication. Tomatoes are among the most broadly consumed vegetables around the world, with emerging nations representing a sizeable level of the worldwide inventory network [27].

Even though dialect cellulosic biomass treating the soil is another waste-to-abundance answer for natural waste administration and the circular economy, it actually has critical ecological blemishes that should be addressed to make it more feasible and solid. This approach examines the significant obstructions experienced while fertilizing the soil with cellulosic waste biomass, as well as the scientific advancements made over time to resolve these issues in a sustainable way. It resolves a significant overall issue: the arrival of ozone-depleting substances during the fertilizer-soil interaction, which restricts its pertinence for a more significant scope. Besides, it dives into how different natural minerals and organic added substances affect the physiochemical and organic properties of manure, which is used to produce eco-accommodating fertilizer with low smell, low ozone-depleting substance emissions, and an ideal C/N proportion. It gives one-of-a-kind bits of knowledge by representing the impact of added substances on microbial proteins and their pathways engaged with the cellulosic biomass breakdown [28].

To advance ecological manageability, waste portrayal has been used in research to make ideas on the best waste treatment and management solutions. The recognizable proof of different waste streams through waste reviews helps with the ID of recoverable things and, surprisingly, the decrease of biodegradable rubbish going to landfills, thus limiting potential worries presented by removal destinations [29]. Be that as it may, because of the cost ramifications of performing such reviews, these reviews are habitually restricted to the characterization of MSW because of the assessment of such wastes in metropolitan regions, without an exhaustive waste review.

While efforts on solid waste arrangement at the home level are well known, those at higher educational institutions (HEIs) have for the most part gone unaddressed. Nonetheless, because HEIs have similar issues regarding successful waste administration as districts and have a significant impact in fostering a supportable society, they may be considered reciprocals to towns and urban communities. The ecological and monetary ramifications of the consideration of improper parts in all legal parts were evaluated utilizing life cycle appraisal and natural life cycle costing approaches in view of compositional examination. For both promotion and soil treatment processes, three unmistakable situations were inspected. Independently gathered food waste was reused by actually utilizing modern anaerobic digestion and fertilizing the soil. Nonetheless, the consideration of unseemly materials in SC-FW not just proposes specialized issues with promotion and treating the soil but also decreases the nature of the cycles' results. Accordingly, improper materials identified in SC-FW force critical ecological and financial outcomes [30].

This study examines the outflows from civil solid waste transportation, open consuming, treating the soil, reusing, anaerobic absorption, cremation, and land filling under a few municipal solid waste management situations in Phnom Penh. The examination likewise considers the diminished emissions as an advantage of reusing and power generation from cremation and promotion. The viability of source isolation in greenhouse gas (GHG) relief is assessed utilizing different waste division rates. The results demonstrate that scenario 5 accomplishes the most net waste reduction, avoiding approximately 1.15 M kg CO_2eq/day. This is achieved by treating 389 t/day of organic waste, 714 t/day of mixed recyclables, 777 t/day of edible food waste, and 1,280 t/day of combined waste through soil fertilization, recycling, promotion, and combustion, respectively [31].

Treating the soil is the controlled transformation of biodegradable natural wastes and items into stable products by utilizing microorganisms. Fertilizing the soil is a long-utilized process, although it has various defects that cut off its inescapable use and effectiveness. Microorganism location, low supplement status, long soil fertilization time, long mineralization time, and smell creation are some of the disadvantages. After some time, these issues have expanded public attention to the utilization of compound manures delivered by the Haber-Bosch process as a choice to compost.

Compound composts make supplements accessible to plants, yet their disadvantages outweigh their advantages. Substances, for instance, add to ozone-depleting substances outflows, natural contamination, annihilation of soil living beings and marine life, ozone layer consumption, and human affliction. Thus, ranchers are turning around to fertilizing the soil to recuperate soil richness. Treating the soil is a significant agrarian practice that guides in the reusing of homestead waste. Fertilizing the soil takes an extensive stretch as a result of the presence of components that take more time to compost, particularly during co-fertilizing the soil. This survey looks at waste management through fertilizing the soil, different soil fertilizing techniques, fertilizing soil parts, long-term treatment of soil, system fundamentals, current fertilizing soil patterns, and future possibilities [32].

The respondents' degree of mindfulness and degree of consistency with the arrangements of the solid waste management programme regarding isolation, 3Rs (reduce, reuse, and recycle), and removal were researched. The relationship between respondents' mindfulness and consistency with the arrangements of the solid waste management program was likewise noticed. A sum of 367 respondents participated in the overview, which included teachers, staff, and students who finished their Natural Science certification. Most of the respondents were students, as per the outcomes.

This study discovered that respondents did not practice or screen solid waste management programs. Respondents know about the stock of solid waste management in view of isolation or partition, as well as the 3Rs incorporating removal. In light of the discoveries, it is exhorted that the program be generally carried out, as the discoveries exhibit that respondents are just somewhat mindful of and never conform to the solid waste management program's removal guidelines. A subsequent survey of respondents' information and consistency ought to be performed to identify decreases in maintenance and execution. Respondents should also know about isolation, reusing, and treating the soil standards and cycles [33].

Water serves various capabilities in the papermaking process. During the beginning phase of the papermaking system, it goes about as a suspending medium and an expanding specialist for the filaments, circulating and moulding them into a uniform sheet. It likewise goes about as a dissolvable for various synthetic substances and added substances that are used to further develop item quality. In the mash and paper area, water recovery has been a basic obligation for quite some time. Ecological regulations, wastewater release costs, and the significant expense of freshwater drive the reception of interaction water and wastewater treatment innovation. Ongoing headways have empowered diminished water use and natural effects, yet the recuperation of treated water and significant parts, for example, strands, makes water reusing frameworks financially savvy. Subsequently, the financial suitability of these advances has been basic in their application [34,35]. Figure 3.3 describes recycling and composting waste in waste manufacturing in the concept of zero waste.

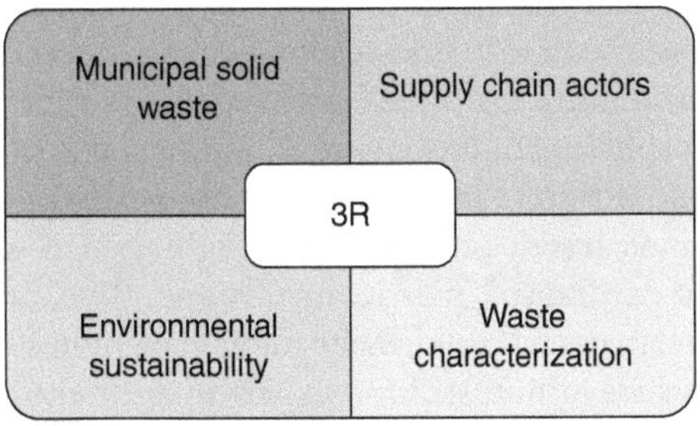

Figure 3.3 Recycling and composting waste in waste manufacturing.

3.4 WASTE-TO-ENERGY

Waste-to-energy is the most common way of converting waste into energy as power, heat, or biofuels. Waste energy innovations are fit for handling many waste streams, including civil solid waste, modern waste, and rural waste. Waste energy innovations are becoming prevalent as a way to diminish waste and create sustainable power. By 2021, there will be roughly 2,400 waste energy plants in operation all over the planet, with a complete limit of more than 400 terawatt hours of energy each year.

With the ascent of metropolitan populaces, the developing pace of civil and modern wastes has turned into a significant worldwide worry in recent years. Therefore, executing imaginative and effective waste administration and taking care of solutions are essential. The exemplary waste-to-energy idea has been set up for quite a while to handle this test; notwithstanding, waste-to-material is a somewhat late way to deal with an eco-accommodating development that is more helpful to the decrease of greenhouse gas emissions. These wastes can be changed into useable materials, for example, underlying folio and cementitious composites for innovative applications utilizing the geopolymerization innovation. This examination work will explain how the complex geopolymer definition is being utilized to exhibit waste use and epitome [36].

Nonetheless, the Worldwide Strong Waste Affiliation reports that multiple billion tonnes of MSW are generated universally every year. As per the standard overview, only 19% of this MSW will be recyclable or handled for mechanical and natural treatment in 2020, with the excess 70% being discarded in dumpsites and clean landfills. This pattern will probably continue as interest in petroleum derivatives rises. The energy area has moved its concentration from customary fills to sustainable power in light of asset consumption and the adverse consequences of utilizing non-sustainable power sources, which has prompted environmental change. Food waste management, explicitly solid waste management and metropolitan solid waste management, is a fundamental part of a circular economy, which calls for us to reduce waste and boost asset use by developing and contracting material cycles and following waste info and results to make financial stream inventories. The blend and nature of MSW are impacted by the social and monetary setting, atmospheric conditions, reusing rates, assortment recurrence, socioeconomics, and different factors. A past report characterized the MSW stream into six kinds based on actual highlights, such as food and yard garbage, paper and cardboard, plastic, metals and glass, and dormant [37].

The zero-waste term in civil solid waste management has been the idealistic objective of every waste administration expert in agricultural nations' urban areas, regardless of how it accompanies many, frequently wrong thoughts. Individuals cannot make no trash except if they do not consume. The expression "zero waste" does not infer that we produce no trash, yet rather that we dump no waste at a landfill. No one needs to live next to a contiguous landfill since it hurts more. The waste administration authority tries to adapt to the intricacies of civil solid waste management by restoring each piece of the waste administration partners to cooperate to manage waste. Individual families and networks, without which waste management's executives will fizzle, were situated as the waste management's centre drive [38].

Waste creation is an undeniable result of espresso creation and utilization. The husk, what isolates from the beans after they are cooked, is one of the results associated with espresso creation. The waste from espresso drinking is spent coffee beans created during the preparing system. The review looked to discover the attainability of changing over espresso creation and utilization waste in the anaerobic absorption process. The trial was done under anaerobic assimilation conditions as per the Racket 38414/S8 standard. The espresso husk had an all-out solid content of 93.37% and a natural matter substance of 93.34%. As far as new matter, the biogas proficiency was 329.50 m3Mg1. The complete solids content of the three ground tests analysed went from 41.27 to 45.72%. The high unpredictable solids fixation, which went from 97.91 to 98.41%, affirmed the biogas potential. The essential waste management headings connected with the creation and utilization of espresso are their stockpiling. Landfilling biodegradable trash is certainly not a helpful option since it discharges contaminants into the environment. One of the most famous purposes of their work is preparation [39].

Subsequently, coal keeps on being the world's most noteworthy wellspring of power age. Making zero-emission arrangements, for example, bioenergy with carbon catch and use is a possible way. The main focus of this investigation is waste biomass oxy burning with a focus on methane framework. A proton trade layer electrolysis fuelled by photovoltaic sun-based energy is remembered to make hydrogen and oxygen for oxy-fuel ignition in a subcritical steam power cycle. Contingent on the activity procedure, a specific measure of extra O_2 created by a cryogenic refining air partition interaction would be essential. The almost unadulterated CO_2 stream leaving the kettle is additionally cleaned, packed, and moved to the methanation reactor. To mimic different working conditions and assess reconciliation productivity, a semi-fixed model of the total plant is constructed. The outcomes uncover that the whole plant is very much coordinated, with a yearly typical productivity punishment connected with CO_2 stayed away from 6% of the places. Since waste biomass is utilized as fuel, the framework discharges 610 kg CO_2/MWh not as much as biomass establishes that do not ingest CO_2 [40].

Economical improvement objectives at an overall gathering for better implementation of SDG goals. SDG 11 is especially connected with the idea of laying out urban areas and networks that give a great life by accomplishing net-zero release and self-maintainability. Bio-roast has arisen as a possible arrangement in accordance with the worldwide local area's endeavours because of its capacity to change over waste into benefit. Bio-scorch is being investigated for use as an adsorbent, a co-impetus to speed up modern grade responses, and a feed for energy units, among different purposes. Moreover, the utilization of bio-roast as a dirt improvement specialist advances the utilization of closed-loop nutrient cycles. Subsequently, to pick biomass feedstock relying on the application, it is essential to have an intensive handle of the biomass properties, aqueous therapy, and interaction boundaries to be utilized for single creation [41].

Notwithstanding the way that food shortage exists in many regions of the planet, food waste yield keeps on increasing because of social and behavioural changes, as well as mechanical viewpoints. Food waste, then again, has a huge amount of natural material and biodegradable parts, demonstrating that it may be a potential hotspot for assembling hydrogen to meet the significant objectives of enhancing energy supply and limiting ecological contamination. Fast industrialization and population growth cause an Earth-wide temperature boost and environmental change by exhausting fossil fuels, and increasing energy demand, and affecting human existence. Besides, most CO_2 discharges into the air are caused by the utilization of petroleum products in energy and transportation areas. Thus, bioenergy and sustainable power are imperative in bringing down the utilization of petroleum derivatives and furnishing an energy-efficient society with a carbon-free climate [42].

Worldwide uncertainties about mistaken waste are developing because of filter feed tainting in groundwater and soil. Be that as it may, with the appearance of the circular economy, supporting reusing rates, diminishing area filling rates, and delivering side-effects, for example, power and fertilizer from garbage have become progressively fundamental. A power plant, a district, a fertilizer office, and a gatherer are all seen in this review as players in a government mediated Stackelberg game. For sure, the public authority gives appropriations to residents, the manure plant, and the power plant to keep costs stable and furthermore forces fines for ozone-depleting substance (GHG) emissions and leachate interruption. The gatherer gathers waste through two contending channels: one from the city's garbage cans and the other by buying recyclable waste from residents, each with its own assortment capability. Expanding waste collection to work on residents' social government assistance, likewise, is an ideal dynamic to boost the benefit of every individual [43].

Nonetheless, with regards to standard cement, there will clearly be various issues. From one perspective, creating substantial materials is exorbitant and energy-escalated. Nonetheless, due to the widespread use of non-biodegradable materials, many waste products are either discarded or costly to dispose of, as they are often non-reusable and pose a significant risk of environmental contamination. In the meantime, combined with the strength needs of the substantial, the energy emissions of geopolymer concrete have become huge as of late. It was found that delivering NaOH beds required more energy. Accordingly, choosing the fitting forerunner can assist with decreasing overall energy use. Due to their nearby accessibility and waste material, agrarian results can be utilized as antecedents to defeat this escalated energy [44]. Waste is currently considered as a sign of inefficiency. Waste creation is a human activity, not a natural one. Currently, land filling and incinerating wastes are prevalent waste management practices; however, in addition to losing raw resources, these methods cause environmental, water, soil, and air damage. Waste is regarded as a useful resource in the new "zero waste" concept. A critical component of the methodology is the development and management of products and procedures that limit waste volume and toxicity while also preserving and recovering all resources rather than burning or burying them. Unlike a linear system where waste is generated from product consumption, the end of one product becomes the beginning of another with zero waste. A scientific treatment approach, resource recovery, and reverse logistics may enable municipal, industrial, agricultural, biomedical, construction, or demolition trash to become a raw material for another process [45].

Global climate change and resource depletion are critical challenges for modern societies. Furthermore, population growth and fast industrialization raise energy demand. In order to meet ongoing energy demand, developed and developing countries must implement sustainable waste management practices. Waste-to-energy technology, in addition to avoidance and recycling, might undoubtedly be advantageous. Biofuels can be produced from trash in addition to heat and energy generation. The aviation industry is growing, and its greenhouse gas emissions account for around 2% of total global emissions. Higher altitudes have more persistent aircraft emissions, which have a greater environmental impact. As a result, decarbonization of the sector is a continuing problem. Sustainable aviation fuels appear to be a promising alternative in the coming years. The use of waste materials will help to increase the overall production of bio-jet fuels. Currently, the American Society for Testing Materials standards have validated eight paths for blending limitations of up to 50% with conventional jet fuels [46]. Figure 3.4 shows the perception of waste-to-energy in waste manufacturing in the concept of zero waste.

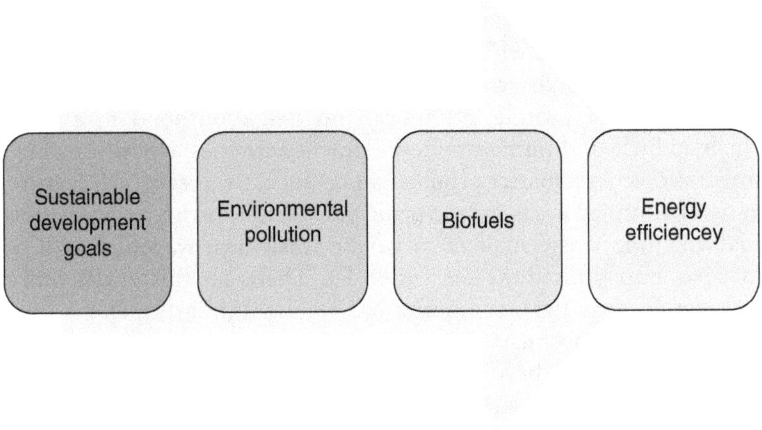

Figure 3.4 Perception of waste-to-energy in waste manufacturing.

3.5 WASTE MANAGEMENT STRATEGIES

Through a contextual investigation project in China, the fossil fuel by-products of development and destruction waste created by building redesign are inspected utilizing a day-to-day existence cycle evaluation approach. Three waste administration situations were made for a Suzhou building remodel project. The main situation relates to the standard waste management methods in China; the subsequent situation is grounded in the unconditional 3Rs technique, stressing the waste's downstream effect; and the third situation considers the waste's consolidated upstream and downstream impacts. According to the findings, construction debris has a distinct form when compared to debris from development and demolition initiatives. The restoration material phase of the life cycle of building repair projects produces the most notable and substantial fossil fuel by-products, in comparison to the disassembly, restoration development, and repair material completion of life phases of the waste board [47].

Another effect of urban overcrowding is that city officials must pay greater attention to concerns like easy parking, quick and clean sewage and garbage disposal, and cheaper electricity and water costs. These are just a few things that are made easier by efficient transport without lengthy traffic jams. They are supposed to provide the essential necessities of city dwellers, including water, housing, transportation, education, and healthcare, while making the most efficient use of the natural and financial resources already in place and causing the least amount of environmental damage feasible. It is therefore more important than ever to employ cutting-edge technical technologies to provide municipal services more quickly and effectively [48].

It is impossible to conduct a comprehensive analysis of construction and demolition waste management without considering specific local-scale factors and characteristics, which has hindered the optimal integration of circular economy principles in the construction industry. This study shows the significance of putting resources into neighbourhood arrangements that include little and miniature development activities as well as state-run administrations. As a matter of some importance, the discoveries feature the meaning of confined areas for the principal stockpiling under district ward.

This will urge waste makers to isolate trash nearby, which will bring down costs and limitations for networks. Then, on this scale, ordinary administrative activities at building locales are similarly pivotal because they advance improvement as far as elevating adherence to legitimately required systems and best practices for the administration of development-related waste. Notwithstanding, executing direct nearby techniques is easier for upgrading practice than depending exclusively on regulatory legitimate necessities [49,50].

Both biology and general wellbeing are harmed by solid waste. Uncollected fluid and solid garbage bin defile the air, water, and soil and lead to irresistible illnesses. Squander the board (WM) is one of the primary markers of supportability, and scholars from different business schools presently observe WM to be a critical subject. Moreover, WM should utilize suitable administration procedures to prevent the adverse consequences of waste on the environment and human wellbeing [51].

It is guessed that expanded industrialization, interests in sustainable power, and social improvement will make waste management in Togo a serious issue. Despite the fact that a great deal of variables influence asset and waste management, rules and plan systems are crucial. The legitimate prerequisites and administrative structures relating to waste management, along with the lawful execution process itself, should be progressively exhaustive to meet the mounting need for ecological insurance. The measures implemented in Togo to work on the fuse of additional feasible modern cycles include limitations and guidelines on the age of civil solid waste (MSW), decentralization of MSW on the board, approaches and motivator frameworks that empower squander waste decrease, reuse, and reusing, further developed requirements through the examination and treatment of infringement, and consolation of large-scale socio-economies in the administration of MSW [52].

Human wellbeing, the climate, and the far and wide ordinary utilization of plastic are profoundly impacting the environment. Subsequently, worldwide global cooperation is exemplified in decreasing the ecological dangers related to plastic utilization, encompassing something other than the execution of techniques and arrangements. This study aims to offer explicit knowledge into potential worldwide cooperation drives for the post-treatment and the board of plastic. Initially, a complete life cycle examination of the energy, ecological, and monetary necessities is done from birth to death for different plastic polymers [53].

A compelling MSW board technique could profit from both hard and soft measures. Be that as it may, for metropolitan experts in developing countries, putting resources into hard infrastructure, for example, sewage, and waste frameworks is an expensive undertaking. Minimal-expense social and vital mediations can assist with working on the adequacy of the ongoing foundation in these sorts of circumstances. Observational exploration exhibits how civil waste that is discarded is impacted by family inclinations about the booking and recurrence of waste assortment as well as admittance to road waste containers. The requirement for cleaner urban areas is additionally powered by expanded public information on the need of effective waste disposal, which is critical for upgrading MSW the board [54].

Some outsider co-product makers have recently created and carried out coproduction innovation. Green customers are profoundly drawn to co-products fabricated from surplus materials from customary creation, and they are in any event able to pay a premium for ecological protection. Since the co-product reduces the deals of their customary items, unique gear producers might be hesitant to embrace coproduction advances.

In many assembling areas, misuse of unrefined components persists in spite of upgrades in efficiency and asset proficiency, especially in those that have high requirements for normal assets, such as cowhide, wood, and stone. Numerous products just utilize natural substances whose quality outperforms a specific standard due to explicit models for surface, variety, and thickness; the excess bad quality parts are essentially disposed of. Thus, clinical solid waste result from homes, medical services offices, and senior living offices is separated into hazardous and non-hazardous classes. This is in accordance with WHO rules, which express that 15% of clinical solid waste ought to be hazardous and 85% non-hazardous [55,56].

Garbage from building and destruction tasks represents a lot of the solid waste filling landfills universally. Development and destruction trash is a huge part of the solid waste filling landfills around the world. Reusing building junk might assist with decreasing the amount of waste that is shipped to landfills as well as the demand for energy and other natural resources. An administration progressive system that starts with re-examine, reduce, upgrade, revamp, reuse, incinerator, treating the soil, reusing, and at last removal is probably going to be useful in diminishing development waste [57].

Home-grown waste creation has received huge consideration in recent years. Dry waste represents a lot of civil solid waste and it is taking care of and handling such waste that presents specialized concerns while utilizing a significant portion of urban resources. In order to implement long-term MSW policies, some metropolitan areas, districts, and neighbouring states need an accurate evaluation, prediction, and categorization of pressing issues. Subsequently, it has acquired significant review consideration, especially in a long time with the quick development of regions in emerging countries because of globalization. The increase of urbanization as well as population development has brought about colossal expansions in MSW creation [58].

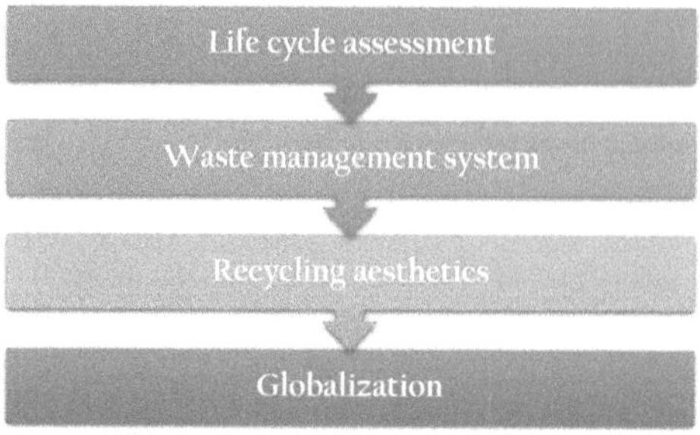

Figure 3.5 Insight of waste-to-energy strategies in waste manufacturing.

The development business is one of the world's driving wellsprings of solid waste. Truth be told, it is assessed that development and destruction waste represents no less than 30% of all strong rubbish produced universally. Lately, there has been an expansion in the quantity of examinations in the writing that cover the board choices. The development business makes the most. Given the potential results of an unfortunate administration and taking care of the board in Community India (CI) means a step back. Thus, CI all over the planet is progressing from a direct economy to the new worldview of the circular economy [59].

Commonly, the executive's waste management strategies are focused on in a progressive order, with prevention and minimization at the top and removal at the bottom. This is because anticipation and minimization are the best ways to deal with diminishing waste's natural effect. The board arrangements are the basic for defending both human wellbeing and the climate. We can add to a more supportable future by limiting how much waste we produce and discard. Figure 3.5 shows an insight of waste-to-energy strategies in waste manufacturing in the concept of zero waste.

3.6 CONCLUSIONS

Zero-waste production is essential for a sustainable future. Businesses can reduce their environmental impact, improve their financial performance, and boost their brand reputation by using zero-waste ideas and techniques. While zero-waste production is difficult to achieve, it is doable with a commitment to ongoing development and investment in new technology. Government programs, industry associations, and consulting firms are among the options available to assist businesses on their path to zero waste.

The following conclusions were drawn from the above chapter discussion:

- In green manufacturing, source reduction is the technique of eliminating waste before it is created. This can be accomplished in a variety of ways. Source reduction is a critical component of green manufacturing.

Manufacturers can decrease their environmental effect, enhance efficiency, and save money by employing source reduction methods.

- Strategies for coordinating AI into creation support acknowledge new methodologies; save time, energy, and assets; and diminish waste. However, various methodologies and apparatuses are utilized in the assembling area for PCs; ongoing advances, for example, AI and information mining procedures, are useful in addressing troublesome modern research challenges.
- The waste administration authority looks to adapt to the intricacies of civil solid waste management by resuscitating each piece of the waste administration partners to cooperate to manage waste. Individual families and networks, without which the executives will fizzle, were situated as the waste management's centre drive.
- Human health, the environment, and the widespread everyday use of plastic are all highly vulnerable. Thus, global international collaboration is exemplified in reducing the environmental risks associated with plastic consumption, encompassing more than just executing strategies and policies.
- Typically, waste management solutions are prioritized in a hierarchical order, with prevention and minimization at the top and disposal at the bottom. This is because prevention and minimization are the most effective approaches to reduce waste's environmental impact. Waste management solutions are critical for safeguarding both human health and the environment.

REFERENCES

1. Yang, G., Zhang, Q., Zhao, Z., & Zhou, C. (2023). How does the "Zero-waste City" strategy contribute to carbon footprint reduction in China? *Waste Management*, *156*, 227–235.
2. Ahmed, F., Hasan, S., Rana, M. S., & Sharmin, N. (2023). A conceptual framework for zero waste management in Bangladesh. *International Journal of Environmental Science and Technology*, *20*(2), 1887–1904.
3. Baba-Nalikant, M., Syed-Mohamad, S. M., Husin, M. H., Abdullah, N. A., Mohamad Saleh, M. S., & Abdul Rahim, A. (2023). A zero-waste campus framework: perceptions and practices of university campus community in Malaysia. *Recycling*, *8*(1), 21.
4. John, E. P., & Mishra, U. (2023). A sustainable three-layer circular economic model with controllable waste, emission, and wastewater from the textile and fashion industry. *Journal of Cleaner Production*, *388*, 135642.
5. Khalid, M. Y., Arif, Z. U., Hossain, M., & Umer, R. (2023). Recycling of wind turbine blade through modern recycling technologies: road to zero waste. *Renewable Energy Focus, 44*, 373–389.
6. Sarangi, P. K., Singh, A. K., Srivastava, R. K., & Gupta, V. K. (2023). Recent progress and future perspectives for zero agriculture waste technologies: pineapple waste as a case study. *Sustainability*, *15*(4), 3575.

7. Khan, S. A. R., Tabish, M., & Yu, Z. (2023). Investigating recycling decisions of internet recyclers: a step towards zero waste economy. *Journal of Environmental Management, 340*, 117968.
8. Kurniawan, T. A., Meidiana, C., Othman, M. H. D., Goh, H. H., & Chew, K. W. (2023). Strengthening waste recycling industry in Malang (Indonesia): lessons from waste management in the era of Industry 4.0. *Journal of Cleaner Production, 382*, 135296.
9. Okorie, O., Russell, J., Cherrington, R., Fisher, O., & Charnley, F. (2023). Digital transformation and the circular economy: creating a competitive advantage from the transition towards Net Zero Manufacturing. *Resources, Conservation and Recycling, 189*, 106756.
10. Leberruyer, N., Bruch, J., Ahlskog, M., & Afshar, S. (2023). Toward Zero Defect Manufacturing with the support of Artificial Intelligence—Insights from an industrial application. *Computers in Industry, 147*, 103877.
11. Alkaraan, F., Elmarzouky, M., Hussainey, K., & Venkatesh, V. G. (2023). Sustainable strategic investment decision-making practices in UK companies: the influence of governance mechanisms on synergy between industry 4.0 and circular economy. *Technological Forecasting and Social Change, 187*, 122187.
12. Yaro, N. S. A., Sutanto, M. H., Baloo, L., Habib, N. Z., Usman, A., Yousafzai, A. K., & Noor, A. (2023). A comprehensive overview of the utilization of recycled waste materials and technologies in asphalt pavements: towards environmental and sustainable low-carbon roads. *Processes*, 11(7), 2095.
13. Oluleye, B. I., Chan, D. W., Olawumi, T. O., & Saka, A. B. (2023). Assessment of symmetries and asymmetries on barriers to circular economy adoption in the construction industry towards zero waste: a survey of international experts. *Building and Environment, 228*, 109885.
14. Saira, G. C., & Shanthakumar, S. (2023). Zero waste discharge in tannery industries – an achievable reality? A recent review. *Journal of Environmental Management, 335*, 117508.
15. Ding, Z., Awasthi, S. K., Kumar, M., Kumar, V., Dregulo, A. M., Yadav, V., ... & Awasthi, M. K. (2023). A thermo-chemical and biotechnological approaches for bamboo waste recycling and conversion to value added product: towards a zero-waste biorefinery and circular bioeconomy. *Fuel, 333*, 126469.
16. Xu, S., Nupur, R., Kannan, D., Sharma, R., Sharma, P., Kumar, S.,... & Bai, C. (2023). An integrated fuzzy MCDM approach for manufacturing process improvement in MSMEs. *Annals of Operations Research, 322*(2), 1037–1073.
17. Veeman, D., & Palaniyappan, S. (2023). Process optimisation on the compressive strength property for the 3D printing of PLA/almond shell composite. *Journal of Thermoplastic Composite Materials, 36*(6), 2435–2458.
18. Isaja, M., Nguyen, P., Goknil, A., Sen, S., Husom, E. J., Tverdal, S., & Lamplmair, P. (2023). A blockchain-based framework for trusted quality data sharing towards zero-defect manufacturing. *Computers in Industry, 146*, 103853.
19. Kumar, S., Gopi, T., Harikeerthana, N., Gupta, M. K., Gaur, V., Krolczyk, G. M., & Wu, C. (2023). Machine learning techniques in additive manufacturing: a state of the art review on design, processes and production control. *Journal of Intelligent Manufacturing, 34*(1), 21–55.

20. Sohns, T. M., Aysolmaz, B., Figge, L., & Joshi, A. (2023). Green business process management for business sustainability: a case study of manufacturing small and medium-sized enterprises (SMEs) from Germany. *Journal of Cleaner Production*, *401*, 136667.
21. Akinwande, A. A., Adesina, O. S., Adediran, A. A., Balogun, O. A., Mukuro, D., Balogun, O. P.,... & Kumar, M. S. (2023). Microstructure, process optimization, and strength response modelling of green-aluminium-6061 composite as automobile material. *Ceramics*, *6*(1), 386–415.
22. Jafari, M., Shafaie, S. Z., Abdollahi, H., & Entezari-Zarandi, A. (2023). A green approach for selective ionometallurgical separation of lithium from spent Li-ion batteries by deep eutectic solvent (DES): Process optimization and kinetics modeling. *Mineral Processing and Extractive Metallurgy Review*, *44*(3), 218–230.
23. Nureen, N., Xin, Y., Irfan, M., & Fahad, S. (2023). Going green: how do green supply chain management and green training influence firm performance? Evidence from a developing country. *Environmental Science and Pollution Research*, *30*(20), 57448–57459.
24. Motla, R., Kumar, A., Saxena, N., & Sana, S. S. (2023). Inventory optimization in a green environment with two warehouses. *Innovation and Green Development*, *2*(4), 100087.
25. Ruidas, S., Seikh, M. R., Nayak, P. K., & Tseng, M. L. (2023). An interval-valued green production inventory model under controllable carbon emissions and green subsidy via particle swarm optimization. *Soft Computing*, *27*(14), 9709–9733.
26. Lino, F. A., Ismail, K. A., & Castañeda-Ayarza, J. A. (2023). Municipal solid waste treatment in Brazil: a comprehensive review. *Energy Nexus*, *11*, 100232.
27. Türkten, H. (2023). Impact of revenue generated via composting and recycling of wastes produced in the greenhouse tomato supply chain on reducing income inequality: a case study of Türkiye. *Sustainability*, *15*(18), 13801.
28. Ansari, S. A., Shakeel, A., Sawarkar, R., Maddalwar, S., Khan, D., & Singh, L. (2023). Additive facilitated co-composting of lignocellulosic biomass waste, approach towards minimizing greenhouse gas emissions: an up to date review. *Environmental Research*, *224*, 115529.
29. Mbama, C. A., Otegbulu, A., Beverland, I., & Beattie, T. K. (2023). Solid waste recycling within higher education in developing countries: a case study of the University of Lagos. *Journal of Material Cycles and Waste Management*, *25*(2), 886–898.
30. Le Pera, A., Sellaro, M., Sicilia, F., Ciccoli, R., Sceberras, B., Freda, C.,... & Cornacchia, G. (2023). Environmental and economic impacts of improper materials in the recycling of separated collected food waste through anaerobic digestion and composting. *Science of the Total Environment*, *880*, 163240.
31. Kasmuri, N., Razak, S. N. A., Yaacob, Z., Miskon, M. F., Ramli, N. H., & Zaini, N. (2023). Waste segregation through recycle and composting activities in urban and suburban areas. In *IOP Conference Series: Earth and Environmental Science* (Vol. 1135, No. 1, p. 012059). IOP Publishing.
32. Pheakdey, D. V., Van Quan, N., & Xuan, T. D. (2023). Potential greenhouse gas emissions and reduction from municipal solid waste management in Phnom Penh municipality. *Greenhouse Gases: Science and Technology*, *13*(5), 620–633.
33. Hamroyevna, M. M. (2023). Integration of waste recycling, composting and reduction strategies in Surkhandarya. *Multidisciplinary Journal of Science and Technology*, *3*(2), 47–49.

34. Bona, C. A. D., & Manuel, M. S. (2023). Awareness and compliance of the students, faculty, and, staff of Kalinga State University on the solid waste management program. *International Journal of English Literature and Social Sciences (IJELS)*, 8(3), 62–77.
35. Murugesan, S., Ramasamy, V., Varatharajan, R., Ilanchelian, S., & Chandrasekar, S. (2023, October). Recycling of waste paper and treatment of wastewater using organic filters – a review. In *AIP Conference Proceedings* (Vol. 2912, No. 1). AIP Publishing.
36. Rahman, A., & Amritphale, S. (2023). Advanced geopolymer: Utilizing industrial waste to material to achieve zero waste. In Todd Hoare (Ed.), *Advanced Materials from Recycled Waste* (pp. 255–272). Elsevier.
37. Rezania, S., Oryani, B., Nasrollahi, V. R., Darajeh, N., Lotfi Ghahroud, M., & Mehranzamir, K. (2023). Review on waste-to-energy approaches toward a circular economy in developed and developing countries. *Processes, 11*(9), 2566.
38. Permana, A. S., As'ad, S., & Potipituk, C. (2023). The multipronged approach of solid waste management toward zero waste to landfill site: an Indonesia and Thailand experience. In Ting, D.S.-K. & Stagner, J.A. (Eds.), *Pragmatic Engineering and Lifestyle: Responsible Engineering for a Sustainable Future* (pp. 113–129). Emerald Publishing Limited.
39. Czekała, W., Łukomska, A., Pulka, J., Bojarski, W., Pochwatka, P., Kowalczyk-Juśko, A.,... & Dach, J. (2023). Waste-to-energy: biogas potential of waste from coffee production and consumption. *Energy*, *276*, 127604.
40. García-Luna, S., & Ortiz, C. (2023). Conceptual assessment of sustainable methane production from oxycombustion CO_2 capture in waste-to-energy power plants. *Energy Conversion and Management*, *292*, 117348.
41. Ganesapillai, M., Mehta, R., Tiwari, A., Sinha, A., Bakshi, H. S., Chellappa, V., & Drewnowski, J. (2023). Waste to energy: a review of biochar production with emphasis on mathematical modelling and its applications. *Heliyon*, *9*(4), e14873.
42. Koshariya, A. K., Krishnan, M. S., Jaisankar, S., Loganathan, G. B., Sathish, T., Ağbulut, Ü., ... & Pham, N. D. K. (2023). Waste to energy: an experimental study on hydrogen production from food waste gasification. *International Journal of Hydrogen Energy, 54, 1–12.*
43. Ghozatfar, A., Yaghoubi, S., & Bahrami, H. (2023). A novel game-theoretic model for waste management with waste-to-energy and compost production under government intervention: a case study. *Process Safety and Environmental Protection*, *173*, 729–746.
44. Raju, G. K., Nagaraju, T. V., Jagadeep, K., Rao, M. V., & Varma, V. C. (2023). Waste-to-energy agricultural wastes in development of sustainable geopolymer concrete. *Materials Today:* Proceedings (Scopus, Q2).
45. Yazdani, S., & Lakzian, E. (2023). Conservation; waste reduction/zero waste. In Ting, D.S.-K. & Stagner, J.A. (Eds.), *Pragmatic Engineering and Lifestyle: Responsible Engineering for a Sustainable Future* (pp. 131–152). Emerald Publishing Limited.
46. Kokkinos, N. C., & Emmanouilidou, E. (2023). Waste-to-energy: applications and perspectives on sustainable aviation fuel production. In Wu, C. (Ed.), *Renewable Fuels for Sustainable Mobility* (pp. 265–286). Springer Nature Singapore.

47. Cook, N., Goodwin, D., Porter, J., & Collins, J. (2023). Food and food-related waste management strategies in hospital food services: a systematic review. *Nutrition & Dietetics*, *80*(2), 116–142.
48. Ma, W., Hao, J. L., Zhang, C., Di Sarno, L., & Mannis, A. (2023). Evaluating carbon emissions of China's waste management strategies for building refurbishment projects: contributing to a circular economy. *Environmental Science and Pollution Research*, *30*(4), 8657–8671.
49. Demircan, B. G., & Yetilmezsoy, K. (2023). A hybrid fuzzy AHP-TOPSIS approach for implementation of smart sustainable waste management strategies. *Sustainability*, *15*(8), 6526.
50. Ramos, M., Martinho, G., & Pina, J. (2023). Strategies to promote construction and demolition waste management in the context of local dynamics. *Waste Management*, *162*, 102–112.
51. Derhab, N., & Elkhwesky, Z. (2023). A systematic and critical review of waste management in micro, small and medium-sized enterprises: future directions for theory and practice. *Environmental Science and Pollution Research*, *30*(6), 13920–13944.
52. Beguedou, E., Narra, S., Agboka, K., Kongnine, D. M., & Afrakoma Armoo, E. (2023, July). Review of togolese policies and institutional framework for industrial and sustainable waste management. *Waste*, *1*(3), 654–671.
53. Loy, A. C. M., Lim, J. Y., How, B. S., Yiin, C. L., Lock, S. S. M., Lim, L. G.,... & Yoo, C. (2023). Rethinking of the future sustainable paradigm roadmap for plastic waste management: a multi-nation scale outlook compendium. *Science of the Total Environment*, *881*, 163458.
54. Nepal, M., Karki Nepal, A., Khadayat, M. S., Rai, R. K., Shyamsundar, P., & Somanathan, E. (2023). Low-cost strategies to improve municipal solid waste management in developing countries: experimental evidence from Nepal. *Environmental and Resource Economics*, *84*(3), 729–752.
55. Jin, M., Li, B., Xiong, Y., Chakraborty, R., & Zhou, Y. (2023). Implications of coproduction technology on waste management: who can benefit from the coproduct made of leftover materials? *European Journal of Operational Research*, *307*(3), 1248–1259.
56. Takunda, S., & Steven, J. (2023). Medical solid waste management status in Zimbabwe. *Journal of Material Cycles and Waste Management*, *25*(2), 717–732.
57. Lu, Y., Ge, Y., Zhang, G., Abdulwahab, A., Salameh, A. A., Ali, H. E., & Le, B. N. (2023). Evaluation of waste management and energy saving for sustainable green building through analytic hierarchy process and artificial neural network model. *Chemosphere*, *318*, 137708.
58. Lakhouit, A., Shaban, M., Alatawi, A., Abbas, S. Y., Asiri, E., Al Juhni, T., & Elsawy, M. (2023). Machine-learning approaches in geo-environmental engineering: exploring smart solid waste management. *Journal of Environmental Management*, *330*, 117174.
59. Soto-Paz, J., Arroyo, O., Torres-Guevara, L. E., Parra-Orobio, B. A., & Casallas-Ojeda, M. (2023). The circular economy in the construction and demolition waste management: a comparative analysis in emerging and developed countries. *Journal of Building Engineering*, *78*, 107724.

Chapter 4

Implementation and analysis of clean energy

Arulmurugan Gnanasekaran, Marimuthu Rengasamy, Ragunath Lakshmanan, Rajesh Nasinathan, and Kamatchi Rajaram

4.1 INTRODUCTION

The significant increase in energy demand caused by rapid industrialisation resulted in fossil fuel burning, accounting for 74% of worldwide anthropogenic greenhouse gas emissions in 2019 [1]. Hence, efforts to reduce carbon emissions from the energy system focus on decarbonising power generation and shifting end-use industries to electricity. For those reasons, governments throughout the globe are enacting legislation and policies to promote renewable energy (RE) use, conservation measures, and technical innovation [2]. RE adoption must increase dramatically by 2050 in order to attain the Sustainable Development Goals of the United Nations (UN) and limit global temperature rise to 1.5°C [3].

RE has become a prominent solution to address the carbon footprint. It is forecasted that RE will remain the predominant energy source until 2040 [4]. RE includes a range of energy forms like biomass, solar energy, geothermal power, mini-hydro, and biogas. Many countries have revised their energy strategies to increase the proportion of RE in their energy mix due to the pressing need to reduce the harmful effects of fossil fuels and growing global concerns [5].

Solar power is a readily available and abundant source of RE that can be harnessed at various capacities, ranging from small-scale applications to large-scale power generation in the megawatt range. The continent of Asia, in particular, holds great potential for solar power generation due to its high levels of solar radiation [6]. The urgent need for sustainable energy, reduction of carbon emissions, implementation and development of RE technologies has been emphasised at important global conferences such as the Paris Agreement Conference and COP26 in Glasgow [7]. These conferences have seen countries come together and commit to reducing their carbon emissions in order to address the issue of global warming. The significance of these efforts has been further underscored by the recent report from the Intergovernmental Panel on Climate Change.

DOI: 10.1201/9781003470342-4

Asian countries have the advantage of generating a relatively high amount of electricity from rooftop PV due to the extended sunlight hours and high solar radiation [8]. This not only makes homes energy self-sufficient and independent from the grid but also provides resilience against frequent power outages caused by weak or poor grids. Integrating rooftop PV systems with energy storage makes residential buildings more resilient to utility grid outages. Additionally, grid-connected rooftop PV systems can export excess electricity to the utility network, thereby increasing the capacity of the local grid and making the investment more appealing for building owners [9]. Hence, recently, Asian governments are implementing more policies of solar rooftop PV for electricity shortage and to reduce CO_2 emissions. This chapter reveals the government policies and regulations for the implementation of solar rooftop PV systems in various Asian countries like India, China, Indonesia, the Philippines, Vietnam, and the United Arab Emirates (UAE). Furthermore, this chapter explores CO_2 reduction in the above countries by implementing solar rooftop PV with existing energy systems or alone.

4.2 IMPLEMENTATION OF GOVERNMENT POLICIES FOR SOLAR ROOFTOP PV IN VARIOUS ASIAN COUNTRIES

4.2.1 India

Solar energy-based power, harnessed from the sun's rays, holds the potential to emerge as India's primary energy source, ensuring a reliable power supply and enhancing energy security. The direct normal irradiation (DNI) across India is shown in Figure 4.1. It is seen that the DNI increases year by year. Hence, it is the right time to concentrate and trap energy from solar power. Recently, the Indian solar industry has been gradually expanding since the launch of the National Solar Mission (NSM) by the Ministry of New & Renewable Energy (MNRE), Government of India, with the aim of reaching 100 GW capacity by 2022 [10]. The primary purpose of this initiative is to lower the long-term cost of solar power production, promote widespread deployment, and contribute to global warming mitigation. Figure 4.2 depicts the total installed capacity of solar power in different states of India. As a result, solar energy has been recognised as the best choice of RE in India, with minimal environmental impact.

India possesses a significant abundance of solar radiation, receiving an ample amount each year, surpassing 5,000 trillion kilowatts [13]. This surpasses the energy demand within the country and can be utilised effectively. To alleviate the strain on traditional energy sources and minimise carbon emissions, a range of solar technologies are accessible for both urban and rural areas. As part of the NSM, the Indian government has proposed

Figure 4.1 Direct normal radiation in India [11].

introducing grid-connected rooftop PV systems and small-scale solar systems. A budget of 5,000 INR crores has been earmarked over 5 years until 2019–2020 for this initiative. The MNRE is actively overseeing the implementation of this scheme and aiming to install grid-connected solar rooftop PV systems in suitable places.

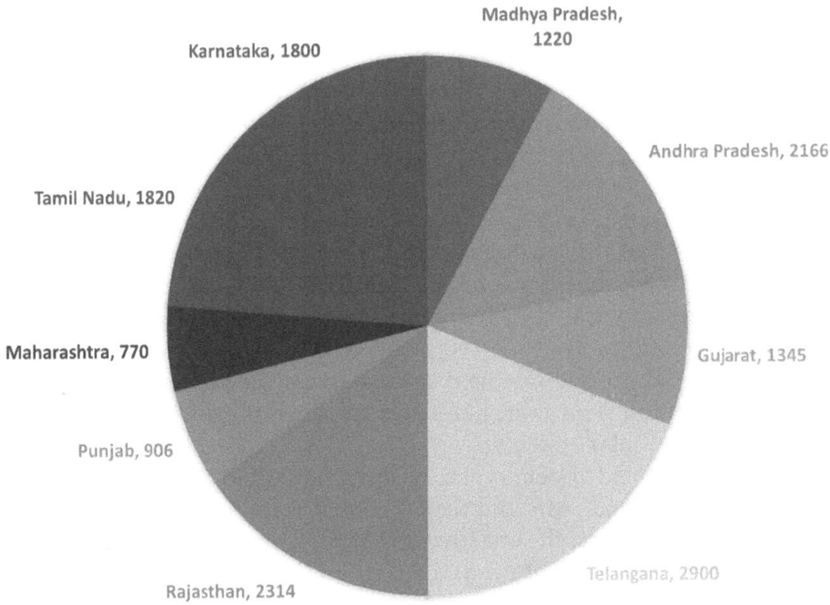

Figure 4.2 Installed capacity of solar power in different states of India [12].

The growing demand for electricity poses a considerable challenge for both government bodies and electricity utilities. In response to this challenge, the scheme called "Development of Solar Cities" has been introduced to aid cities in transitioning to solar energy. A total of 60 cities have been selected for this initiative across 11 schemes, with support from the ministry. In 2010, the Prime Minister of India inaugurated the Jawaharlal Nehru National Solar Mission, initially aiming to install 20 GW of grid-connected solar power by 2022 [14]. This objective was later amplified to 100 GW in 2015, subsequent to approval from the cabinet. To encourage RE and facilitate its implementation, the government has introduced various policy measures, such as net-metering policies, the establishment of the Suryamitra scheme, and the International Solar Alliance. Furthermore, state nodal agencies have been established at both state and central levels to promote the widespread adoption of efficient solar energy usage. Solar energy systems are expected to become the primary electricity source by 2050 [15].

According to infrastructure researchers in India, coal-fired power plants contribute to 76% of the total power production, while the remaining 24% comes from other sources [16]. Coal-fired power plants had a collective installed capacity of 198 GW in 2018, with their share of total installed capacity increasing from 2012 to 2016 and then steadily falling. This reduction is mainly due to the concerns about CO_2 and the integration of RE sources into the grid. Coal-fired power plants are known to generate pollutants that contribute to acid rain and air pollution. Despite technological advances that have resulted in newer plants emitting less pollution, removing impurities from coal remains difficult due to its chemical composition. Nevertheless, emissions from coal-based plants continue to exceed normal levels by a significant margin.

A consortium of ministries and institutions involved in research and development oversees a variety of projects. These entities have established suitable frameworks for conducting projects related to solar energy. Advisors cum specialists will evaluate the submitted projects before presenting them to the ministry for approval. Ground-mounted solar systems generate power which is then transferred to other places via grid connections. As a result, rooftop solar systems generate power in a distributed method, whereas ground-mounted solar systems generate the power in a centralised method. However, numerous considerations favour the use of ground-mounted solar systems.

A ground-mounted system typically costs more than a solar rooftop PV system, owing to land costs and the additional resources required for grid connection. In contrast, the effectiveness of rooftop solar systems depends on the condition of the roof on which they are installed. When the roof is in poor condition, the ability to generate electricity is reduced. Rooftop solar energy systems are often more expensive than coal-fired power plants. Nonetheless, advancements in technology and the implementation of various schemes have substantially reduced these costs.

Developing countries are currently focusing their efforts on increasing their solar energy capacity, particularly by building rooftop systems. India, in particular, has made significant progress in expanding its rooftop installations as part of its efforts to reduce costs. Nonetheless, the execution of huge installations has accidentally created barriers to the deployment of rooftop systems. India has set a goal of 40 GW of rooftop PV installations in 2022, with industrial and commercial customers accounting for more than 70% of these. In contrast, household users account for <20% of the whole capacity [17].

In India, there is a significant demand for electricity throughout the year, and thus, converting all the coal-fired power plants into RE-based sources is much more difficult due to the intermittent nature. Nonetheless, given technical improvements that have improved the efficiency and dependability of solar power systems, it is critical to take measures to replace certain coal-fired plants with solar energy-based systems. Although replacing all coal-based facilities is difficult, a portion can be converted into a solar-powered energy source to reduce carbon emissions.

According to Table 4.1, almost 50% of coal-based power plants have been replaced by solar-powered ones. In 2012, the replacement of 373 TWh of coal-based power units by solar production resulted in a 130 Mt CO_2 reduction in carbon emissions. Similarly, in 2017, 2022, 2027, 2037, 2042,

Table 4.1 Replacement of 50% of coal-based power plants by solar PV [12]

Particulars	*Unit*	*2012*	*2017*	*2022*	*2027*	*2032*	*2037*	*2042*	*2047*
Solar generation replacing coal generation	TWh	373	512	550	643	712	781	856	950
New coal generation	TWh	373	512	550	643	712	781	856	950
Emission reduction	Mt CO_2 Eq.	−130	−178	−191	−223	−247	−271	−297	−330

and 2047, carbon emissions were reduced by 178 Mt CO_2, 191 Mt CO_2, 223 Mt CO_2, 47 Mt CO_2, 271 Mt CO_2, 297 Mt CO_2, and 330 Mt CO_2, respectively. These numbers show a significant drop in carbon emissions. The Indian government has launched a number of initiatives and actively encouraged people to adopt solar energy-based equipment [12].

4.2.2 China

China, as one of the emerging East Asian countries, has experienced a notable shift in energy consumption patterns in recent decades. Moving away from heavy reliance on fossil fuels and coal, the country has made significant strides in utilising RE sources. This transition has propelled China toward a greener energy future, with initiatives dating back to the early 2000s in response to environmental concerns stemming from rapid economic development. Per capita electricity generation by source in China, 2022, as shown in Figure 4.3.

China took a major stride toward green energy with the introduction of the Renewable Energy Law in 2005. This legislation established goals for RE generation and required grid companies to purchase RE at a set rate, offering financial incentives to RE providers. The enactment of the RE Law facilitated China's swift growth in RE output. From 2005 to 2013, China's RE capacity surged significantly, driven by the expansion of wind, hydro, and solar power [18].

China has developed a solar photovoltaic (PV) poverty alleviation (PVPA) strategy with over 0.24 million PVPA power plants planned between 2014 and 2020 [19]. However, our present understanding of its consequences, which include not just direct poverty alleviation but also secondary aims like CO_2 reduction, is fairly limited.

China's PVPA programme has significantly improved socio-economic and environmental conditions throughout the country between 2014 and 2020. The government has allotted almost $3.52 trillion to establish

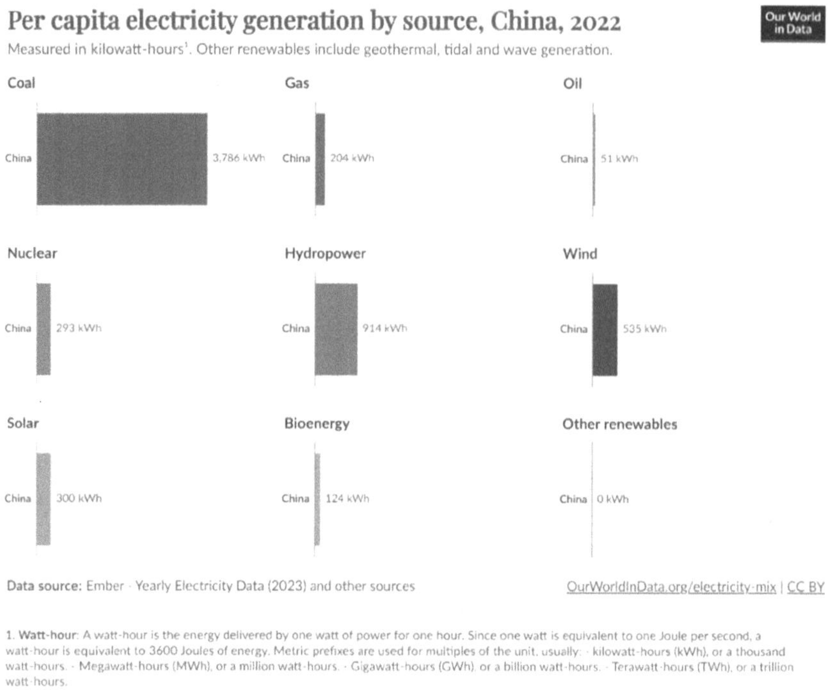

Figure 4.3 Per capita electricity generation by source in China, 2022 [20].

around 0.24 million PVPA plants in 1,224 cities with a capacity of 24 GW (Figure 4.4a and b). Two types of PVPA power plants have been implemented by China: village collective plants and rooftop household plants. The village-level plants, which numbered 81,765 and were put into service after January 2018, were mostly constructed in Shandong (10.20%), Hebei (9.35%), Jiangxi (11.89%), Anhui (16.47%), and Henan (14.72%). With a total of 157,476 solar rooftop PV household plants installed by the end of 2017, the majority of them were built in Anhui (37.75%), Hebei (12.78%), and Jiangxi (25.76%). The mean building cost for PVPA plants was 7.3 CNY/W, with the majority at around 85%, ranging between 5 and 8 CNY/W. PVPA plants had an average construction cost of 7.3 CNY/W, with 85% of those costs ranging between 5 and 8 CNY/W. It is noteworthy that construction prices were greater in the south than in the north, with the centre region spending more than the western regions (Figure 4.4c).

Although the village-level installations were more cost-effective, the building costs for solar rooftop PV systems (priced at 7.47 CNY/W) were somewhat higher. The solar payback period (SPP) of 64.5% of PVPA plants nationwide ranged between 6 and 8 years, with an average SPP of 5.95 years.

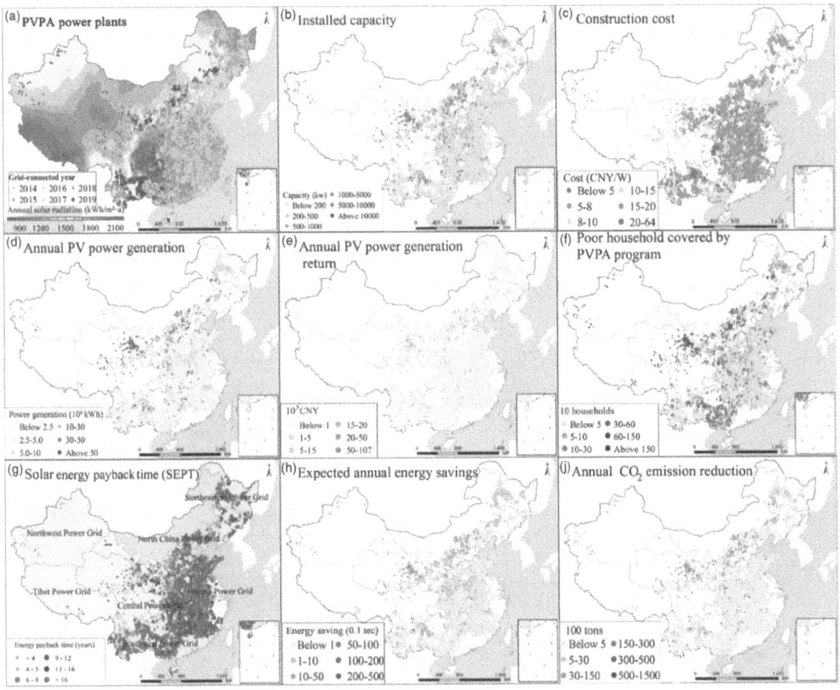

Figure 4.4 China's PVPA power plants with energy-saving and reduction in emission [21]. (a) PVPA power plant installation and annual solar irradiation. (b, c) Installed capacity and construction cost. (d, e) Generation of average annual power and revenue generated by individual plant. (f) Poor households covered by the PVPA programme. (g) Details of SPP. (h, i) Annual energy savings and reduction of CO_2 emission, respectively.

Figure 4.4g depicts the detailed SPP in China. Village-level installations had a shorter SPP of 5.53 years compared to residential rooftop installations, which was mostly because of higher generating energy capacity and easier access to government subsidies. In comparison to the northern area, the southern area's SPP for PVPA power plants was longer. Furthermore, there was a negative correlation found between PVPA power plants' SPP and sunlight hours, suggesting that plants with higher sunlight exposure had shorter SPP. PVPA power plants have the ability to save energy while also lowering pollution. By 2020, 31.75 million tonnes of CO_2 emissions were reduced by China's PVPA plants, which generated 34.35 billion kWh of power, worth CNY 31.9 billion (about 4.6 billion USD).

It is projected that China's PVPA plants will run for 25 years. They are expected to generate 774 billion kWh of power throughout this time, saving 278.65 million tonnes of energy and reducing CO_2 emissions by 715.75 million tonnes between 2020 and 2045. Figure 4.4d, e, h, and i demonstrates

Chinese PVPA plants' enormous potential to combat climate change by depicting the geographical distribution of average annual PV power generation, annual PV power generation return, annual energy savings, and annual emission reduction from 2020 to 2045, respectively. The maps reveal that Central and North China are poised to experience the most significant impact in terms of power generation income and emission reduction.

4.2.3 Indonesia

Indonesia's per capita energy consumption is 1.131 MWh, while the country's overall consumption was 0.99 tonnes of electricity (toe) in 2022. From 2013 to 2019, total energy use increased by 3.4% yearly; however, in 2020, it decreased by almost 5% [22]. Figure 4.5 illustrates the major source of energy in Indonesia. The government has consistently raised electricity

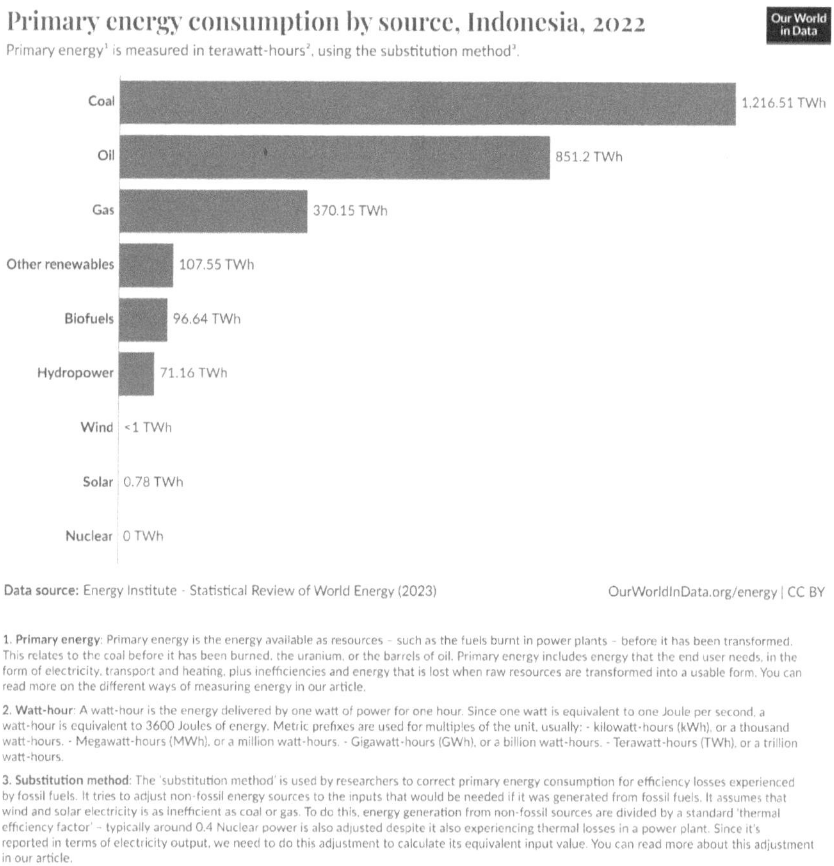

Figure 4.5 The installed different sources of energy in 2022, Indonesia [27].

tariffs in recent times, resulting in a corresponding rise in per capita energy consumption.

Ministry of Energy and Mineral Resources (MEMR), Indonesia, is developing various plans to support solar rooftop PV systems, which aim to support the country's solar PV industry and fulfil its international obligations to minimise the emission level [23]. Additionally, the report emphasises that commercial, residential, and industrial PV installations can send back the excess power into the grid via a net-metering scheme. The government expects that the net energy metering (NEM) project would result in the installation of around 1 GW of solar PV in another 3 years, allowing PV system owners to save up to 30% on their energy costs. The goal of the NEM programme is to reduce the quantity of electricity exported to Perusahaan Listrik Negara, the utility provider, by emphasising PV installations with high self-consumption rates [24]. The government has also implemented the MEMR Regulation No. 4 of 2020 as an additional step to strengthen these regulations. The regulations aim to promote more industrial clients to install solar panels and reduce energy expenses [25].

The Indonesian government emphasised the need to develop green technology, economy, and goods in order for the country to compete effectively in the global market. The planned Super Grid, which will begin in 2025, promises to provide electricity availability to local communities across Indonesia. Regulatory changes, such as power wheeling, will make it easier to transmit electricity directly from renewable sources to operational facilities, driving the Green Grid forward even further. Indonesia has set ambitious goals for 2025, including obtaining a 23% share of RE, lowering emissions in accordance with nationally decided contribution targets, and achieving net zero emissions by 2060 with international assistance [26].

The MEMR highlighted the need for financial support in addition to the government-instituted regulatory framework. This framework comprises projects such as the Electric Power Supply Business Plan (RUPTL 2021–2030), legislation governing the carbon economy, and direction to net zero emissions in 2060. The Indonesian government's efforts include eliminating the export energy discount, extending the set-off period, expediting licencing procedures, and enacting comprehensive RE laws and plans. As a result, customers will become very interested in developing rooftop solar PV installations under the NEM programme.

Indonesia has revised its national determined contributions (NDC) for 2022. The country has marginally elevated its unconditional emission reduction target from 29% in 2030 to 31.89% when compared to the business-as-usual (BAU) scenario. Likewise, the conditional emission reduction targets have been increased from 41% to 43.2%. These conditional targets are contingent upon the achievement of global agreements, including bilateral cooperations, technology development and transfer, and financial assistance.

4.2.4 The Philippines

Fossil fuels have been integrated into the energy generation sector of the Philippines, with coal emerging as the primary sector of the national energy mix, representing 30.9%, followed closely by oil at 29.9% and 4.6% of natural gas. Conversely, RE sources have contributed 34.6% to the country's energy mix out of it, solar and wind contribute 15.4%, biofuel and waste account for 17.8%, and hydropower contributes 1.4%, as depicted in Figure 4.6. Over the years, the Philippines has experienced consistent growth in the utilisation of solar PV, reflecting the government's focus on its development.

The Renewable Energy Act of 2008 was approved by the government to transition from fossil fuel-based power generation to renewable one [28]. Considering that most of the power utilised in the Philippines, coal and oil-fired power plants play an essential role. This Act also aims to increase the nation's RE investments' competitiveness. It has created a net-metering policy framework and an Energy Regulatory Commission (ERC)-managed incentive programme. Customers are encouraged to actively engage in the production of RE through this project, which also allows them to export excess power to a distribution utility (DU).

The solar energy capacity of the Philippines has increased significantly during the last 10 years; it is expected to increase from 2 MW in 2011 to 1,048 MW by 2020. The country's ability to generate RE and move away from conventional energy sources has been greatly enhanced by the Renewable Energy Act of 2008. Any excess power generated by consumers is routed back into the DU's distribution network through the NEM programme. After subtracting any further production adjustments, the DU then credits the user's electric account with an amount equal to the DU's average production cost in pesos; these credits are not included in the bill. Per kilowatt-hour (kWh), customers are charged P9.985 for electricity that comes from the grid. Additionally, clients may sell extra power produced by their rooftop solar systems for as low as P5.42/kWh.

NEM programmes in the Philippines are aggressively promoting consumer participation in the production of RE, hence promoting an environment with minimal carbon emissions [29]. Additionally, by 2022, the Philippines government wants to increase the amount of solar PV systems

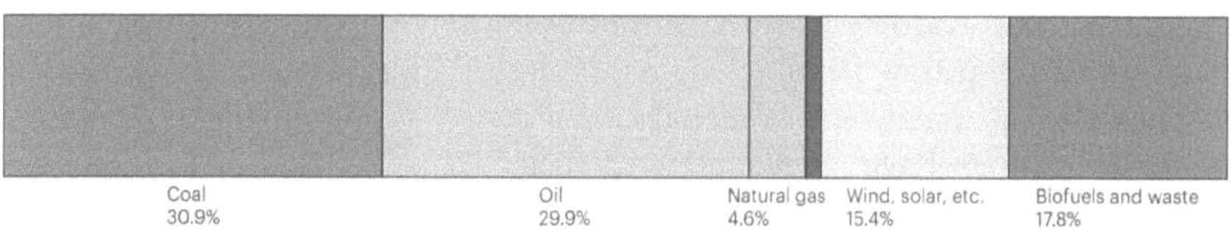

Figure 4.6 Total energy supply in the Philippines, 2021 [31].

installed to 3 GW. Moreover, it is projected that by the end of 2030, the total installed solar capacity would attain 8.7 GW, with solar rooftop PV systems constituting 35% of the installations. The Philippines' solar energy industry has expanded more quickly through NEM programmes. The ERC modified its NEM regulations in 2020 in an effort to encourage more people to participate in a programme that allows energy end users to feed power into the grid [30]. Changes were made to the ERC's net-metering programme for RE, which permits ordinary electricity consumers to become "prosumers" who generate electricity for their own purposes and sell the excess to DU. According to the ERC, DUs must complete the procedure within 20 days of receiving the letter of interest, assuming all necessary licences and permissions are obtained from the appropriate authorities. With the use of affordable, cutting-edge, and renewable technologies, the ERC's Amended Net-Metering Rules seek to empower consumers by enabling their own energy generation and, consequently, reducing their utility expenses.

In economically developed countries, per capita energy-related CO_2 emissions generally exhibit higher levels. However, these emissions can significantly differ based on the economic structure and energy system of each country. For example, countries that rely largely on carbon-intensive means of transportation (such as driving and flying) have a higher percentage of energy-intensive industries (such as steel or chemicals) or rely heavily on fossil fuels for power generation and would have higher per capita emissions. As depicted in Figure 4.7, the CO_2 emission per capita in the Philippines until 2021 amounts to 1.161 tCO_2/capita. After the implementation of solar rooftop PV systems, which are replaced with existing energy plants, CO_2 emission is reduced to 1.143 million tonnes CO_2eq [32].

4.2.5 Vietnam

Electricity generation in Vietnam, 2021, is depicted in Figure 4.8. Recently, Vietnam has emerged as the Association of Southeast Asian Nations (ASEAN) region's leading adopter of RE sources such as solar and wind [34].

Figure 4.7 CO_2 emissions per capita, the Philippines [33].

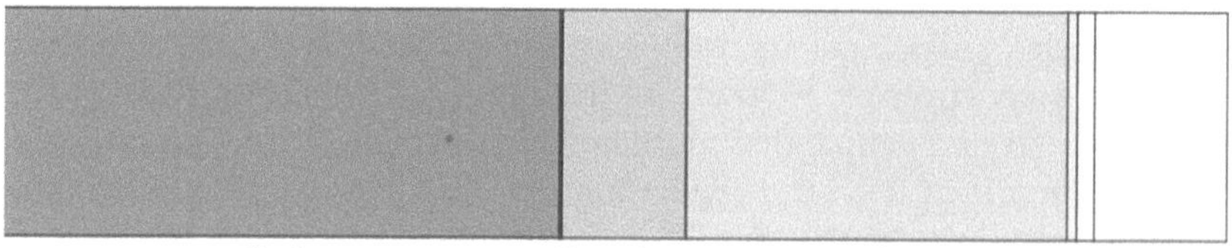

Figure 4.8 Electricity generation in Vietnam, 2021 [37].

Vietnam outperformed Thailand in terms of total solar and wind power potential in 2019. Vietnam's total solar PV capacity is expected to reach 16,500 MW by the end of 2020. This accomplishment surpasses the initial target of 850 MW set for 2020 (Government of Vietnam, 2016), bringing Vietnam closer to the provisional goal of installing a solar power capacity of 18.6 GW by 2030, as outlined in Vietnam's Power Development Plan 8 (Government of Vietnam, 2016).

Vietnam installed more than 1 lakh solar rooftop PV systems in 2019 and 2020, a notable accomplishment [35]. Vietnam broke all previous records for solar capacity in 2020, with 9.3 GW coming from solar rooftop PV systems that power the nation's electrical grid—despite the obstacles posed by the COVID-19 outbreak. Vietnam's solar rooftop capacity is expected to rise eightfold by December 2020 from approximately 378 MW in 2019. Through Decision No. 11/2017/QD-TTg, the Vietnamese government has made significant contributions to the expansion of solar power and advanced the development of RE in the nation [36].

The yearly purchase and offering prices for solar rooftop grid-connected PV systems will be set by the Ministry of Trade and Industry based on the current exchange rate between VND/USD. The payment for excess electricity is set at 9.36 US cents/kWh (2,087 VND) and is adjusted annually based on the exchange rate between VND/USD. Any extra power generated in one payment cycle will be rolled over to the next one. Surplus power will be sold to power Vietnam at the feed-in-tariff price upon the expiry of the Power Purchase Agreement ("PPA").

Greenhouse gas emissions are the main contributor to climate change and are mostly produced by burning fossil fuels. Global CO_2 emissions are still trending far higher than what is needed to prevent the worst effects of climate change despite attempts to reduce them. In Vietnam, CO_2 emissions from fuel combustion in the energy plant is 284.75 million tonnes, which is shown in Figure 4.9. After the implementation of a solar rooftop PV system that replaced 50% of the existing energy plant, the CO_2 emission is avoided by 14.32 million tonnes CO_2eq [32].

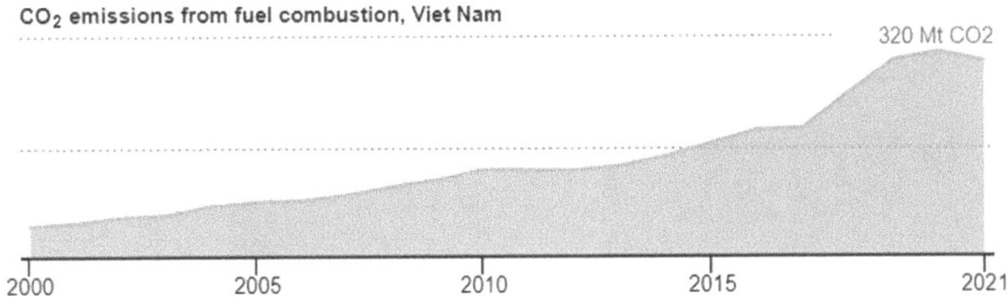

Figure 4.9 CO_2 emission from fuel combustion, Vietnam [38].

4.2.6 The UAE

The trend of the UAE's percentage of renewable power generation from 1990 to 2022 is shown in Figure 4.10. Renewable power generation was very low before 2015, but new legislative changes and technology improvements have caused a sharp rise that will see it reach 10% by 2022 [39]. It is crucial to emphasise that the UAE's main RE source is solar energy.

Figure 4.11 shows a significant increase in the UAE's installed solar energy capacity. Until 2013, solar energy installations were nearly non-existent. But after 2018, there was a remarkable increase, with the installed capacity surpassing 2,500 MW by 2020.

The solar rooftop plan (SRP) in Abu Dhabi takes the incentive to promote the installation of solar PV systems on the rooftops of commercial buildings. Masdar and the Abu Dhabi Water and Electricity Authority (ADWEA) are leading this programme, which aims to produce 500 MW of power from the solar roofs over the next 20 years. The SRP in the UAE began in 2017, with an initial installed capacity of approximately 4 MW [41].

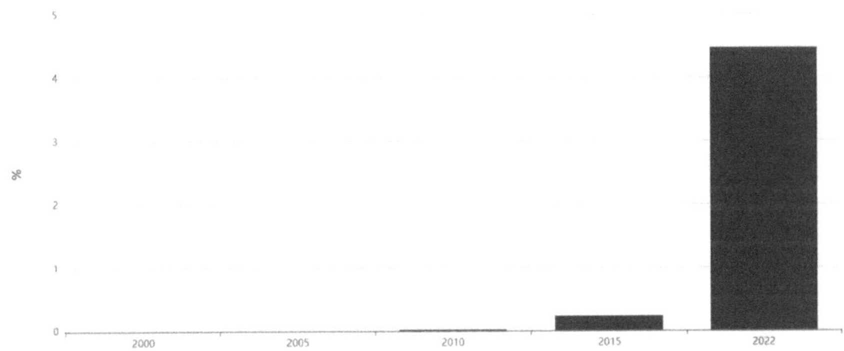

Figure 4.10 RE-based power generation in percentage between 1990 and 2022 in the UAE [40].

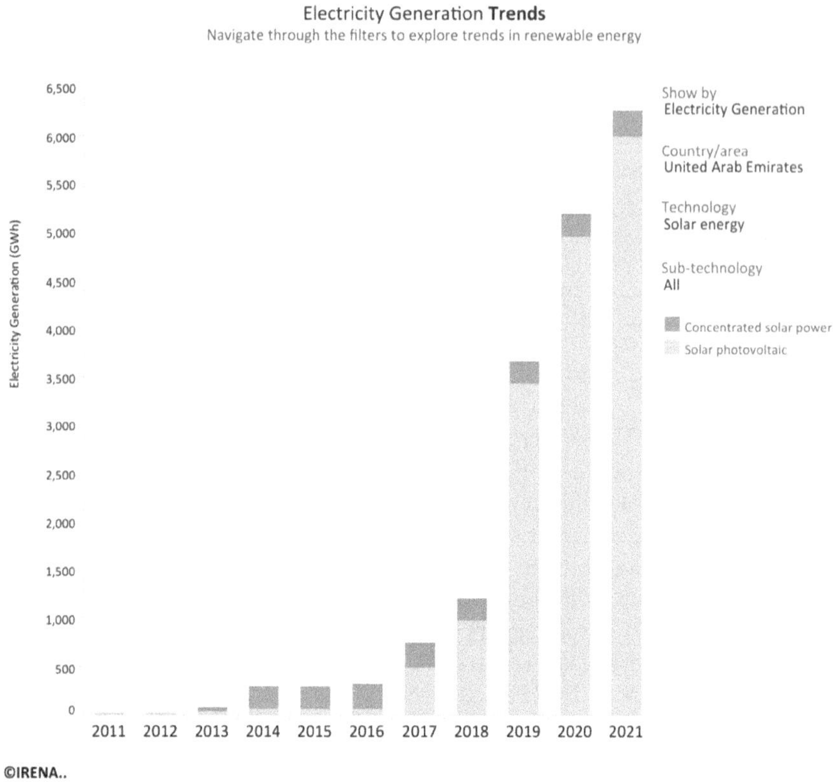

Figure 4.11 Solar power capacity in the UAE [42].

By the end of 2017, the capacity had grown to 20 MW. As part of the programme, rooftop PV panels with a 2.4-MW capacity have been installed on 11 government buildings, producing 4.024 GWh of power yearly. This reduces 3,220 tonnes of CO_2 emissions. Owing to the special environmental circumstances in Abu Dhabi, a pilot study is required for evaluating the effectiveness of PV panel facilities and providing an appropriate financial incentive scheme. The SRP is expected to replace existing electricity facilities, which is avoiding 2.326 million tonnes of CO_2eq emissions [32].

4.3 CONCLUSION

The depletion of conventional energy resources and severe climate change have made the transition to sustainable and clean energy solutions imperative. Among all RE sources, the solar rooftop PV system plays a major role in sustainable energy production. This chapter details how different Asian countries have taken the initiative to implement the solar rooftop PV system in order to reduce CO_2 emissions. Furthermore, this chapter reveals successful government policies and regulations of the solar rooftop PV system that have significantly improved the energy efficiency with the existing system or alone in various Asian countries like India, China, Indonesia, the Philippines, Vietnam, and the UAE. The said Asian countries had participated in the Paris Agreement Conference and COP26 in Glasgow and discussed to reduce CO_2 emission by implementing solar rooftop PV system with the existing energy system partially or fully. In the future, solar rooftop PV system adoption will increase dramatically by 2050 in order to attain the SDG and limit global temperature rise to 1.5°C.

REFERENCES

1. Olivier, J.G., Schure, K.M. and Peters, J.A.H.W., 2017. Trends in global CO_2 and total greenhouse gas emissions. *PBL Netherlands Environmental Assessment Agency*, 5, 1–11.
2. Zhou, Q., Wu, J., Imran, M., Nassani, A.A., Binsaeed, R.H. and Zaman, K., 2023. Examining the trade-offs in clean energy provision: Focusing on the relationship between technology transfer, renewable energy, industrial growth, and carbon footprint reduction. *Heliyon*, 9(10), e20271.
3. Azzuni, A., Aghahosseini, A., Ram, M., Bogdanov, D., Caldera, U. and Breyer, C., 2020. Energy security analysis for a 100% renewable energy transition in Jordan by 2050. *Sustainability*, 12(12), 4921.
4. Xiaosan, Z., Qingquan, J., Iqbal, K.S., Manzoor, A. and Ur, R.Z., 2021. Achieving sustainability and energy efficiency goals: Assessing the impact of hydroelectric and renewable electricity generation on carbon dioxide emission in China. *Energy Policy*, 155, 112332.
5. Saygin, D., Kempener, R., Wagner, N., Ayuso, M. and Gielen, D., 2015. The implications for renewable energy innovation of doubling the share of renewables in the global energy mix between 2010 and 2030. *Energies*, 8(6), 5828–5865.
6. Prăvălie, R., Patriche, C. and Bandoc, G., 2019. Spatial assessment of solar energy potential at global scale. A geographical approach. *Journal of Cleaner Production*, 209, 692–721.
7. Carey, E. and Yang, X., 2023. From Paris to Glasgow and beyond: What future for clean energy technology deployment under Article 6? In: Considine, J., Cote, S., Cooke, D., and Wood, G. (Eds.), *A Research Agenda for Energy Politics* (127–154). Edward Elgar Publishing.
8. Rababah, H.E., Ghazali, A. and Mohd Isa, M.H., 2021. Building integrated photovoltaic (BIPV) in Southeast Asian countries: Review of effects and challenges. *Sustainability*, 13(23), 12952.
9. Ayompe, L.M. and Duffy, A., 2013. Feed-in tariff design for domestic scale grid-connected PV systems using high resolution household electricity demand data. *Energy Policy*, 61, 619–627.

10. Behuria, P., 2020. The politics of late late development in renewable energy sectors: Dependency and contradictory tensions in India's National Solar Mission. *World Development,* 126, 104726.
11. Direct normal radiation in India. https://globalsolaratlas.info/download/india
12. Akhtar, I., Kirmani, S., Jameel, M. and Alam, F., 2021. Feasibility analysis of solar technology implementation in restructured power sector with reduced carbon footprints. *IEEE Access,* 9, 30306–30320.
13. Pandey, A., Pandey, P. and Tumuluru, J.S., 2022. Solar energy production in India and commonly used technologies—An overview. *Energies,* 15(2), 500.
14. Sharma, A., Srivastava, K. and Kar, S.K., 2015. Jawaharlal Nehru national solar mission in India. In: Sharma, A., and Kar, S.K. (Eds.), *Energy Sustainability through Green Energy* (pp. 47–67). Springer.
15. Breyer, C., Bogdanov, D., Khalili, S. and Keiner, D., 2021. Solar photovoltaics in 100% renewable energy systems. In: Meyers, R.A. and Jäger-Waldau, A. (Eds.), *Encyclopedia of Sustainability Science and Technology* (p. 1). Springer.
16. Cropper, M.L., Guttikunda, S., Jawahar, P., Malik, K. and Partridge, I., 2018. Costs and benefits of installing flue-gas desulfurization units at coal-fired power plants in India. In: Mock, C.N., Nugent, R., Kobusingye, O, et al., (Eds.), Injury Prevention and Environmental Health, 3rd ed. Washington (DC): The International Bank for Reconstruction and Development / The World Bank. https://www.ncbi.nlm.nih.gov/books/NBK525204
17. Chowdhury, S.A., 2020. National solar energy roadmap, 2021–2041. Sustainable and Renewable Energy Development Authority (SREDA). https://www.scribd.com/document/531525711/Draft-National-Solar-Energy-Roadmap-2021-2041
18. Yang, X.J., Hu, H., Tan, T. and Li, J., 2016. China's renewable energy goals by 2050. *Environmental Development,* 20, 83–90.
19. Wang, Y., Cui, X. and Huang, H., 2023. Spatial patterns and environmental benefits of photovoltaic poverty alleviation programs in China. *Environmental Impact Assessment Review,* 103, 107272.
20. Per capita electricity generation by source in China, 2022 [online]. Available, https://ourworldindata.org/energy-mix
21. Zhou, Y., Wang, H., Liu, Z., Bhattarai, N., Paudel, J. and Qiu, H., 2023. Can solar photovoltaic poverty alleviation policies reduce carbon emissions and increase income in China? *Environmental Science & Technology,* 57(49), 20583–20594.
22. Indonesia Energy Information, 2022 [Online]. Available, https://www.enerdata.net/estore/energy-market/indonesia
23. Indonesia: New MEMR Regulation on Rooftop Solar. [Online]. Available, https://insightplus.bakermckenzie.com/bm/projects/indonesia-new-memr-regulation-on-rooftop-solar-generating-interest-in-indonesian-rooftop-solar
24. Tongsopit, S., Amatayakul, W., Saculsan, P.G., Nghia, V.H., Tirpornvitoon, C., Favre, R. and Afanador, A., 2017. Designing renewable energy incentives and auctions: Lessons for ASEAN. USAID Clean Power Asia.

25. Maulia, R.F. and Siregar, A.A., 2020. Unlocking renewable energy potential in Indonesia: Assessment on project viability. Institute for Economic and Social Research. https://lpem.org/wp-content/uploads/2020/07/WP-LPEM-052-Unlocking_Renewable_Energy_Potential_in_Indonesia_Rev.pdf
26. Arkantyo, M.D., Susilawati, N., Aditya, I.B.P. and Krishnamurti, I., The partnership between Indonesia and United Kingdom to achieve the net-zero emission goal on behalf of sustainable development action. https://www.researchgate.net/profile/Aditya-Ida-Bagus-2/publication/369926205_The_Partnership_between_Indonesia_and_United_Kingdom_to_Achieve_the_NetZero_Emission_Goal_on_Behalf_of_Sustainable_Development_Action/links/643501d3609c170a130b55fc/The-Partnership-between-Indonesia-and-United-Kingdom-to-Achieve-the-NetZero-Emission-Goal-on-Behalf-of-Sustainable-Development-Action.pdf
27. The installed different sources of energy in 2022, Indonesia [online]. Available, https://ourworldindata.org/energy-mix
28. Renewable energy Act of 2008, [online]. Available, https://www.iea.org/policies/4754-renewable-energy-act
29. Dahlan, N.Y., Ahmad, N., Ilham, N.I. and Yusoff, S.H., 2022. Energy security: Role of renewable and low-carbon technologies. In: Asif, M. (Ed.), *Handbook of Energy and Environmental Security* (pp. 39–60). Academic Press.
30. Rudnick, H. and Velasquez, C., 2019. Learning from developing country power market experiences: The case of the Philippines. World Bank Policy Research Working Paper, (8721).
31. Total energy supply in Philippines, 2021 [Online]. Available, https://www.iea.org/countries/philippines/energy-mix
32. Avoided Emissions Calculator [Online]. Available, https://www.irena.org/Data/View-data-by-topic/Climate-Change/Avoided-Emissions-Calculator
33. CO_2 emissions per capita, Philippines 2021 [Online]. Available, https://www.iea.org/countries/philippines/emissions
34. Pandey, A.K., Kalidasan, B., Reji Kumar, R., Rahman, S., Tyagi, V.V., Krismadinata, Said, Z., Salam, P.A., Juanico, D.E., Ahamed, J.U. and Sharma, K., 2022. Solar energy utilization techniques, policies, potentials, progresses, challenges and recommendations in ASEAN countries. *Sustainability,* 14(18), 11193.
35. Do, T.N., Burke, P.J., Nguyen, H.N., Overland, I., Suryadi, B., Swandaru, A. and Yurnaidi, Z., 2021. Vietnam's solar and wind power success: Policy implications for the other ASEAN countries. *Energy for Sustainable Development,* 65, 1–11.
36. Dung, T.V., 2020. Legal and policy framework for renewable energy and energy efficiency development in Vietnam. *Vietnamese Journal of Legal Sciences,* 1(1), 33–47.
37. Electricity generation in Viet Nam, 2021 [Online]. Available, https://www.iea.org/countries/viet-nam/energy-mix
38. CO_2 emission from the fuel combustion, Vietnam [Online]. Available, https://www.iea.org/countries/viet-nam/emissions

39. Share of renewables in electricity production, [Online]. Available, https://yearbook.enerdata.net/renewables/renewable-in-electricity-production-share.html
40. Renewable power generation percentage trend between 1990 and 2022 in the UAE, [Online]. Available, https://yearbook.enerdata.net/renewables/wind-solar-share-electricity-production.html
41. Salimi, M., Hosseinpour, M. and N. Borhani, T., 2022. Analysis of solar energy development strategies for a successful energy transition in the UAE. *Processes,* 10(7), 1338.
42. Installed capacity of solar energy in the UAE [Online]. Available, https://www.irena.org/Energy-Transition/Technology/Solar-energy

Chapter 5

Circular economy and waste management

Venkata Siva Prasanth Yendeti, Raajasekar Mathiyalagan, and Jayakrishna Kandasamy

5.1 INTRODUCTION

5.1.1 Green manufacturing

The main goal of green manufacturing is to reduce the impact throughout the entire manufacturing process. Achieving green manufacturing involves implementing sustainable methods to replace current ones. The primary aim is to reduce pollution by reducing waste and emissions by effectively utilizing resources and adopting cleaner technologies. The main goal is to reduce energy consumption, limit material usage, and reduce the impact of manufacturing activities through pollution control. Green manufacturing integrates eco methods into production such as by using renewable energy resources, encouraging the use of recycled materials, and adopting optimization strategies to lower emissions and waste generation (Singh et al., 2018; Zhidebekkyzy et al., 2024). Green manufacturing also meets the criteria targeting environmental needs, supporting the growth of the customer needs and for environmentally friendly products usage. Green manufacturing implementation increases the essence of creativity by encouraging the development of technologies and practices which are environment friendly. The role of green manufacturing in shaping the future of the manufacturing industry by reducing dependency on resources and improving accountability is vastly improving (Almadhi et al., 2023).

5.1.2 Waste management

When it comes to a conventional way of open dumping, or even merely solo incineration, this is better. One of the main differences between a linear and circular economy (CE) is that the CE seeks to keep products, components, and materials for circulation instead of wasting them (Agovino et al., 2024) – generating the maximum value from resources while they are being used and then recovering them as their end of life happens (Di Foggia and Beccarello, 2024). The CE implementation has many challenges, and this

DOI: 10.1201/9781003470342-5

drives a path for CE with the deployment of digital solutions along with cutting-edge waste management (Costa et al., 2024). The waste management sector needs to come up, and probably, metropolitan cities may not have an inch of landfills shortly, raising the scope for new growth areas within the economy and giving us hands-on job creation, as well as CE and waste management (Mandpe et al., 2022). This chapter explores these dimensions by demonstrating how a CE can transform waste management from a linear into an entirely new pattern, underlining the impact of policies, digital innovation, and environmentally sound practices that bring societal transition closer.

5.1.3 Circular manufacturing

Circular manufacturing (CM) constructs a closed-loop system by which materials are used as long as possible. CM is sustainable to the environment, which is an industrial technique to prolong the lifespan of resources in the manufacturing sector. CM describes the need of retaining the materials within the production and consumption cycle as long as possible, reducing the waste and the demand for new resources (Dahmani et al., 2021). Implementing recycling, remanufacturing, and repurposing activities within a closed-loop system allows for the continuous reusage of products and materials, which is the main aim of CM (Dumitrescu-Popa et al., 2024). CM encourages diminishing the environmental impact and promotes resource efficiency, thereby transforming the waste into a beneficial input to the future production cycles (Rumo, 2023). Thus, CM contributes towards sustainability by adapting a therapeutic economic framework.

The aim of CM is to integrate the CE principles into manufacturing processes for resource efficiency and sustainability. Implementing CE methodologies such as recycling, remanufacturing, and industrial symbiosis at different levels: micro (individual enterprises and products), meso (networks of firms), and macro (regions and nations). Artificial intelligence (AI), which is a recent booming technology, improves the ability for tracking and monitoring the data, and for making the decisions which support the execution of CM methods. AI makes the process streamlined by analyzing the data in real time and by increasing the efficiency for the utilization of resources and makes it easy for the shift toward a regenerative system. Since AI has been tremendous in the progress of its application at the micro level, the potential at the meso and macro levels remains at the basic level (Abideen et al., 2023), suggesting that there are substantial opportunities for future research in CM.

The transition to CE has diverse and essential benefits for achieving sustainability and resource efficiency in different types of sectors (Köpman and Majava, 2024). The benefit of using CE principles is keeping products and materials in productive service for a prolonged period and effectively maintaining the value of these kinds of resources (Golinska-Dawson et al., 2024). This extends the life cycle of the products. The life cycle advances and the value of material are maximized for the resource input as good emissions/energy consumption minimization. Additive manufacturing (AM), the essence of being an additive process, entails CE because it allows for on-demand production, cuts warehousing costs, and curtails material waste (Suzanne et al., 2020). AM contributes to environmental sustainability by enabling patient-specific treatments by improving the design and to produce biodegradable medical equipment and implants, contributing to environmental sustainability.

By this, the need to apply CE principles is essential, especially in the healthcare industry because the industry is an enormous consumer of resources that generates massive amounts of waste. Implementing CE practices also reveals that implementing it may save costs by diminishing material contractions and waste disposal. And, it has emerged as one of the investment opportunities in recent times, allowing new streams for business transformation through recycling or remanufacturing innovation. CE decreases the reliance on raw materials, contributing to supply chain robustness and reducing the risks linked with resource scarcity. Moving to CE aids in economic and social benefits directly toward environmental protection, enabling the transition for a sustainable growth path in addition to enhancing overall resilience (Hidalgo-Crespo et al., 2024).

5.1.4 The connection

The basis of manufacturing comes from green manufacturing practices that aim to reduce waste and pollution in an industry and lays a foundation for implementing a circular system, whereas by the reuse of materials and reducing reliance on new resources, CM strategies have a significant role in achieving manufacturing objectives. Green manufacturing and CM are inter-related for driving sustainability efforts. Green manufacturing minimizes harm by reducing waste and emissions by integrating innovative methods. Both aim for adopting cleaner technologies to reduce pollution, which happens by ensuring that initial production steps in the manufacturing industry are possible by using eco-friendly methods, such as energy sources, recycling materials, and resource use.

The products are designed for durability, adaptability, and easy disassembly for a prolonged material life cycle through a closed-loop system in CM. CM introduces remanufacturing, recycling, and repurposing of used goods by decreasing environmental pollution. CM boosts resource efficiency, which remains sustainable throughout its lifecycle by maximizing the duration of material use (Acerbi, Forterre, & Taisch, 2021). Green manufacturing and CM strategies are very crucial to sustainability targets, and the long-term benefits are very promising.

Finally, the synergy between reductions in waste and pollution of production and maintenance materials in use, recycling, and remanufacturing leads to a net impact on the environment. This also brings about a better sustainable, robust manufacturing atmosphere into form. For this cause, an interconnected strategy is needed to steer the transition toward the future, balancing economic growth with responsibility. The role of professionals and students is important for this transition.

5.1.5 Analogy

For the sake of better understanding, let us consider a factory to have similarities with that in a kitchen. Green manufacturing will be like having a kitchen with energy-efficient appliances and environmentally friendly appliances and supplies, whereas CM is akin to meal planning that eliminates kitchen baggies and soft cheese containers from your takeout. While both of those lead to a kitchen being more sustainable, CM really came down to holistic resource management. Green manufacturing is "doing things right" in a linear system, while CM creates the design of closed-loop systems to produce almost zero waste.

Let's say a kitchen that has appliances like refrigerators and ovens that are used for energy efficiency, the green manufacturing principal application is what lowers electricity consumption and increases carbon footprints, effective reduction of hazardous air pollution levels and water contamination happens with the least toxic supplies. The methodologies of green manufacturing are consistent, aiming for greater sustainable extraction of raw materials and the minimization of disposal emissions from cooking. However, CM is all about operations inside a kitchen, like developing methodologies such as reuse of the resources by minimizing waste. For instance, perfecting meal planning, which optimizes material usage, also converts the leftovers to other types of meals, and the food waste is composted and converted, which can be used for the fertility incremental process of the farming lands. Adopting an end-to-end approach in the circular loop inside the kitchen operations ensures the longevity of materials, reduces the new resources, and also management practices that might help to generate optimal waste. A similar design implementation of the products in the manufacturing sector can last a long time. It can be easily repaired as well as fully recyclable or manufacturable overall to the end of their lifecycle. Through green and CM combination, we improve efficiency, and the sustainability of production processes can be expanded today and flow into a regenerative system that can create long-term environmental objectives. This kind of combined system is necessary, and awareness needs to be created about CE and its principles. It can reach more than just what happens at the local authority level whenever needed; it can also reach all the different stages of stakeholders, right from design to the end of life, to make sure everything recyclable has to be recycled back to manufacturing.

5.2 THE LINEAR PROBLEM: FROM EXTRACTION TO LANDFILL

In general, almost all the economic models follow a linear path: extract-manufacture-consume-discard. Virgin resources that are pulled from the Earth are transformed into products, used for a limited time, and discarded at last. The linear process begins with extracting pristine materials from the Earth, which are converted into finished goods subsequently. These items will have a short lifespan, and then, they are disposed of, which causes huge and noticeable environmental issues. When those items end their life cycle, they will either be disposed of in landfills or destroyed. These kinds of approaches will have significant effects on the environment and human wellbeing. Landfills will generate methane, a greenhouse gas with harmful chemicals that can contaminate groundwater, while incineration might release harmful substances into the air. This pattern results in the depletion of resources and waste generation, leading to the need to shift toward environmentally friendly practices to avoid these impacts.

5.2.1 Drawbacks in the linear system

5.2.1.1 Resource depletion

Relying on virgin materials had a lot of immense pressure on finite resources. The rates of minerals, fossil fuels, and even abundant resources like wood are increasing. Irresponsible extraction of minerals and consuming approximately 100 billion tons of minerals produces harmful emissions like carbon dioxide which pollutes the ecosystem. At the same time, perceived abundant resources are also extracted unsustainably, e.g., wood. This relentless extraction threatens the availability of these essential materials for future generations, undermining ecological balance and sustainability efforts. The fewer these resources, the worse this environmental impact becomes, resulting in deforestation, loss of habitat, carbon emissions boarding on, and an explosion in biodiversity loss. Much more is required to reverse this trend and strive for the transition toward a sustainable, CE (an economic market model that prioritizes systems that decouple growth from resources), so addressing this depletion is necessary (Serrano-Arévalo et al., 2024).

5.2.1.2 Environmental degradation

The creation of waste will result in the contamination of the air, water, and land. Toxic substances will seep into the ground from landfills, and burning waste will release greenhouse gases and plastic waste that can cause disturbances in our present ecosystems (Tamasiga et al., 2022). Linear waste management systems will have impacts on environmental degradation, and it is notably significant, especially when the world is moving toward the CE.

Classic waste management, which is a linear creation model (extract raw materials—make products—use products and then dump them), will lead to catastrophic environmental problems. This type of linear (take-make-dispose) model will result in the generation of a huge amount of waste that can pollute our world's air, water, and soil, starting with the process of disposing of waste in landfills which can result in the release of toxic chemicals from contamination to the act of burning that emits the greenhouse gases like CO_2 causing disturbances to the ecosystem. Eliminating these environmental costs by reducing waste production through recycling and reusing is the main aim and objective of the CE. As shown by the life cycle assessment (LCA), the combination of mechanical recycling and pyrolysis will have the most significant potential to decrease the impact of plastic waste management. For example, the research study on Saudi Arabia's waste management system concludes that a combination of alternative recycling methods can be able to yield high-quality post-consumer recycled products and can lower their overall impact potential, which is solid evidence for CE practices over a linear way of approach.

5.2.1.3 *Economic inefficiency*

The discarded materials are a loss in economic value. The energy and resources invested in their production are simply thrown away. Waste materials will generate a massive loss in the economy when it follows a linear pattern of economy (Julianelli et al., 2020). Most of the energy and resources are consumed in making these materials. This implies that they should go into the pit and never be seen again. By making wastes and inputs scarce through inefficient use, the costs of environmental remediation are increased. This linear economy will create a cycle of inefficiency and loss, ultimately resulting in the depletion of finite resources needed for long-term economic sustainability and growth. By doing so, we lose a whole series of useful properties that could have been put to use elsewhere by salvaging valuable raw materials and keeping their use for new products. This forms the downside that ought to emphasize why there should be advancement toward a CE—that is, the re-utilization of materials, retention of economic value, and minimization of environmental costs.

5.2.2 Strategies to implement green manufacturing

Implementing green manufacturing takes time and investment, but with the right strategies, it can be implemented. Some of the strategies are increasing energy efficiency, reducing material consumption, pollution control, sustainable sourcing, design for longevity, product-as-a-service (PaaS), remanufacturing, recycling, and industrial symbiosis (Zhang et al., 2020; Xia et al., 2023).

5.3 THE CIRCULAR SOLUTION: A CLOSED-LOOP SYSTEM

When developed from nature, the concepts lead to a closed-loop system in which waste from one process is the raw material for another. This will dramatically reduce the volume of waste generated in the first place. Companies can establish closed-loop systems through a variety of strategies or innovations that imitate nature's own design methods and ensure both material and energy are recycled in perpetuity. This promotes environmentally conscious practices, resource efficiency, and sustainability. Natural systems all ran according to the tenets of a CE: no portion of the output could be called a waste as all of it was accounted for. According to Eid and Al-Abdallah, biomimicry principles, when adopted in industry processes, could influence the development of more sustainable technologies such as those that rely on biodegradable materials, systems that use energy in a more efficient way, and products that are disassembled easily for recycling. These ideas also support several circular economics ideas, plans, and practices, at least those that work to guarantee a less damaging process of production and utilization. A particularly disruptive idea is a CE. It suggests a model where all resources are placed in the system.

5.3.1 Circular strategies for waste management in manufacturing

Manufacturing is one of the main contributors to global waste issues. Yet, the CE provides a model for changing how manufacturers approach waste management and sustainability. This chapter focuses on circular strategies for waste management in manufacturing with the aid of CE, and its effect throughout various industries in a supply chain. Manufacturers can reduce waste and toxic emissions by practicing CE strategies like reducing, reusing, and recycling the materials with improved efficiency to attain sustainability. Proper waste management in manufacturing helps reduce negative environmental impact and improve resource efficiency and profitability. These circular strategies can help manufacturers also to support a more sustainable economic model where very little waste is generated, and resources are used at their optimum to make the industrial sector much more resilient as well as green (Sahadevan & Mishra, 2024). This is important for waste management in manufacturing.

Moving to CE, design for durability and repairability (along with ease of disposal) is key. Furthermore, by designing products to be long-lasting and components replaceable rather than the entire product replaced every 3–5 years once will reduce e-waste. This makes sure that the products have a longer life and are more maintainable and repairable, which in turn ensures better recycling or reuse of these goods ultimately. Products have to be authored in a manner that their durability and repairability make them last long enough for feathers to eventually come into existence. This is a significant part of the lifecycle which are designed as closed-loop systems where materials flow through cycles being used over and again. It is fully CE compliant, which means that in the spirit of a true zero-waste cycle, all products and components are designed to be used forever with perfect reusability.

5.3.2 Product design

5.3.2.1 Durability and reusability

If we are to design things that consider longevity and repairability, circular design is a crucial principle to keep in mind. Longevity and repairability are important to delay the disposal of a product or material. It allows products to last longer and be easily upgraded, with the possibility of modular design in which components can be replaced separately rather than throwing away the whole product; hence, it can significantly cut the waste associated with electronic and electrical equipment. It shall make products robust and long-lasting, with the related advantage in perspective. Durability and repair must be designed into products to function over the longest possible time, closing the loop more and cycling materials through use over a far more extensive time frame. An idea that goes along with the principles of a CE is to keep products, components, and materials at their maximum utility value for the longest possible time without waste.

5.3.2.2 Material selection

According to material selection, use only materials that are either recycled or bio-based or easily recyclable. These would treat the Earth better and give it a chance to heal in the future. Choices of materials are one of the most important parts of a CE, and the shapes of the products affect how well trash is handled. Businesses will be more materially aware if they choose a solution with a high content of recycled material, is sustainable, or is one where materials can be easily returned (Tenhunen-Lunkka et al., 2024). This shall help in reducing the damage and speeding up the pace of healing. The use of such materials during the making of commodities might make reuse and recycling even better, as put forward by a circle economy. This way of doing things does not support reducing or decreasing waste. It is also resource-efficient in that it is made from materials that are used and reused continuously without requiring new inputs. Thus, choosing materials is paramount in attaining our goals of a CE for healthy growth and eco-friendly production.

5.3.2.3 *Design for disassembly*

Easily disassemble goods so valuable parts can be separated and used in new products. Product design includes the design for disassembly feature in waste management techniques for production. The focus will be on easier disassembly of goods so that the materials will be separated into reusable materials for new products. Things consequently last longer and create less waste, and resources are used more efficiently. Manufacturers can recycle parts and materials if they produce goods that can be easily disassembled at the end of their service life. This reduces resource usage and its associated impacts. This helps the CE in manufacturing by designing products with their whole lifetime in mind so that they can be recycled, fixed up, and remade again more efficiently. Design for disassembly, part of product development, supports these ways of making things. The other aspect also concerns global programs that aim to change the industrial model as a contributing factor toward resource efficiency and environmental friendliness (Ostapska et al., 2024).

5.3.3 Optimizing production processes

5.3.3.1 *Lean manufacturing*

Lean concepts optimize cutting procedures, boost the energy economy and eliminate mistakes to cut the wastes produced at the source. Lean manufacturing focuses on reducing real waste generated from the start in order to increase the production efficiency of waste management solutions. This plan will entail putting simple cutting, waste material minimization, energy efficiency, and mistake reduction into practice. Industrial operations have significantly improved by utilizing manufacturing methods that have received scientific validation. Businesses may monitor production processes and make quick choices on interventions to increase efficiency by using real-time data and sophisticated analytics. In this sense, it contributes to waste reduction while enabling a decrease in manufacturing costs while maintaining product quality. Manufacturing is encouraged when it is integrated into the CE framework, which also ensures that it uses resources efficiently and reduces the negative effects of industrial operations on the environment (Taher & Bashar 2024).

5.3.3.2 *Closed-loop manufacturing systems*

Develop internal systems for treating and re-introducing process waste streams. An example is the treatment of water used in one process for reuse in another process. The use of closed-loop manufacturing systems supports the optimization of production processes and waste management techniques in manufacturing. This comes under their fundamental goal of developing systems that effectively recycle process waste, treat and reuse resources, including water, for use in various processes in the bid to reduce the levels of waste and encourage sustainability. Results from the case study show that closed-loop systems can effectively reduce waste and improve resource efficiency. Manufacturers can also reduce material dependency and mitigate their environmental impact through the use of closed-loop systems, which allow them to recover resources from the waste stream. Such processes ease the production process and enable reductions in cost through the reduction of material and disposal fees. Closed-loop manufacturing systems are one of the applications of CE-related concepts that offer economic benefits and enhance developing a more sustainable and resilient manufacturing sector (Dehshiri & Amiri 2024).

5.3.4 Post-production strategies

5.3.4.1 Reuse and repurpose

Reusing old things after the first time they were used. Fixing something, making it look new again, or often, creatively upcycling is what it is. The idea of reusing and repurposing is at the heart of a faster Sustainable Development Goals in managing trash in a CE. As already mentioned, they both save resources and give materials and goods a second chance, which means that fewer of them end up in waste streams. Fixing things or keeping them in good shape is one way to do this. Another way is to find clever ways to reuse things that would have been thrown away otherwise. For instance, a waste segregation system in the building industry makes it possible to reuse materials while sorting them for recycling. This is paired with on-site recycling centers, adding a whole new level of sustainability when recycled materials from new building projects are used. The other part concerns digital methods in real estate administration with building information modeling to improve quality and accuracy in data on building materials toward better choices of their use. Moreover, it could help create an eco-friendly closed-loop system, which would foster the materials to be reused and recycled in future works, supporting customers' consumption in the long run.

5.3.4.2 Remanufacturing

This includes the disassembly of used products, cleaning them, refurbishing the components, and reassembling them into new, functional products. In the circular and waste management approach, remanufacturing is an aspect of the production strategy for manufacturing. This involves taking used products, cleaning and refurbishing their parts, and putting them together to create new functional items. It also helps to save resources and reduce waste to the optimized level. The results of a case study show that remanufacturing is effective in bringing environmental advantages. By reusing materials and components, manufacturers can reduce the use of materials and decrease the cost of production and reduce the impact. Remanufacturing always supports the economy by preserving the value of products and materials in circulation that is within a closed loop, promoting sustainability as well as resource efficiency in manufacturing. This approach is crucial for developing a model because it values resource use and regeneration.

5.3.4.3 Recycling

Consequently, this will allow the realization of efficient sorting and recycling systems in order to recover valuable materials from production waste and end-of-life products. Establishing partnerships with specialized recycling companies is a strategy of building CM in a closed loop: The concept of recycling involves waste management strategies in manufacturing; it aims to recover materials from production wastes and used products. This is demonstrated in a study on marble quarry waste management where new recycling technologies are implemented, for instance, application at the Carrara Marble Basin to treat and recover sludge and waste obtained after the processes of marble extraction. Mineralogical examination revealed that the mineral properties of the waste showed its potential in a variety of industries, including paper, plastics, and pharmaceuticals. Working with those recycling companies ensures adequate resource recovery and reuse without impacting resources as much and increasing resource productivity. This further indicates how recycling helps an economy, showing the possible environmental and economic benefits of proper waste management (Tazzini et al., 2024).

The CE can harness synergy between recycling and composting, two inherently integrated aspects of closing the loop in waste management systems (Victar and Waidyasekara, 2024). Wastewater recycling and recovery reclaim items from waste streams: for example, virgin materials for use in products decrease the need for virgin resources and environmental damages. Keeping stuff in the loop and never allowing it to cycle out is the concept of CE. Advanced recycling and composting systems have helped to reduce landfill waste, cut down on resource demand, and facilitate sustainable urban development. For examplc, in Ghana, interventions that improve recycling and composting have turned out well and have been effective in integrating the informal sector within formal waste management systems and delivering environmental and economic benefits. Finally, the materials are recovered from these waste streams and transformed back into usable resources. The embedding of these methods in an authoritative waste management scheme enables a city to complement robust environmental permitting systems, fostering sustainable growth and thus positively affecting greenhouse gas emissions while indirectly supporting CE objectives (Oviedo-Ocaña et al., 2023).

5.3.4.4 Industrial symbiosis

Industrial symbiosis essentially connects companies and allows them to partner with each other and share the use of each other's waste streams as valuable input in their own production process. One of the ways industrial symbiosis influences manufacturing practices is through the association between companies in recycling waste materials by using them as useful inputs in the manufacturing process. Through this connectivity, one industry's by-products become an input material for another, promoting resource efficiency in a manner that generates less waste. Another business opportunity that businesses can enjoy from the advantages their environment provides through industrial symbiosis is seen in the paper industry. This allows for material, energy, and waste exchange to reduce resource dependency and production costs and minimize the footprint. A very good case in point is the torrefaction of pulp industry sludge with digestion; this disposes of waste from papermaking to form bio-methane and volatile fatty acids. They epitomize the potential of industrial symbiosis in growing an economy. This approach also increases innovation and competition in the manufacturing industry apart from sustainability (Doddapaneni et al., 2022).

5.3.4.5 Collaboration and transparency-extended producer responsibility (EPR)

Again, it is the manufacturer's greater responsibility in collecting, reusing, and recycling the products they have produced at the end of their life by bringing them into compliance with designing for circularity. EPR does promote collaboration and transparency among all practices of waste management undertaken after production. This program requires manufacturers to work the line to collect, reuse, and recycle their products once they have reached the end of their life. This responsibility, therefore, encourages the production of products that can come back full circle, easily recyclable or even reused. A clear example of the reality of EPR practice is in Quebec where a study regarding how municipalities are handling their e-waste provides some insight. Most of them have taken it upon themselves to work hand in hand with these EPRs, though not under the regulation, and are actively involved in the actual collection and recycling processes. The study showed that the drivers for participation are financial incentives offered through EPR programs, perceived legitimacy, and operational efficiencies. However, challenges are mostly on transparency, lack of support for jobs, and recycling rather than emphasis on reuse. These findings therefore call for the need to define the roles, measures for transparency, and collaboration between stakeholders while implementing the EPR program: a step toward promoting the CE.

5.3.4.6 Supply chain collaboration

Partnering with many suppliers from the industries may ensure that they use recycled materials and lead to the implementation of sustainable practices. The engagement of the supply chain with respect to suppliers ensures a transparent and cohesive partnership with regarding to waste management. According to findings in the research on Industry 4.0 technologies, it is evident that the primary objective of implementing measures across the supply chain is to boost sustainability. Partnership with the suppliers ensures that organizations adopt eco practices such as using recycled materials and sustainable technologies. This combined effort streamlines the supply chain performance and enhances the innovation and also the adaptability of the particular industries in the circular loop, helping firms meet their sustainability targets in a changing market. Some studies have proven that when green supply chain practices are integrated with Industry 4.0 tools—such as the Internet of Things, blockchain, and big data analytics—the effectiveness of these collaborations increases considerably, leading to more optimal resource use and lesser environmental impact (Matarneh et al., 2024).

5.3.4.7 Product-as-a-service (PaaS)

Move away from product sales toward leasing or renting services: it rewards companies for developing strong, durable products and then taking responsibility at the end of a product's life. PaaS involves the concept of a strategy for handling waste in manufacturing. The shifting strategies of renting or leasing in product design shift from selling products to offering services. This type of modeling incentivizes the manufacturers to create better products and bear the burden of their management till their end of life. A manufacturer is pushed toward a goal to make a product that lasts, is easy to repair and can be recycled or disposed of correctly when the time comes because it deals with keeping products' ownership and thus any lack of need to minimize resource use. It also leads to sustainability in reducing waste and achieving a CE. PaaS models enable the pooling of the product among multiple customers, therefore reducing impact and further enabling repair and refurbishing. Adopting PaaS helps in promoting the advanced manufacturing industry by aligning incentives with environmental stewardship (Hidalgo-Crespo et al., 2024).

5.3.5 Benefits of circular waste management

This waste generation is managed by reducing environmental impact and avoiding landfills, in addition to saving on the cost of raw materials, such as using recycled content and avoiding the cost of disposing of the waste. The dependency on virgin resources is decreasing, and the supply of such natural resources is guaranteed for future generations. Brand reputation – this builds brand reputation since the stakeholders realize commitment toward sustainability, therefore attracting environmentally sensitive consumers.

5.4 USE CASES – CIRCULAR CHAMPIONS

5.4.1 Waste management examples across industries

Interest in a CE is accelerating in all sectors as innovative companies in these business segments adopt strategies to turn waste into useful resources. A few examples are as follows:

5.4.1.1 Automotive industry

5.4.1.1.1 Renault

Electric vehicle batteries that are given a "second life" for use in stationary storage applications extend the battery's lifespan and reduce the burden on new materials being sourced for manufacture. Another firm bringing the concept of economy into practice is Renault, which has a 'second life' scheme for electric vehicle batteries. This recycles electric vehicle batteries until their end-of-life service, thus extending functionality and reducing the material demand. Results will be shown of how Renault saves resources and contributes to a sustainable automotive industry rather than generating waste. It makes strategies in the CE economically beneficial and gives a real-life application of waste management associated with manufacturing.

5.4.1.1.2 BMW

To reduce the environmental footprint and promote the closed-loop mechanism, BMW introduced steel and aluminum into their vehicles. This has brought a tremendous change in the economic practices within the company. This strategy not only made the circular loop method in using these materials but also impacted the manufacturing process. With this process, BMW reduced the dependency of those materials impacting emissions that are linked to the processing of those materials and saving lots of energy. This process flow shows how sustainability is integrated into the car industry in their production models by saving the environment along with waste reduction. The initiatives and impacts made by BMW highlight the principles of CE in the field of manufacturing (Mihaela Andreea Mitiu, 2024).

5.4.1.2 Electronics industry

5.4.1.2.1 Dell

Dell introduced a program to recover the plastics and metals from used electronics which can be recycled and used in new products. This program was introduced by the company to collect and process the precious metals and plastics from electronics, which are used and recycled and can be reused in the new products. With this process, Dell reintegrates the recycled materials from the used old electronic products into the supply chain, thereby closing the supply chain loop. This program shows how CE strategies impacted electronic companies to improve resource efficiency and sustainability (Leitão et al., 2023).

5.4.1.2.2 Fairphone

Making it easier to upgrade and repair by reducing e-waste and with more lifespan, Fairphone designed smartphones with modular components. The Dutch company started creating phones, leading to the smartphone production in the country with parts that can be easily upgraded and repaired, extending the phone's period and reducing waste. Thus, the CE principles showcase the refurbishment of phones, which drives a significant impact in the electronics industry and is an effective model for other manufacturing industries by improving sustainability (Prabhu & Majhi, 2022).

5.4.1.3 Construction industry

5.4.1.3.1 St Gobain

The initiatives made by the Building Materials company by implementing effective programs for reducing the virgin gypsum and using recycled materials made a reluctant impact. A member of the materials sector, Saint Gobain launched a program that recycles used materials by collecting and recycling plasterboard waste, where the company reduces the construction's dependence on gypsum. This brought a significant impact, preventing waste from entering the landfill and making new products that promote environmentally friendly practices. This program demonstrates how manufacturers can productively manage materials and optimize resources (Patel, 2006).

5.4.1.3.2 Urban Miners

A sustainable alternative introduced by Urban Miners where the scrap of construction materials from demolished sites is to be reused and recycled for reducing waste. The company stands at the forefront of impacting environmental management by providing an alternative solution to reusing materials and reducing wastage in landfills. The company follows the CE path by recycling materials such as metals, alloys, and wood by transforming them.

This states the feasibility and environmental benefits of reducing the waste in landfills.

These initiatives of waste management practices show how industries can integrate and bring a positive impact across various fields. The producers across all sectors must recognize waste as a resource, rethink production process with different processes and emphasize on reusing and recycling. This changes the negative impact made by the industry on the environment into positive by adopting innovation and collaboration with its partners and thereby promoting sustainability.

Following the CE pathway requires efforts from individuals, policymakers, and companies to make progress where waste is no longer seen as a burden but as a tool for development and a creative solution. These steps promote development by increasing the welfare of future generations and circulating in the global economy.

5.4.1.4 Other examples

5.4.1.4.1 HP's Closed-Loop Ink Cartridges

A cartridge system developed by HP that collects and refills them with ink with a locking mechanism reduces the wastage of plastic and extends the cartridge's lifetime. This mechanism process was created to reduce the electronic waste problem by HP and has revolutionized waste management. This process helped in lowering the production costs with the reduction of plastic waste and extending the lifespan, creating a sustainable cycle of cartridges. This closed-loop process is considered from the CE pathway, significantly reducing electronic wastage, solving the environmental problem with an impact on the global economy.

Furthermore, by going through this business flow, industries save a large amount in production costs, which makes it environmentally friendly where extraction of raw materials from nature becomes less. These highlight how the CE impacts the environment along with industrial waste management.

5.5 CHALLENGES AND THE WAY FORWARD

Not only are there advantages, but some concerns take place when transforming into the system of CE circular. One such thing is reorganizing the infrastructure of the various sectors in the industry, which leads to upfront costs. It may also require educating the customers on product reusing and recycling thereby making them to understand the importance of sustainability. This might be harder for customers to understand but it is important than they process it. Clothing and fashion industries where the waste can be recycled prominently think about doing so as they feel consumers may not be interested in dealing with such products.

Although the CE principle is not widely followed in many fields, local governance plays a vital role in implementing sustainable developments. Furthermore, industrialization is crucial in promoting this, as many small-scale businesses are trying to fix this closed-loop mechanism. Despite all these, implementing CE principles to create a sustainable and environmentally friendly solution offers a prominent way to industrial waste management.

5.6 CONCLUSION

Future can be envisioned where implementing the CE leads to the development in collaboration between policymakers and companies to turn industrial waste into resources. CE also helps to have a stable global economy and a pollution-less world for the future generations.

Adopting a CE emphasizes innovation in processing, which reduces waste and conserves rare materials, helping ecosystems. These strategies also benefit in concerns from prevention of pollution and climatic problems. It also creates new business opportunities in evolving new organizational structure in various sectors. Products that are made in the way of this sustainable development need to make sure of durability and design and create supportive guides that will lead to resource research.

In conclusion, adopting CE principles improves the economy while conserving resources with reduced risks, which makes groundbreaking strategies that strive for sustainability. This change and efforts are to be made in all industries, businesses, and decision-makers to strive for an environmentally friendly future.

REFERENCES

Abideen, A. Z., Sorooshian, S., Sundram, V. P. K., & Mohammed, A. (2023). Collaborative insights on horizontal logistics to integrate supply chain planning and transportation logistics planning–A systematic review and thematic mapping. *Journal of Open Innovation: Technology, Market, and Complexity*, 9(2), 100066. https://doi.org/10.1016/j.joitmc.2023.100066

Acerbi, F., Forterre, D. A., & Taisch, M. (2021). Role of artificial intelligence in circular manufacturing: A systematic literature review. *IFAC Papers Online*, 54(1), 367–372. https://doi.org/10.1016/j.ifacol.2021.08.040

Agovino, M., Cerciello, M., Musella, G., & Garofalo, A. (2024). European waste management regulations and the transition towards circular economy. *Journal of Environmental Management*, 354, 120423. https://doi.org/10.1016/j.jenvman.2024.120423

Almadhi, A., Abdelhadi, A., & Alyamani, R. (2023). Moving from linear to circular economy in Saudi Arabia: Life-cycle assessment on plastic waste management. *Sustainability*, 15(10450). https://doi.org/10.3390/su151310450

Costa, S., Donner, M., Duquennoi, C., & Savary, V. (2024). Biological valorization of urban solid biowaste: A study among circular bioeconomy start-ups in France. *Sustainable Chemistry and Pharmacy*, 39, 101545. https://doi.org/10.1016/j.scp.2024.101545

Dahmani, N., Benhida, K., Belhadi, A., Kamble, S., Elfezazi, S., & Jauhar, S. K. (2021). Smart circular product design strategies towards eco-effective production systems: A lean eco-design Industry 4.0 framework. *Journal of Cleaner Production*, 320, 128847. https://doi.org/10.1016/j.jclepro.2021.128847

Dehshiri, S. J. H., & Amiri, M. (2024). Considering the circular economy for designing closed-loop supply chain under hybrid uncertainty: A robust scenario-based possibilistic-stochastic programming. *Expert Systems with Applications*, 238, 121745. https://doi.org/10.1016/j.eswa.2023.121745

Di Foggia, G., & Beccarello, M. (2024). Designing circular economy-compliant municipal solid waste management charging schemes. *Utilities Policy*, 81, 101506. https://doi.org/10.1016/j.jup.2023.101506

Doddapaneni, T. R. K. C., Cahyanti, M. N., Orupõld, K., & Kikas, T. (2022). Integrating torrefaction of pulp industry sludge with anaerobic digestion to produce biomethane and volatile fatty acids: An example of industrial symbiosis for circular bioeconomy. *Fermentation*, 8(9), 453. https://doi.org/10.3390/fermentation8090453

Dumitrescu-Popa, I.-S., Chita, E.-I., Craciun, M. A., & Panait, M. (2024). Challenges of the circular economy: Inequality and economic behavior. *Proceedings of the 18th International Conference on Business Excellence,* 18, 798–810. https://doi.org/10.2478/picbe-2024-0070

Golinska-Dawson, P., Zysnarska, Z., & Pender, A. (2024). Assessment of the maturity of product-as-a-service business models for household appliances from the perspective of R strategies in circular economy. *Procedia CIRP*, 122, 1083–1088. https://doi.org/10.1016/j.procir.2024.02.038

Hidalgo-Crespo, J., Riel, A., Vogt Duberg, J., Bunodiere, A., & Golinska-Dawson, P. (2024). An exploratory study for product-as-a-service (PaaS) offers development for electrical and electronic equipment. *Procedia CIRP*, 122, 521–526. https://doi.org/10.1016/j.procir.2024.01.076

Julianelli, V., Caiado, R. G. G., Scavarda, L. F., & Cruz, S. P. de M. F. (2020). Interplay between reverse logistics and circular economy: Critical success factors-based taxonomy and framework. *Resources, Conservation & Recycling*, 158, 104784. https://doi.org/10.1016/j.resconrec.2020.104784

Katsanakis, N., Ibn-Mohammed, T., Moradlou, H., & Godsell, J. (2023). Circular economy strategies for life cycle management of returnable transport items. *Sustainable Production and Consumption*. https://doi.org/10.1016/j.spc.2023.11.016

Köpman, J., & Majava, J. (2024). The role of product design in advancing the circular economy of electric and electronic equipment. *Resources, Conservation & Recycling Advances*, 21, 200207. https://doi.org/10.1016/j.rcradv.2024.200207

Leitão, F. O., Martins, T. S., Guarnieri, P., & Ouro-Salim, O. (2023). Transition from linear to circular economy of electrical and electronic equipment: A review. *Business Strategy and Development*, 6, 430–446. https://doi.org/10.1002/bsd2.249

Mandpe, A., Paliya, S., Gedam, V. V., Patel, S., Tyagi, L., & Kumar, S. (2022). Circular economy approach for sustainable solid waste management: A developing economy perspective. *Waste Management & Research*, 41(3), 499–511. https://doi.org/10.1177/0734242X221126718

Matarneh, S., Piprani, A. Z., Ellahi, R. M., Nguyen, D. N., Le, T. M., & Nazir, S. (2024). Industry 4.0 technologies and circular economy synergies: Enhancing corporate sustainability through sustainable supply chain integration and flexibility. *Environmental Technology & Innovation*, 35, 103723. https://doi.org/10.1016/j.eti.2024.103723

Mitiu, M. A. (2024). Eco-innovation and eco-design in the current automotive industry. *International Journal of Mechatronics and Applied Mechanics*, 15, 135–145.

Ostapska, K., Rüther, P., Loli, A., & Gradeci, K. (2024). Design for disassembly: A systematic scoping review and analysis of built structures designed for disassembly. *Sustainable Production and Consumption*, 48, 377–395. https://doi.org/10.1016/j.spc.2024.05.014

Oviedo-Ocaña, E. R., Abendroth, C., Domínguez, I. C., Sánchez, A., & Dornack, C. (2023). Life cycle assessment of biowaste and green waste composting systems: A review of applications and implementation challenges. *Waste Management*, 171, 350–364. https://doi.org/10.1016/j.wasman.2023.09.004

Patel, N. (2006). A strategy for the management of municipal waste in Wakefield. WRAP

Prabhu, S. N., & Majhi, R. (2022). Disposal of obsolete mobile phones: A review on replacement, disposal methods, in-use lifespan, reuse, and recycling. *Waste Management & Research*, 41(1), 18–36. https://doi.org/10.1177/0734242X221105429

Rumo, D. (2023). Forgotten dust: Following plasterboard for non-destructive circular economies. *Frontiers in Sustainability*, 4, 994452. https://doi.org/10.3389/frsus.2023.994452

Sahadevan, V. K., & Mishra, U. (2024). A green manufacturer–retailer circular economic production model with carbon emissions and waste control. *Environment, Development and Sustainability*, 26, 12367–12406. https://doi.org/10.1007/s10668-023-03805-w

Serrano-Arévalo, T. I., Díaz-Alvarado, F. A., Tovar-Facio, J., & Ponce-Ortega, J. M. (2024). The impact of circular economy indicators in the optimal planning of energy systems. *Sustainable Production and Consumption*, 44, 234–249. https://doi.org/10.1016/j.spc.2023.12.018

Singh, A., Philip, D., Ramkumar, J., & Das, M. (2018). A simulation based approach to realize green factory from unit green manufacturing processes. *Journal of Cleaner Production*, 182, 67–81. https://doi.org/10.1016/j.jclepro.2018.02.025

Suzanne, E., Absi, N., & Borodin, V. (2020). Towards circular economy in production planning: Challenges and opportunities. *European Journal of Operational Research*, 287, 168–190. https://doi.org/10.1016/j.ejor.2020.04.043

Taher, M. A., & Bashar, M. A. (2024). The impact of lean manufacturing concepts on industrial processes' efficiency and waste reduction. *International Journal of Progressive Research in Engineering Management and Science*, 4(6), 338–349.

Tamasiga, P., Miri, T., Onyeaka, H., & Hart, A. (2022). Food waste and circular economy: Challenges and opportunities. *Sustainability*, 14(9896). https://doi.org/10.3390/su14169896

Tazzini, A., Gambino, F., Casale, M., & Dino, G. A. (2024). Managing marble quarry waste: Opportunities and challenges for circular economy implementation. *Sustainability*, 16(3056). https://doi.org/10.3390/su16073056

Tenhunen-Lunkka, A., Balatsas-Lekkas, A., Mouazan, E., Palola, S., Ngo, T., Salo, M., Hylkilä, E., Sundqvist, H., Luomala, H., Pennanen, K., Sorvari, K., Petänen, P., & Lahtinen, J. H. (2024). Implementing a circular business model for reusable packaging: Multidisciplinary learnings from reusable pizza packaging. *Sustainable Production and Consumption*, 48, 62–83. https://doi.org/10.1016/j.spc.2024.05.006

Victar, H. C., & Waidyasekara, A. S. (2024). Optimising construction waste management in Sri Lanka through circular economy strategies: A focus on construction and renovation and use and operate stages. *Engineering, Construction and Architectural Management*. https://doi.org/10.1108/ECAM-06-2023-0608

Xia, T., An, X., Yang, H., Jiang, Y., Xu, Y., Zheng, M., & Pan, E. (2023). Efficient energy use in manufacturing systems—Modeling, assessment, and management strategy. *Energies*, 16(3), 1095. https://doi.org/10.3390/en16031095

Zhang, Y., Song, Y., & Zou, H. (2020). Transformation of pollution control and green development: Evidence from China's chemical industry. *Journal of Environmental Management*, 275, 111246. https://doi.org/10.1016/j.jenvman.2020.111246

Zhidebekkyzy, A., Kalmakova, D., & Kotásková, A. (2024). Responsible production: A systematic review and future research directions. *Eurasian Journal of Economic and Business Studies*. https://doi.org/10.47703/ejebs.v68i2.380

Chapter 6

Nature-inspired bio-materials

An alternative resource

Velmurugan Ganesan, Siva Shankar Vinothan, Natrayan Lakshmaiya, Manikandan Thangaraj, and Narayanasamy Pandiarajan

6.1 INTRODUCTION

The fusion of nature-inspired architecture and materials research represents a cutting edge of innovative approaches in an era of intensifying environmental difficulties and a pressing worldwide push for environmentally friendly methods. This chapter takes the reader on a broad study of "Nature-Inspired Bio-Materials: An Alternative Resource," an investigation aimed at revealing the transformational possibilities woven throughout the rich fabric of the natural world. It is vital that we rethink how we handle purchasing materials and architecture when we see previously unheard-of ecological transformations and resource limits. Nature is an abundant source of ideas for creating alternatives that are both practical and ecologically sound because of its complex designs and adaptive methods that have been refined throughout aeons of development (Chen 2024; Settlage 1994). The discipline of mimicking the trends and functions of the natural world, or biological mimicry, serves as the organising principle for this investigation. Nature exhibits a wide range of creative solutions, from the minuscule to the microscopic, which have enabled life to flourish in a variety of settings. The universe of nature is rife with findings that have yet to be put into practice, such as the environmentally friendly procedures used in the photosynthesis of plants, the durability and adaptability of spider cotton, and the self-repairing qualities that characterise biological materials. The current section aims to solve the enigmas surrounding biomimicry and demonstrate how these principles from the natural world's manual might be used to transform the field of biological materials. Figure 6.1 shows the nature product converted into inspired materials (Ferreira et al. 2023).

We begin our adventure by venturing into the centre of nature's ecosystems, wherein the complex ballet of plants and animals takes place. In this context, diversification serves as a living example of the durability and adaptability inherent in systems of living things. Nature's creations are a monument to effectiveness and long-term viability, from the intricate structures of foliage that optimise sunlight absorption to the symbiosis

DOI: 10.1201/9781003470342-6

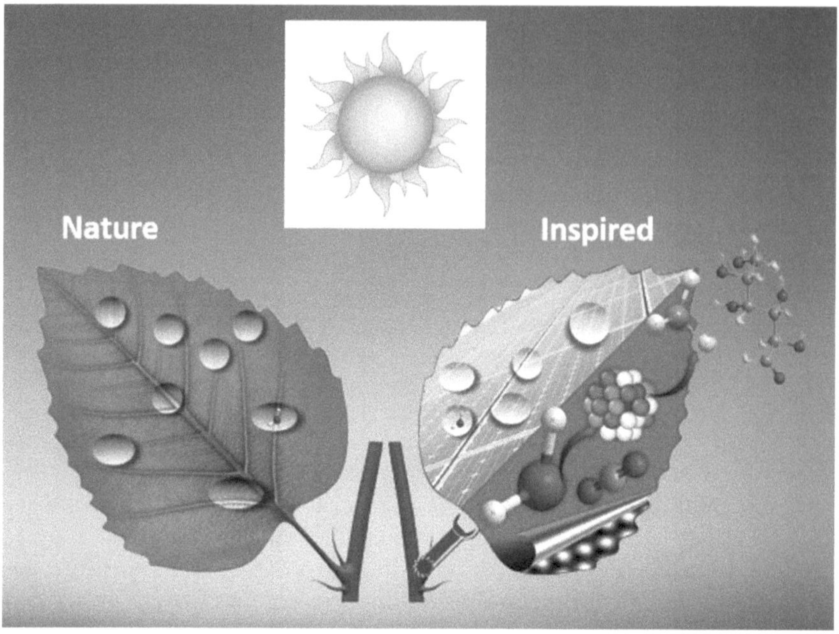

Figure 6.1 The nature product converted into inspired materials.

partnerships that support the metabolism of nutrients in the soil (Cosma and Brighenti 2023). These insights, gleaned from the complex web of ecosystems, provide context for comprehending how nature-inspired biological materials might transform our thinking about how we use resources. The story switches to the most recent developments in material research as we move from fauna-rich environments to the lab. In this section, investigators and scholars begin the fascinating task of translating the language of nature by exploring the structure of molecules and biochemical mechanisms behind the extraordinary qualities of materials from nature. The research centre transforms into an innovative playground wherein biomimicry's tenets serve as guidelines for creating materials that resemble and, in certain instances, even outperform their biological counterparts. A wide range of bio-materials develop in the laboratory environment (Ganesan et al. 2024; Velmurugan, Suresh Kumar, et al. 2023a). The combination of material sciences and imitation sparks an explosion of innovation, resulting in building blocks modelled after the robustness of seashells and bio-inspired polymers with healing capabilities. The section aims to shed light on the numerous options, demonstrating how such biological materials may replace conventional resources with innovations that mirror the effectiveness and beauty of the surroundings.

As we explore the intricacies of biological material synthesis, the conversation broadens to include ecological concerns. Biodegradable and environmentally friendly, nature-inspired bio-materials are strong competitors in the search for environmentally acceptable alternatives. The chapter examines how these materials are used in a variety of areas, including wrapping, medical care, and power. In addition, the discussion touches on the idea of a circular economy in addition to tangible attributes. The cyclical cycles of development, degradation, and regeneration found in nature offer a model for a more environmentally friendly utilisation of resources. The section looks at the ways that bio-materials, which draw inspiration from the circularity seen in ecosystems, might help reduce waste and promote a regeneration connection to the natural world (Nadar et al. 2023). As a result of this investigation, it is clear that biological materials influenced by the environment are not just innovative scientific discoveries but also concrete steps towards a fundamental change in how humans interact with the environment. Taking cues from the environment's resiliency and inventiveness, this section aims to emphasise how urgent it is to adopt different assets and approaches. As the pages turn, the story intertwines the themes of long-term viability, the science of matter, and biomimicry, showing how the combination of these fields offers the possibility of a more tranquil and harmonious cohabitation in the natural world. This chapter attempts to promote a greater awareness of the transformational potential inherent in the notion of "Nature-Inspired Bio-Materials: An Alternative Resource" via a lens that is academic, environmentally friendly, and philosophic. The investigation that results invites us to envision a time when environmental responsibility and inventiveness coexist, where materials draw inspiration from environmental knowledge, and where long-term viability is not just a theoretical concept but a real-world reality that we are creating right now (Taneja et al. 2023).

By examining six different subdivisions, the goal of this book chapter is to thoroughly examine the transformational potential of "Nature-Inspired Bio-Materials: An Alternative Resource." First of all, "Nature's Blueprint: Inspiring Sustainable Bio-Materials" looks at how the complex patterns seen in the natural world might serve as inspiration for the creation of sustainable materials. After that, "Biomimicry Unleashed: Crafting Alternatives from Nature" explores how material science applies biomimicry concepts and demonstrates how nature's techniques might be liberated to create inventive and eco-friendly alternatives. By emphasising useful applications, "Eco-Friendly Solutions: Harnessing Nature's Bio-Materials" expands on the investigation by showing how bio-materials originating from nature may offer environmentally acceptable answers to pressing problems. Through a series of focused lessons taken from nature's many offerings, "Innovative Materials: Lessons from the Natural World" demonstrates how these lessons may guide the design and synthesis of materials that outperform their conventional equivalents. The book "Green by Design: Exploring Nature-Inspired Bio-Materials" goes one step further in this inquiry by looking at how bio-materials that draw inspiration from nature may be purposefully developed with sustainability in mind. Lastly, "Application and Future Scope: A Sustainable Path Forward" offers a comprehensive viewpoint with the goal of directing biological mimicry into the creation and use of biological materials in order to steer towards a future that is environmentally conscious. All together, these subtopics seek to provide a thorough grasp of the complex interrelationships among nature, biomimicry, and bio-materials, emphasising the latter's potential to act as ecologically sound replacement resources across a range of sectors and uses (Wang et al. 2023).

6.2 NATURE'S BLUEPRINT: INSPIRING SUSTAINABLE BIO-MATERIALS

Nature's pattern is an unrivalled opportunity for motivation in the complex choreography of the environment, wherein organisms flourish and life unfolds in a perfect symphony. The significant topic of "Nature's Blueprint: Inspired Environmentally Friendly Biological Materials" opens this chapter's investigation. Nature has evolved over centuries to create solutions that combine form and function in an effortless manner to achieve maximum durability and efficiency (Velmurugan and Natrayan 2023). The fundamental concepts of the environment's designs entice us to discover their mysteries, from the complicated designs of leaves that maximise sunlight utilisation to the capacity that allows living structures to mend themselves. The main goal is to explore the intricate intricacies of these blueprints, interpret the language expressed in the structures of molecules, and demonstrate how the creativity of nature may serve as an inspiration for the creation of sustainable bio-materials. As we set out on our adventure, the pages open up to show how the patterns found in nature may act as a beacon of hope, heralding in an entirely novel generation of substances that not just replicate but also magnify the ecological qualities seen in the environment (Iñiguez-Moreno et al. 2023).

6.2.1 Exploring nature's designs for eco-friendly materials innovation

Researchers and scientists have begun to look at the natural world as an alternative source of motivation for material development in their search for sustainable and environmentally friendly alternatives. Over thousands of years, nature has adjusted designs to be both ecologically benign and efficient. Making new, environmentally friendly substances by mimicking the shapes and behaviours found in nature is a technique known as biological mimicry. The investigation of the environment's designs has the potential to transform a number of industries, including manufacturing, packing, and power production, in addition to the building and manufacturing sectors (Shukla et al. 2022).

6.2.1.1 *Biomimicry in material science*

Studying biological systems—plants, animals, and microorganisms—in order to comprehend their distinctive architecture and behaviours is known as biomimicry. Subsequently, scientists use these ideas to create materials that emulate the effectiveness of nature. For example, the water-repellent surface of lotus leaves has served as an inspiration for the development of self-cleaning materials. Researchers have created coverings that stop filth and contaminants from building up on surfaces by studying the tiny structures seen on lotus leaves. Another source of influence is tarantula cotton, one of the most resilient and flexible substances found in nature (Velmurugan 2023). Scientists are investigating methods to imitate the atomic makeup of spider silk for application in building materials and textiles. Given its strength and pliability, spider silk is a viable option for creating long-lasting and environmentally friendly textiles. Plant cell walls include a substance called cellulose, which is a multifunctional biological material that has encouraged the creation of sustainable containers. Researchers are developing biodegradable packaging supplies that may substitute regular plastics by imitating the structure of cellulose, which would reduce garbage and pollutants in the surroundings (Jing et al. 2021).

6.2.1.2 *Eco-friendly materials in construction*

Although the building sector contributes significantly to ecosystem deterioration, biomimicry provides ways to reduce its environmental imprint. Investigating termite mound layouts for environmentally friendly structure circulation is one instance. Scientists are combining airflow from nature, made possible by the intricate arrangement of termite hills, into construction plans in an effort to lessen the demand for synthetic air conditioning. Furthermore, scientists are looking at the adaptability of some plant structures—like bamboo—to get ideas for sustainable construction products. Bamboo is renowned for its endurance, adaptability, and quick development. Manufacturers are creating material composites by mimicking these characteristics, which may provide ecologically sound substitutes for conventional construction resources, including steel and concrete. Figure 6.2 shows some samples of sustainable materials-based construction (Ganesan and Kaliyamoorthy 2020).

6.2.1.3 *Applications in energy production*

Investigators are using the effective generation of energy methods that nature has created to provide renewable energy options. For instance, the photosynthesis process in plants served as inspiration for the development of artificial photosynthetic devices. Scientists are developing tools to more effectively capture solar energy by modelling the way plants transform sunlight into electricity. The wings of some insects and birds have

Figure 6.2 Sustainable materials-based constructions.

also influenced improvements in wind turbine construction (Lakshmaiya, Kaliappan, et al. 2022). By analysing the aerodynamic properties of these innate fliers, scientists are creating smaller and more effective turbines, which can help provide greener energy. Biomimicry offers enormous possibilities for investigating the designs found in nature to develop sustainable resources for a variety of sectors. Scientists are drawing influence from the long-term viability and effectiveness of natural systems in a variety of fields, including the study of materials, building, and energy generation. We can build an even more harmonic and mutually beneficial connection between human creativity and the surroundings by taking inspiration from nature's designs. This will help us build an ecologically sound and cleaner environment (Abedanzadeh et al. 2021).

6.2.2 Harvesting inspiration: sustainable bio-materials from the natural world

The effects of conventional production techniques and supplies on the ecosystem have come to the fore in recent years. Consequently, there has been a significant increase in demand for environmentally friendly substitutes, with an emphasis on biological materials that are sourced from renewable resources. This article examines the exciting field of using natural inspirations to produce ecological bio-materials, looking at the creative methods researchers and academics are using to provide environmentally friendly options for a range of sectors (Sanjeevi et al. 2021).

6.2.2.1 *Biomimicry as a source of inspiration*

In biomimicry, problems that humans face are addressed by taking cues from the natural world. Throughout thousands of years of development, nature has refined elements and designs, offering a wealth of inspiration for environmentally friendly alternatives. For example, the cleaning technique of lotus leaves has influenced the creation of repellents for water, which lessens the demand for toxic chemical treatments (Kim 2021).

6.2.2.2 *Spider silk: nature's high-performance fibre*

Spider silk is an amazing substance that is biodegradable, flexible, and strong. Genetic engineering is being used by academics to synthesise spider silk protein molecules, opening the door to environmentally friendly replacements for traditional fabrics and surgical devices.

6.2.2.3 *Mycelium-based materials*

Fungi's mycelium, or root system, is a potential biological material with a wide range of uses. It is an environmentally beneficial substitute for plastics and bubbles as it can be developed into a variety of shapes and forms. Businesses are looking at mycelium-based building materials and packaging as environmentally friendly substitutes.

6.2.2.4 *Plant-based polymers and bioplastics*

Conventional plastics made from petroleum and coal are a major source of pollution. As ecological substitutes, scientists are looking into bioplastics and polymers derived from plants. Plastics that decompose may be made from starch from maize, sugarcane, and various other plant-based resources, which lessens the adverse effects of wrapping and single-use goods on the ecosystem (Mohan et al. 2020).

6.2.2.5 *Inspiration from marine life*

Materials with special qualities have been developed as a result of inspiration from marine animals like coral and sponges from the sea. Scientists are looking into ways to create environmentally friendly building supplies by imitating the tenacity and adaptability of reef fish. Furthermore, bio-based adhesives for a variety of uses are being created, drawing inspiration from the sticky proteins found in mussels (Ganesan et al. 2021).

An encouraging approach to solving environmental problems is to use natural inspiration to produce renewable biological materials. The combination of biological imitation, material science, and technological advancement presents industry with intriguing opportunities to shift to more environmentally friendly methods. We may anticipate a move towards a healthier tomorrow wherein substances are not just useful but also ecologically conscious, as long as investigation in this area is carried out (Biswal et al. 2020).

6.2.3 From flora to fabric: unveiling nature's blueprint for green materials evolution

Natural substances are becoming more and more sought after by companies as well as by scholars searching for sustainable and environmentally friendly materials. Nature has long offered guidelines for the creation of green materials that are not only ecologically benign but also novel in their composition and characteristics, from flora to textiles. The study of the environment's aesthetic principles has led to the development of a new area called biological mimicry, in which scientists and engineers use the natural environment's innate knowledge as motivation to develop new materials that will satisfy the needs of a future without fossil fuels.

6.2.3.1 Biomimicry in materials science

In order to address human issues, biomimicry, also known as biomimetics, mimics the structures, functions, and procedures found in the environment. This method has produced materials in the field of material science that are modelled after the extraordinary forms and functions of plants, microbes, and mammals. The focus is on developing materials with distinct qualities drawn from the intrinsic qualities of their surroundings, in addition to being environmentally benign (Sekar et al. 2020).

6.2.3.2 Plant-based fibres: nature's textiles

The manufacture of textiles from plant-based fibres is ingrained in human cultures' traditions and historical fibre. Plants have consistently served as a vital source of fibre for the production of textiles, from prehistoric societies to contemporary uses in industry. The use of plant-based fibres as a green substitute for traditional textiles made from resources that are not renewable has increased recently as ecological responsibility has come into greater focus. This investigation into natural textiles celebrates the renewable and biodegradable qualities of these materials while showcasing their adaptability and promise to completely transform the textile sector. Plant-based fibres have been used to make textiles for thousands of years. Archaeological sites throughout Asia, Africa, and North America have shown traces of cotton farming. Cotton is a mainstay in the production of textiles and is obtained from the fluffy fibres around the embryos of the cotton vine. Hemp fibres are processed into fabrics for clothes, among other uses, whereas flax, which is extracted from the bulbous part of the plant

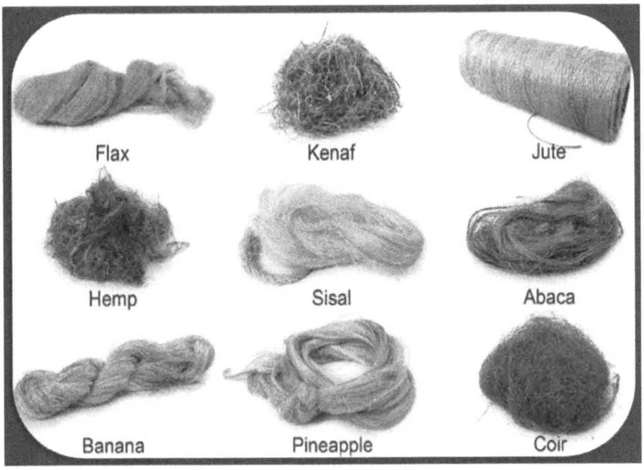

Figure 6.3 Plant-based fibre materials.

that produces flax, is used to make linen. These fibres have been around for a long time, which speaks to their adaptability, convenience, and longevity in a variety of settings (Silva 2019; Spencer 2019). Figure 6.3 shows the examples of plant-based fibre.

The intrinsic permanence of plant-based fibres is one of their main benefits. Plant-based fibres may be grown in agriculture and are sources of renewable energy, in contrast to synthetic fibres made from petrochemical products. For instance, cotton is a crop that can be cultivated every year, and advances in agricultural techniques have centred on increasing output while reducing its ecological impact. In addition, plant-based fibres have an organic end-of-life cycle and are recyclable, which sets them apart from artificial substances that leave a lasting ecological footprint. Technological developments in the processing industry have increased the range of possible uses for plant-based fibres (Velmurugan, Suresh Kumar, et al. 2023b). Recent advances were added to traditional techniques of plant fibre extraction to increase efficiency and decrease waste. For example, the calibre of the fibres has increased and the environmental effect has decreased when enzymes are used in the regeneration process, which separates the fibres from the stem. Furthermore, advancements in spinning and loom weaving techniques have made it possible to produce superior fabrics using plant-based fibres that satisfy commercial and modern fashion requirements (du Plessis and Broeckhoven 2019).

Using waste from agriculture for extracting fibres is a new trend in plant-based fibres. Traditionally regarded as trash, crop residues, including rice, bananas, and pineapple, have been recycled for the production of textiles. By doing this, waste is reduced and farming processes are enhanced.

The creation of these fibres supports a circular economy, which minimises wastage and maximises the use of resources in accordance with the concepts of sustainable growth. Plant-based fibres stick out as a naturally conceived response to the problems facing the clothing sector in the process of turning plants into fabric. Such fibres, which have a long history, are environmentally friendly, sustainable, and have a wide range of uses. With the ongoing advancements in processing methods and the use of agricultural garbage, plant-based fibres have the potential to significantly influence the textile sector's transition towards a more ecologically mindful and sustainable future (Perera and Coppens 2019).

6.2.3.3 Biofabrication and 3D printing

In the sphere of the development of sustainable resources, biological manufacturing and three-dimensional printing are emerging as revolutionary innovations that provide creative and environmentally friendly methods of material development and production. These methods, which take their cue from the intricacy of nature, make it possible to create intricate structures and useful substances with a degree of accuracy and efficiency that was previously impossible (Alagumalai et al. 2021). This section delves into the concepts, uses, and possible consequences of biological manufacturing and three-dimensional printing in relation to the creation of eco-friendly materials. Using biological substances, or cells that are alive, to create functioning tissues or substances is known as biological manufacturing. Biofabrication has become well-known in the field of green materials because of its capacity to produce environmentally friendly alternatives with specific characteristics. In order to improve functionality, this procedure frequently makes use of materials that are biocompatible, like biopolymers or fibres from nature. It may even involve the addition of live cells (Coppens and Perera 2018).

The fundamental idea behind biological manufacturing is to replicate the natural processes involved in tissue growth and restoration. Scientists can create materials that mimic or improve cellular behaviour and tissue creation by comprehending the molecular mechanics behind these activities. Using biological materials to create an environment that is conducive to cell development is one example of this. The creation of environmentally friendly supplies for a range of sectors has led to their use in biological manufacturing. For example, by utilising natural polymers generated from plant or algal cells, scientists have investigated the use of biofabrication to construct scaffolds that degrade for tissue engineering purposes. Furthermore, the use of biological manufacturing processes can result in naturally degradable materials for packaging that have less of an impact on the environment than ordinary plastics. According to the fundamentals of 3D printing, elements are deposited layer by layer under the direction of electronic models or computer-aided design (CAD) files. Layer-by-layer construction enables the construction of customised structures and intricate geometries.

When it comes to environmentally friendly substances, 3D printing may make use of sustainable resources, including plant-based materials, polymers that decompose, and reused plastics. There are several uses for 3D printing in the creation of environmentally friendly products. Recycled materials may be used to 3D print architectural elements for sustainable construction. Moreover, 3D printing enables on-demand production, which reduces waste in the production of personalised goods for consumers. 3D printing is used in the biomedical industry to produce customised medical implants made of biodegradable substances (Oguntona and Aigbavboa 2017; Nichols 2017).

Blending the biological significance of biological manufacturing with the accuracy of 3D printing, the combination of these two technologies creates a potent synergy. With the use of this method, sophisticated, useful, and sustainable materials may be produced for use in everything from medical devices to customised consumer products. The substances utilised in biological manufacturing are occasionally referred to as "bioinks." Such bioinks might be made of live cells mixed with bio-materials. Hybrid manufacturing techniques that combine biological manufacturing and 3D printing enable the creation of complex structures with biological components, opening up new opportunities in the fields of regenerative medicine and sustainable material production. Although 3D printing and biological fabrication have great potential, there are obstacles to be addressed. Widespread use of these technologies depends on guaranteeing their scaling, improving the characteristics of bioinks, and overcoming regulatory obstacles. Future advancements can concentrate on the incorporation of multifunctional and other environmentally friendly supplies as well as the investigation of cutting-edge printing methods that improve precision and velocity. At the cutting edge of the development of green materials are biological manufacturing and 3D printing, which provide creative and ecologic alternatives for a range of industries (Barthlott et al. 2016). The combination of these innovations enables the production of substances that adhere to social and ecological responsibility criteria in addition to functional needs. The potential for revolutionary and green material creation is becoming more and more attractive as biological manufacturing and 3D printing innovation develop, opening the door to more robust and environmentally friendly futures (Velmurugan, Siva Shankar, et al. 2023).

6.2.3.4 Case studies: nature-inspired innovations

A number of examples demonstrate how well nature-inspired patterns may be incorporated into the creation of new substances. For instance, substances that can mend themselves when damaged, substances that are robust and light, substances that are motivated by the architecture of plants, and surfaces that repel water and resemble leaves from lotuses. These examples demonstrate how biomimicry may be used to solve a range of problems and spur environmentally friendly innovation. By taking influence from design concepts found in the natural world, the progression of renewable resources from flowers to cloth illustrates the possibilities for innovative sustainability. Plant-based fibres, biological manufacturing, biological mimicry in materials research, and 3D printing are all advancing the creation of substances that satisfy the needs of industry while simultaneously adhering to ecologically conscious requirements. The transition from plants to textiles offers the prospect of a more environmentally sound future as we strive to decipher the environment's pattern (Appleton 2013).

6.3 BIOMIMICRY UNLEASHED: CRAFTING ALTERNATIVES FROM NATURE

The idea of "Biomimicry Unleashed: Crafting Alternatives from Nature" is a vibrant example of how human creativity and the knowledge provided by nature may work together to create innovative solutions. The imitation of biological processes and designs, or "biomimicry," has evolved from a theory to a potent instrument that is ushering in a new age in the fields of material science and engineering. This chapter explores the fascinating topic of biomimicry, showing how the intricate details of biological systems—from the smallest levels to the macroscopic—can provide an endless supply of motivation for creating substances that are alternatives to those often used in construction. The goal of this voyage is to reveal the untapped potential of biological mimicry in creating a creative and environmentally friendly future. The pages are open to show how academics and researchers are using nature's brilliant ideas to create technologies that not only resemble but also outperform their organic equivalents. This provides a window into a future whereby the environment is used as a creative inspiration to create alternatives (Ausubel 2012).

6.3.1 Exploring nature's blueprints: biomimicry's foundations

An innovative multidisciplinary subject called biomimicry uses the complex patterns and mechanisms of nature as inspiration to address difficult human problems. This novel strategy has gained traction as researchers, scientists, and architects learn more about deciphering and mimicking the designs found in nature. We will delve into the intriguing realm of nature-inspired invention in this investigation of biomimicry's roots, looking at its tenets, uses, and revolutionary effects on a range of professions.

6.3.1.1 Understanding the biomimicry

Because wildlife is a living creative laboratory, imitation fundamentally signifies a radical shift in how we approach development and resolve issues. It entails a thorough investigation of the complex mechanisms, procedures, and frameworks that have developed over billions of years throughout the world of nature. By skilfully fusing function with form, biomimicry forces us to recognise the genius of nature's creations and gain understanding from them. Biomimicry is the study of how creatures adapt and how habitats are sustainable in order to find ways to integrate the lessons learned from nature into the problems facing humanity. This methodology not only prioritises effectiveness and grace, but it also cultivates a profound reverence for the interdependence of all species on Earth. In the end, recognising biomimicry is an opportunity to take a cue from the environment, seeing that the intricate framework of life that envelops us could hold the answers to plenty of our issues (Latha et al. 2023).

6.3.1.2 *Principles of biomimicry*

- **Sustainable Solutions:** In addition to providing for immediate requirements, nature has produced environmentally friendly alternatives that support the general wellbeing of ecosystems. Applying these environmentally friendly ideas to human-made systems is the goal of biomimicry.
- **Form and Function:** In nature's creations, form and function coexist harmoniously. Recognising the connection between an organism's functionalities and structural makeup is the main goal of biomimicry, which results in effective and meaningful innovations.
- **Adaptability and Resilience:** Natural systems are excellent at adjusting to shifting conditions and conquering obstacles. The goal of biomimicry is to improve robustness when faced with changing circumstances by incorporating adaptive methods into technology as well as design.

6.3.1.3 *Applications of biomimicry*

- **Architecture and Design:** In an effort to produce structures that not only fit in with the surroundings but also maximise energy efficiency and offer creative approaches to climate management, designers and builders increasingly employ biomimicry in their work.
- **Biomimicry and Biotechnology:** Through taking cues derived from processes that are seen in nature, biomimicry has transformed the medical industry. Biomimicry is bringing new medical concepts to the table, from bio-inspired medical devices to creative delivery methods for drugs.
- **Robotics and Engineering:** The creation of bio-inspired robotics has been influenced by the investigation of animal behaviour and movement. These machines are more agile, adaptable, and efficient because they draw their principles from the natural world.
- **Materials Science:** Biomimicry has impacted the creation of sophisticated materials whose characteristics are modelled after those exhibited by organic materials. It also involves robust, lightweight materials derived from silk spiders and materials with healing properties derived from natural systems (Danks 2010).

6.3.1.4 Case studies

Velcro: The creation of Velcro, which was motivated by the hook-like features on the burdock seeds that stick to fur, is among the most well-known instances of biomimicry. Georges de Mistral, a Swiss technician, imitated this organic bonding process to develop the widely used adhesive substance.

Shinkansen and Kingfisher Beak: The kingfisher's beak inspired the layout of Japan's Tohoku bullet train, which improved aerodynamics and reduced sound. High-speed train transport was revolutionised by its biomimetic architecture.

Lotus Effect and Self-Cleaning Surfaces: The lotus leaf's ability to repel moisture and maintain its cleanliness inspired the development of surfaces that clean themselves in paintings and fabrics, among other things. This feature lessens the requirement for washing and upkeep. The field of biomimicry has a bright future ahead of it, with potential applications in tackling worldwide problems including shortages of resources, global warming, and degradation of the environment. Researchers, scientists, and biology working together will probably produce even more innovations that push the limits of ecological sustainability and creativity. By using biomimicry to investigate nature's schematics, a plethora of opportunities for developing inventive, sustainable, and efficient solutions have become apparent. We can revolutionise industry and tackle urgent global crises by taking inspiration from the natural world's ingenuity. As biological imitation develops, it provides a glimmer of hope for a time when technology will harmoniously integrate with the complex patterns found in the environment (Vollrath and Porter 2009).

6.3.2 Innovative solutions from the wild: unleashing biomimicry's potential

Human ingenuity has always found inspiration in nature. In recent years, researchers, technologists, and architects have come to realise the enormous scope of imitating nature's answers, which has given rise to the idea of biomimicry—the practice of using lessons from nature to address problems faced by humans. This method entails studying, comprehending, and mimicking the clever mechanisms present in the environment in order to develop technologies that are effective and ecological. This section will examine the intriguing topic of biomimicry along with several of the most original applications it has yielded in a variety of domains.

6.3.2.1 Biomimicry in design and architecture

Biomimicry has had a significant impact on design and architecture, resulting in the creation of structures that are both functionally sound and environmentally friendly. Some instances are the Eastgate Centre in Zimbabwe, which imitates the cooling systems of termite hills, and the ecosystems of the Eden Project, which were influenced by the biomechanical properties of bubbles of soap. Designers and architects are creating profitable, environmentally friendly buildings that contribute to a greener economy by investigating nature's effective architecture (Wilson 2007).

6.3.2.2 *Materials science and biomimicry*

Biomimicry has provided avenues for producing new substances influenced by natural processes. The environment is adept at making substances with extraordinary features. For example, scientists have turned to the lotus leaf to develop surfaces with superhydrophobic properties that would result in materials that can clean themselves. Among the most powerful substances found in nature, spider silk has served as an inspiration for the creation of artificial materials that are incredibly malleable yet resilient. The use of synthetic materials is transforming sectors like production, health care, and materials.

6.3.2.3 *Biomimicry in robotics and technology*

The architectural ideas found in nature have a major effect on the subject of robotics. Scientists are developing robots inspired by the mechanics of livestock, such as the cheetah, in order to achieve incredible agility and quickness. Drones that mimic the efficient flying movements of insects and birds are known as simulated robots. Such machines demonstrate enhanced performance, flexibility, and environmental sustainability by imitating solutions that are found in the natural world, opening the door for revolutionary developments across a range of sectors (John et al. 2005).

6.3.2.4 *Medical innovations*

The intricate systems of life that exist in nature have sparked advances in medical research. Research into the natural functions of human tissue has made it possible to create biomimetic devices and prostheses that blend in with their surroundings. In an effort to improve the accuracy and efficiency of prescribing, researchers are additionally looking into the use of biomimicry in systems for drug delivery. This technique draws influence from biological processes. There is a great deal of promise for biomimicry to transform healthcare practices and therapies, and right now, research is showing encouraging outcomes.

6.3.2.5 *Sustainability and environmental conservation*

A key factor in developing viable alternatives for ecological issues is imitation. Innovations in the fields of managing waste, power, and farming are creating environmentally friendly procedures by modelling the reuse, recycling, and reducing waste tactics seen in nature. Biomimicry, which draws inspiration from environments, promotes regenerative design that benefits ecosystems while minimising its negative effects on the natural world.

6.3.2.6 Challenges and future prospects

Biomimicry has a lot of promise, but there are drawbacks as well, such as ethical issues, the requirement for multidisciplinary cooperation, and the difficulty of converting biological ideas into useful applications. It will need consistent work from scientists, engineers, and legislators to overcome these obstacles. As scientists continue to discover the mysteries of the natural world, the field of biomimicry promises to provide even more revolutionary discoveries in the future. From the perspective of biomimicry, creative solutions from the natural world highlight the enormous potential for understanding the patterns found in nature. Biomimicry is affecting a wide range of sectors, from environmentally friendly construction to medical improvements, and it is providing a route towards more effective, profitable, and harmonic advances in technology. Biomimicry is a monument to the genius of nature and our capacity to use it to improve the state of the planet as we keep discovering its mysteries (Velmurugan and Babu 2020).

6.4 ECO-FRIENDLY SOLUTIONS: HARNESSING NATURE'S BIO-MATERIALS

Living sustainably and ethically handling resources have moved the emphasis to environmentally friendly alternatives that take advantage of the biological substances found in the environment. This paradigm change acknowledges the need to lessen our influence on the environment and adopt eco-friendly methods that follow the rules of ecological equilibrium. The book "Eco-Friendly Solutions: Leveraging Mother Nature's Biological Materials" looks at creative methods to use the abundant life-giving substances at our disposal to produce goods, materials, and technology that not only successfully satisfy our own needs but also help to preserve the environment. Discovering the fundamental principles of nature will help to create harmonious relationships among humans and their surroundings (Mohanty et al. 2002).

6.4.1 Natural wonders: exploring eco-friendly materials from the Earth

The search for and use of materials that are sustainable is now essential in a world where sustainable development is becoming more and more important. The fascinating voyage "Naturally Wonders: Discovering Environmentally Friendly Materials Developed on Planet" delves into the world of environmentally friendly resources that come from our planet. It is crucial to maximise the potential of natural resources as mankind works to lessen its ecological impact and tackle the problems posed by climate change. This investigation explores the wide range of components that Earth's crust provides, highlighting their adaptability, robustness, and environmentally beneficial qualities. This expedition aims to reveal the hidden gems that our world contains, encouraging a change towards a more ethical and peaceful connection with nature. It does this by utilising both cutting-edge inventions and traditional techniques.

6.4.1.1 Historical roots of sustainable materials

Examining the complex background of humanity, this part explores the long-standing roots of environmentally friendly substances. Learn how ancient societies cleverly employed resources from the Earth for equipment, clothing, and a place to live, from the clay bricks used in Babylon to the resilient thatched roofs of many cultures. Examining such historical foundations helps us understand the robustness and tried-and-true sustainability of substances that have been around and used for millennia. This historical backdrop provides context to comprehend the current upsurge in the pursuit of ecologically sound methods as we work to reestablish connections with traditional knowledge and incorporate it into our modern-day environment (Suresh Kumar et al. 2021).

6.4.1.2 Renewable energy in material production

The source of power used in the manufacture of substances is a crucial factor to take into account while searching for environmentally friendly materials. The transformational effect of energy from sustainable resources on extracting materials and production procedures is examined in the next section. Discover how using nature's energy may drastically lessen the environmental impact of the material manufacturing process, from wind-powered factories to solar-powered quarries. Through the integration of sustainability and energy efficiency concepts, we are taking a step towards improving the ecological sustainability of supplies while also being in line with larger initiatives to move towards a more sustainable economic environment.

6.4.1.3 Plant-based marvels

An abundant variety of recyclable substances originating from plants may be found in the natural environment. The following subsection highlights the wide variety of plant-based wonders that are now well-known in the field of sustainability. Bamboo is a resource that may be used for a variety of purposes, including building, textiles, and as a plastic alternative. Examine the adaptability of hemp, which is praised for its usage in fabrics and may find use in environmental sectors. See the creative ways that plants are turned into products that offer both utility and ecological responsibility, from cork to soy. By realising the promise of materials made from plants, we open the door to a day when the everyday items we use coexist peacefully with the environment.

6.4.1.4 Challenges and future prospects

As we traverse the terrain of environmentally friendly substances, it becomes more imperative to address the obstacles impeding their broad implementation. Challenges including awareness among consumers, financial viability, and scalability are rigorously examined in this section. Examine the challenges of striking a balance between the need to protect the planet and the needs of humanity as a whole. Investigate the legislative measures and cooperative initiatives that are meant to address these issues as well. Imagine a day in the future when environmentally friendly decisions are an essential part of our society, not merely a substitute. We can all work together to create a future where discovering natural beauties results in sustainable and peaceful living on our planet by recognising the obstacles and charting a path ahead (Lakshmaiya, Ganesan, et al. 2022).

6.4.2 Green innovations: the power of nature in sustainable solutions

A trend towards green technologies has emerged from the convergence between science and the natural world, during a time of increased awareness of environmental issues and the imperative need for equitable growth. This dynamic combination of natural forces and human creativity has become a source of optimism, providing innovative answers to a number of the most urgent problems our world is currently experiencing. The book "Green Innovations: The Power of Nature in Sustainable Solutions" challenges us to investigate the revolutionary possibilities that arise from using the natural world's inherent knowledge. This trip takes shape as we explore ground-breaking ideas where sustainable methods and environmentally friendly technology come together to establish a harmonic harmony between protection of the environment and human advancement.

6.4.2.1 Roots of green innovation

In order to fully understand the importance of green innovations, the first chapter takes us on an engrossing tour of the past foundations that have moulded the present-day paradigm of sustainability. This section sheds light on the origins of ecological awareness by following its lineage back to the earliest ecological movements that arose in reaction to the damaging impacts of industrialisation on the natural world. We explore significant turning points that have brought sustainability to the foreground of public worries, including the founding of the inaugural celebration of Earth Day in 1970, the creation of international ecological accords, as well as other important turning points. Through an analysis of the development of legislation and public understanding, this section offers significant insights into the catalyst events that have shaped the history of environmentalism. The story shows how responsibility for the environment has gradually but dramatically changed, from the formation of ecologically responsible development goals to the acknowledgement of the changing climate as a critical worldwide concern.

6.4.2.2 *Renewable energy revolution*

This subject thrusts us towards the centre of the breakthrough in renewable energies, which is a crucial facet of green technologies. The investigation starts by examining the state of the world's energy and the necessity of moving away from petroleum and coal. When we analyse their revolutionary influence on the production of electricity, energy from renewable sources—such as solar, hydroelectric, wind, and geothermal—takes centre stage. Extensive dialogue revolves around state-of-the-art solar technology, design improvements for wind turbines, energy storage advancements, and intelligent electrical system integration. This part of the book highlights the possibility of building a robust and sustainable energy infrastructure that can fulfil the increasing demands worldwide while illuminating the benefits of solar power from an economic and environmental standpoint (Velmurugan, Natrayan, et al. 2023).

6.4.2.3 *Sustainable agriculture and food security*

This chapter delves into the complex domain of responsible agriculture and food security, examining the convergence between farming and the environment. The chapter emphasises the problems with traditional farming methods and the necessity for a paradigm shift to more environmentally friendly and regenerative methods. Agricultural ecology, precision agriculture, vertical agriculture, and other cutting-edge methods designed to maximise ecological impact and improve food security are presented to viewers. The conversation explores how technology may improve industrial environments' ecosystems, cut waste products, and make the most use of available resources. Through an analysis of the possible convergences between food security and environmentally conscious farming, the chapter presents a story of optimism and opportunity among worldwide obstacles.

6.4.2.4 *Epilogue: the promise of tomorrow*

The epilogue, "The Promise of Tomorrow," aptly named, brings the ideas from the previous sections together to create an outlook for what lies ahead. We consider how profoundly green innovations may influence the future of sustainability and their capacity to revolutionise. The chapter focuses on the role that people, organisations, and government may play in encouraging a shared commitment to responsible environmental care. The story ends by encouraging readers to imagine and actively participate in a world where natural power and technological progress coexist, promising an environmental and restorative society. The epilogue summarises the main idea of the book with an appellation to action and some serious contemplation, instilling in readers a feeling of accountability and hope for the next generation.

6.5 INNOVATIVE MATERIALS: LESSONS FROM THE NATURAL WORLD

Researchers and entrepreneurs in the rapidly changing field of material science are finding more and more motivation in nature. Over thousands of years of development, nature has created a wide variety of extraordinary materials with distinct qualities and uses. The investigation of living features and their emulation by scientists to produce new substances is known as mimicry, a science that arose from the desire to comprehend and mimic nature's clever creations. In addition to offering answers to engineering issues, this investigation into the workings of the outside world has created new opportunities for the production of environmentally benign and durable materials. In this regard, investigating the meeting point of human inventiveness and the wonders of nature provides insightful knowledge and beneficial lessons for the development of new substances that have the potential to transform a number of sectors while contributing to a better tomorrow.

6.5.1 Borrowing from nature: the evolution of innovative materials

There has been a notable change in the intriguing field of material development towards design and engineering that draws inspiration from nature. When we explore the field of biomimicry, the development of novel materials tells a fascinating story of how humankind has benefited from nature's genius. The complex architecture of spider silk, the tenacity of spider silk, and the ability for self-healing in some creatures have all been used as inspiration and models. The emergence of bio-inspired materials deviates from conventional artificial substances as researchers and engineers draw influence from the innate efficiency and environmental sustainability of ecosystems. This chapter tells the tales behind the invention of innovative simulated substances, such as the surfaces that clean themselves, influenced by the leaves of lotus, and the portable, exceptionally strong composite influenced by the makeup of bones. The investigation continues to the cutting edge of nanotechnologies, wherein substances imitate the minuscule characteristics of gecko toes or wings of butterflies.

Transdisciplinary cooperation between biological researchers, designers, and material scientists is necessary to synthesise these advances. We examine the difficult task of converting fundamental biological concepts into adaptable, practical applications, focusing on the fine line that separates copying nature from tailoring it to suit human requirements. This story goes outside the lab to demonstrate the potential effects of bio-inspired substances on a range of sectors, including building, medical care, and design. The section examines the moral and environmental ramifications of human attempts to mimic nature as we see new materials emerge, advocating for an ecologically conscious approach to environmentally conscious development. The nexus of material research and biomimicry demonstrates the creative ways that human creativity is guided by blueprints found in nature, opening up an array of opportunities that rethink the function of content in the years to come and in creating a more resilient and ecologically sound globe.

6.5.2 Eco-inspiration: nature's influence on next-gen materials

The fascinating trip "Eco-Inspiration: Nature's Influence on Next-Gen Materials" explores the complex connection that exists between the environment and the cutting-edge technologies that shape future generations. The first part of the investigation looks at the history of eco-inspired substances, which have their roots in the area of biomimicry, and uses the adaptability and clever designs of nature as a source of inspiration for advances in science and technology. Nature's greatest blueprints, which range from the chemical makeup of spider silk to the intricate design of the wings of butterflies, serve as an inspiration for the creation of environmentally friendly substances with exceptional qualities. The story develops into the next-generation material development stages as we explore the terrain of material science. Researchers and programmers use the durability and adaptability of bones and shells to create biological composites, drawing influence from biological structures. The journey continues into the world of self-healing materials that mimic the ability of organic things to regenerate. We see the development of products that not just mimic the splendour of nature but also outperform traditional materials in terms of functionality and sustainability as they move from the lab to practical applications (Senthil Muthu Kumar et al. 2021).

The investigation goes further under the environmentally friendly characteristics of materials, highlighting the creation procedure's utilisation of reused and natural resources. Recycled materials and polymers derived from plants are integrated as essential parts, demonstrating an industrial innovation's dedication to ecological stewardship. The path of those substances is painstakingly followed in this chapter, starting from their inception in the lab and ending with applications in manufacturing, healthcare, and design. Empirical instances demonstrate the transformational capacity of eco-inspired substances, highlighting their influence on conventional businesses and their function in alleviating environmental issues. The story also explores the moral implications of material progress, touching on issues with the extraction of resources, manufacturing procedures, and end-of-life problems. One major theme that emerges is the thoughtful use of environmentally influenced substances in contemporary technology and facilities, providing a window into an era where creativity and sustainable development coexist harmoniously (Dinesh et al. 2020).

Finally, "Eco-Inspiration: Nature's Impact on Next-Gen Substances" offers a comprehensive investigation, following the path of substances from their organic beginnings to their phases of creation in the lab to their revolutionary uses in a variety of sectors. In the quest for a cleaner and more durable future, this voyage not only demonstrates scientific breakthroughs but also emphasises the moral and ecological advantages of adopting materials influenced by nature. In the end, the story imagines a society in which naturally influenced technologies help to create a harmonious relationship between human creativity and the delicate equilibrium of the environment.

6.5.3 Green alchemy: transforming ideas from the natural world into materials innovation

"Green Alchemy: Converting Concepts in the Environment into Biomaterials Development" delves deeply into the field of biomimicry and provides an in-depth analysis of the ways in which concepts found in the environment are driving innovative material development across a range of industries. For example, the investigation of mounds of termites in the building sector has led to the development of creative ventilation techniques that improve energy usage and lessen the adverse environmental impact of structures. Perched examples of imitation in architecture may be seen in Dubai's Burj Khalifa, which optimised its shade and cooling systems by taking cues from the layout plants best suited to the deserts. Furthermore, in an effort to find more environmentally friendly options, materials made from bacteria like mushroom composites—which are lightweight, strong, and biodegradable construction materials—have been developed using fungus. In the field of conveyance, research on bird flying has influenced the development of more aerodynamic car concepts, which have lowered pollutants and improved fuel economy. Furthermore, the creation of synthetic materials for cars on the outside, including coatings modelled after shark skin, not only lowers drag but also stops pollutants from building up, encouraging the creation of greener and more environmentally friendly transport options (Ramesh et al. 2023).

Developments like synthetic textiles, which mimic leaves' repellent to water qualities and eliminate the demand for chemically derived water-repellent materials, have completely changed the textile business.

Businesses such as Bolt Threads and others have drawn influence from the silk of spiders to create synthetic materials that emulate its tensile power and suppleness, providing a sustainable substitute for conventional fabrics. The cleaning properties of lotus leaves have been mimicked for medical purposes, which has impacted the creation of antibacterial surfaces and helped combat hospital-acquired diseases. The use of biomimetic concepts in the production of medical devices, as shown by the creation of needles modelled after the structure of mosquito's proboscis, demonstrates how the natural effectiveness of things may improve the accuracy and minimally invasiveness of medical procedures. The story is further developed with examples from the field of packaging, wherein mycelium-inspired materials for packaging made of mushrooms are starting to replace traditional containers with more environmentally friendly options. These materials provide a workable answer to the worldwide problem of pollution from plastic, in addition to being biodegradable.

With all of these instances, "Green Chemistry" presents a thorough investigation of how nature's genius can be used to transform material development in a number of industries. The real-time applications highlight how urgent it is to implement environmentally friendly habits and encourage a shared vision of a day when green alchemical concepts would steer technological breakthroughs, striking an appropriate equilibrium between advancement for humans and protecting the environment (Zarrinbakhsh et al. 2016).

6.6 GREEN BY DESIGN: EXPLORING NATURE-INSPIRED BIO-MATERIALS

In an atmosphere in which environmentally friendly methods are becoming more and more important, the search for novel materials has brought us to the fascinating world of nature-inspired biological materials, which blend imitation wisdom with cutting-edge technology in order to revolutionise sectors and lessen environmental impact. According to plant-derived polymers in bioengineered clothing, this excursion into sustainable design provides opportunities that not only correspond with environmentally friendly principles.

6.6.1 Sustainable innovation: harnessing the power of bio-materials

In a time when addressing environmental issues is critically important, the idea of environmentally friendly development rises to the fore as a paradigm-shifting idea in the field of building material research. Particularly, the use of bio-materials is revolutionising sectors and forming a future that is more environmentally friendly, leading to an exponential increase in material development every year. Biological materials are an alternative to conventional materials, which are frequently harmful to the planet. They are made from naturally occurring substances, including fungi, microbes, and vegetation. This comprehensive investigation explores the dynamic and developing field of bio-materials, looking at their special qualities, uses, and ecological advantages. The story proceeds with a thorough analysis of the exponential rise in bio-materials R&D projects, especially within the past 10 years. The collaborative efforts of academics, technologists, and environmental aficionados have led to a continuous growth in novel bio-material remedies, driven by a worldwide understanding of the pressing need for sustainable alternatives. The trend of annual growth, from the introduction of polymers made from plants in the first decade of the 2000s to the most recent developments in advanced manufacturing for bio-compositions, shows a persistent and growing dedication to environmentally friendly methods across a variety of industries (Almgren and Gamstedt 2010). Figure 6.4 shows the sustainable innovation from nature.

Investigations from a range of industries offer concrete illustrations of the substantial advancements bio-materials have made in consumer goods, manufacturing, and construction—particularly during the last 5 years. Novelties like sustainable wrapping, fabrics made from waste from agriculture, and biodegradable plastics highlight the real influence of these products in lessening environmental footprints. Furthermore, the discussion goes beyond the scientific domain to examine the economic and societal aspects of this shift. Adoption of bio-materials is both commercially and ecologically feasible, supporting a more circular and renewable society and bolstering industry resilience generally. But the story gets more complex when we examine the potential and problems that come with the widespread use of biological materials. Scalability, which is cost-effectiveness, and infrastructural modifications turn out to be important aspects of this

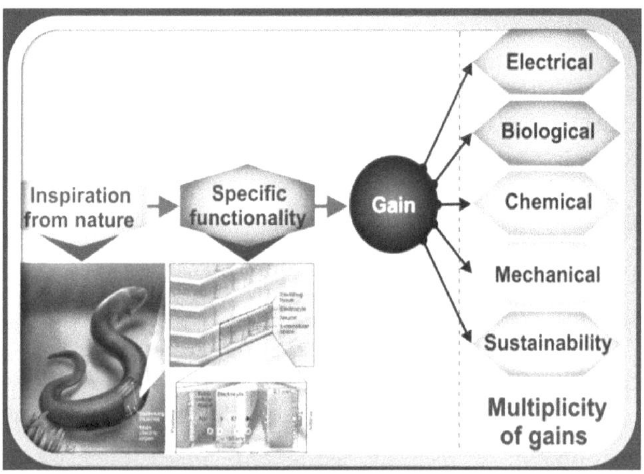

Figure 6.4 Sustainable innovation from nature.

revolutionary path. The intricate nature of an innovative, sustainable environment is highlighted by the careful balancing act between environmental concerns, technical breakthroughs, and social repercussions.

Negotiating this complex landscape makes one thing very evident: sustainable development requires cultural and structural transformation as much as a scientific endeavour, especially in the field of biologic materials. The trend of material innovation's annual increase, particularly over the last 10 years, indicates that we are all committed to redefining our interaction with the natural world. Accepting biological materials is a clear step towards a future that is environmentally friendly, as it paves the path for a more peaceful and regenerative cohabitation with the world. Enmeshed with yearly development structures, this trip into the core of bio-material development demonstrates the transformational potential for sustainability in transforming the very substance of our materialised universe.

6.6.2 Eco-friendly solutions: the future of design inspired by nature

In the discipline of green approach, "Eco-Friendly Options: The Next Generation of Design Influenced by Environment" is considered a masterpiece that offers a comprehensive examination of the revolutionary path towards environmentally friendly products that draw inspiration from nature. The story takes readers on a deep dive into the dangers inherent in conventional design methods, revealing the complex web of damage to the environment created by substances that fuel pollution, forest loss, or an ever-increasing amount of trash. Conventional design materials, which are frequently derived from materials that are not renewable, are criticised for contributing to the worsening of the ecological problem on a worldwide scale. The extensive effects of such substances on wildlife, ecosystems, and climate are carefully examined in this book. The story, which focuses on plastics, illustrates the terrible effects of plastic pollution by portraying it as an ecological enemy that clogs streams, endangers aquatic life, and stays in landfills for generations.

The book advocates for sustainable choices that draw inspiration from nature's genius and promotes a paradigm shift in the field of design thinking as a reaction to these environmental concerns. As a guiding concept, biomimicry arises, encouraging architects to take cues from efficient and environmentally friendly practices that are innate to the environment. As the story progresses, a variety of environmentally beneficial materials are revealed, including the adaptable bamboo, the sturdy cork, and the ground-breaking fungus. Every substance is praised for being renewable, versatile, and having an admirably small environmental impact. These materials are not just abstract ideas; they are actively investigated in a number of sectors to show how they might take the place of ecologically hazardous materials. Bamboo's solidity and adaptability are expected to draw the attention of the building sector, which is sometimes chastised for its resource-intensive methods. In a comparable manner, mycelium-based fabrics are urged to be adopted by fashion designers as an environmentally friendly replacement for traditional textiles, lessening the environmental impact of conventional materials. The book pushes for a future in which making sustainable decisions is not a choice but a necessity, hoping to function as an impetus for a thorough revolution in design processes (Muthusamy et al. 2022). Figure 6.5 shows the future things created by sustainable materials

By carefully examining the issues raised by conventional design materials and providing a detailed explanation of alternatives influenced by the natural world, the story aims to profoundly influence the mindsets of designers, builders, and customers. Modern problems in material science are enormous, and solving them is now necessary for long-term advancement. Traditional substances frequently have a negative influence on human health, the natural world, and the depletion of resources. The trend towards organic components in architecture offers a viable way to get beyond these obstacles and bring in a new age of environmentally responsible creativity.

6.6.2.1 Current problems in material science

- **Depletion of Resources:** A lot of conventional substances, including metals and polymers derived from fossil fuels, are dependent on scarce resources. Such materials' harvesting and use contribute to the gradual degradation of resources, harming the ecosystem over time.

Figure 6.5 The future things created by sustainable materials.

- **Pollution and Waste:** The manufacturing and removal of traditional materials can result in significant waste creation and contamination. Particularly, plastic lingers in the natural world for long periods of time, harming species and ecosystems.
- **Energy Intensity:** Producing some materials requires significant energy inputs, which increases the release of greenhouse gases and exacerbates the effects of global warming.
- **Toxicity:** During production, use, and destruction, human life may be at risk due to the presence of harmful compounds in certain substances. This negative effect puts additional strain on the garbage management procedures.

6.6.2.2 Overcoming problems through the future of design inspired by nature materials

- **Renewable and Biodegradable Materials:** Materials influenced by nature, such as wood and composites made of mycelium, which provide answers to the problem of environmental loss. Compared to their conventional equivalents, these materials need fewer resources in order to grow rapidly and are recyclable. They also frequently biodegrade, which reduces their long-term impact on the environment.
- **Minimal Environmental Impact:** Materials originating from biomimicry and concepts of nature-inspired design sometimes possess a notably reduced impact on the environment. During the whole period, they generate fewer pollutants and waste products because of the use of environmentally responsible production techniques.
- **Energy-Efficient Production:** Environmentally friendly techniques are typically used in the production of goods influenced by the environment. For example, compared to conventional manufacturing techniques, mycelium-based products may be grown with little to no energy input, helping to create a more environmentally friendly and resource-conscious society.
- **Non-toxic Alternatives:** Designers may produce products that are naturally non-toxic by taking cues from the natural world. Bio-based substitutes for hazardous chemicals guarantee that the products are safe for both the planet and individuals.
- **Circular Economy Practices:** The design direction that draws inspiration from nature encourages the application of circular economy principles. The components are made to be readily composted or reusable, promoting a closed-loop process that lowers trash and impacts the environment as a whole.

In conclusion, the use of organic compounds in architecture represents a bright future for solving today's materialistic problems. A kinder, more rejuvenating, and ecologically sensitive future is possible with these societal shifts, which embrace biomimicry, natural resources, and sustainable industrial practices.

6.7 APPLICATION AND FUTURE SCOPE

6.7.1 Application of nature-inspired materials

The development of bio-materials influenced by the environment represents a turning point in the field of materials at a time when environmental problems are causing concern for the entire globe. The organic globe's inherent adaptability and resilience serve as a model for this novel strategy. This investigation explores various uses of biological materials influenced by the natural world, emphasising their adaptability and ability to transform a number of sectors as substitute sources (Velmurugan, Babu, et al. 2023).

- **Foundations of Nature-Inspired Bio-Materials:**
  Investigating biomimicry is the first step towards comprehending the relevance of materials influenced by the natural world. This fundamental feature illustrates how the complex systems and mechanisms found in nature serve as an inspiration for the creation of materials that emulate the effectiveness of the natural world. Inspired by the world of biology, scientists and engineers are currently developing materials that fulfil functional needs while adhering to regeneration and ecological values.
- **Bio-Materials in Medicine and Healthcare:**
  Bio-materials are substances influenced by the environment and have important uses in medical treatment and care. Their biological compatibility, which makes them perfect for tissue science, systems for delivering drugs, and surgical devices, is covered in detail in this part of the chapter. Bio-inspired polymer compounds, scaffolding architectures that look like organic cells, and organs made from synthetic material all show how biologic materials have changed medical technology in revolutionary ways. These materials provide improved outcomes for patients and ethical health care processes.
- **Eco-Friendly Packaging Solutions:**
  Bio-materials inspired by nature provide a sustainable substitute for traditional materials for packaging, which are causing increasing environmental issues. This section examines waste-reducing ideas, including biomimetic designs and plant-based polymers. These alternatives are more than simply substitutes; they have got the power to completely transform the packaging industry, greatly lowering the usage of plastics that are only used once and advancing the ideas of the sustainable economy.

- **Sustainable Construction with Bio-Materials:**
  Using bio-materials drawn from nature transforms the building industry, which is well-known for having a negative impact on the environment, in a sustainable way. Biomimetic architecture layouts and components derived from bamboo and mycelium serve as examples of environmentally friendly building techniques that put the environment and strength first. The building industry is being led by bio-materials in a direction that is more regenerative and ecological in the years to come.
- **Bio-Materials for Renewable Energy:**
  Bio-materials with natural inspiration play a major role in the quest for clean and sustainable energy. This section examines their contribution to the development of technologies like substances used in battery packs and solar cells inspired by biological processes. Biomimicry, which makes use of the design principles found in nature to develop efficient and environmentally friendly solutions, has an impact on innovations in the field of renewable energy. This shift towards a more sustainable economic future is a result.
- **Textile Industry and Wearables:**
  Bio-materials influenced by the environment provide enormous advantages to the textile industry. Plant-based textiles, synthetic patterns influenced by patterns from nature, and garments made of materials that decompose all help to lessen the fashion sector's adverse environmental impacts. Bio-materials are substances that provide a sustainable substitute for traditional textiles, mitigating the issues of contamination and loss of resources related to traditional materials.
- **Water Filtration and Purification:**
  Bio-materials with natural inspiration are essential for tackling water shortages and contamination. Water treatment technologies have demonstrated the efficacy of bio-inspired filtering screens and components sourced from polymers that are naturally occurring. These materials provide effective and long-lasting ways to guarantee clean and safe water supplies by imitating the particular permeability seen in ecosystems.
- **Consumer Electronics and Technology:**
  Bio-materials are substances influenced by the environment that make progress in the field of electronic products for consumers. These substances are bringing in a new era of environmentally friendly internet, from throwaway electronic parts to biomimetic design principles impacting consumer aesthetics. The incorporation of biological materials into electronic equipment lowers waste from electronics and makes technological goods' lifecycles more environmentally friendly.

The investigation of bio-materials influenced by the environment as an alternative resource culminates in highlighting their revolutionary capacity in a wide range of industries. A future that is more ecologically mindful and sustainable is emerging as companies adopt biomimicry and make use of biological materials' adaptability. This voyage emphasises the value of taking cues from nature and the creativity needed to bring human invention and the environment together in a harmonious manner. Bio-materials with a natural theme are a ray of hope, opening doors for a day when resourcefulness and environmental conservation coexist together in many different sectors of the economy (Ng et al. 2022).

6.7.2 Future scope of the research based on natural materials

Future natural materials research has a lot of potential and presents a strong route for environmentally conscious development and growth. The need to investigate, use, and broaden the uses of natural resources in a variety of industries is becoming more and more pressing as we go through a period of increased ecological consciousness. The study area encompasses a wide range of fields, including construction, medical care, material science and technology, and more. This study path aims to capitalise on natural materials' intrinsic capabilities and address important issues, offering several prospects for ground-breaking improvements.

- **Sustainable Construction and Architecture:**
  There is great potential for research on organic substances in design and construction. The way that structures are constructed and designed may be completely changed by looking into the ecological effect, insulating capabilities, and structural integrity of building supplies like mycelium, bamboo, and straw. The findings of this study may open the door to the development of environmentally friendly, energy-efficient constructions that reduce the carbon footprint left by using traditional building supplies.
- **Healthcare and Biomedical Applications:**
  Investigation into the application of organic substances to medical devices is expanding. The creation of surgical devices and technologies with greater biocompatibility and fewer negative consequences is motivated by biomimicry ideas. Investigation of materials such as naturally occurring polymers for pharmaceutical delivery systems, chitosan for wound repair, and cotton for sutures brings up possibilities for disposable and environmentally friendly alternatives in the biomedical sector.
- **Textiles and Fashion Industry:**
  Companies in the fashion and textile industries are looking into organic materials as potential substitutes for synthetic textiles. The production of disposable and sustainable textiles from resources such

as organic cotton, hemp, and pineapple fibres is the major focus of studies in this area of study. Examining eco-friendly methods and organic pigments for the dyeing procedure helps businesses move towards less harmful methods for the environment.

- **Renewable Energy Solutions:**
  Research on natural materials will continue to focus on sustainable energy in the future. Research is needed to fully realise the endless possibilities of substances like organic solar panels, which are modelled after plants' processes of photosynthesis, and to investigate the usage of organic fibres in systems for storing energy. The purpose of the study is to improve both the sustainability and effectiveness of green energy technology.
- **Waste Management and Circular Economy:**
  The idea of economic circularity is greatly advanced by the use of organic elements. The usage of biodegradable substances and biomimicry can be studied in relation to composting, recycling, and reducing waste. Examining natural materials in place of plastic that is only ever used can result in creative ways to reduce the ecological impact of packing as well as throwaway goods.
- **Ecosystem Restoration and Environmental Conservation:**
  Restoration of ecosystems and protecting the environment can greatly benefit from the use of natural resources. This field of study might concentrate on creating materials that decompose for ecological rehabilitation initiatives or employing natural resources to lessen the effects of pollution in water and soil. It will be interesting to investigate how natural materials support resilience to climate change and diversity in the years to come.
- **Digital Fabrication and Advanced Manufacturing:**
  Exciting opportunities arise when natural materials and innovative manufacturing methods like computer-aided manufacturing and 3D printing come together. Research might focus on streamlining procedures for building complex structures out of natural materials, investigating how scalable they are, and pushing the limits of what is practical and aesthetically pleasing.

In short, studies utilising organic substances have wide-ranging and transformational potential in the future. In order to fully explore the untapped potential of the environment, researchers, engineers, creators, and entrepreneurs are encouraged to collaborate across disciplinary boundaries on this research frontier. Studying organic substances is crucial to forming a sustainable, resilient, and peaceful environment for the world and civilisation as we nonetheless unravel the mysteries of nature (Jagadish et al. 2020).

6.8 CONCLUSION

When we look at our research into "Nature-Inspired Bio-Materials: An Alternative Resource," we see a journey through the lens of biomimicry that leads to a lot of eco-friendly alternatives that come from nature. The complex patterns found in the middle of the forest, deciphered from nature, and transformed into state-of-the-art laboratory discoveries provide witness to the boundless possibilities of bio-materials influenced by the biological world's knowledge. Our thorough investigation spanning six different subtopics has yielded insightful information on the transformational possibilities for biological materials influenced by nature. The chapters in their entirety highlight the promise of these materials in bringing about a paradigm shift towards a more sustainable and environmentally friendly future, from deciphering the blueprints of nature to creating substitutes by means of the concepts of biological mimicry, and from demonstrating practical uses to investigating novel materials. It sheds light on the real-world uses of bio-materials and closes the gap between their potential and their actual implementation.

This book chapter provides a path for an environmentally responsible and ecologically sensitive strategy by illustrating how nature-inspired design ideas may be integrated into the field of material science. Bio-materials, which have their roots in the concepts of biological mimicry, show promise as a substitute resource and as a monument to human ingenuity, bringing in an era of environmentally benign and cyclical material processes. As we make our way through the creative teachings found in nature, we see a day in which biological materials that are influenced by nature are purposefully created with durability as a key component. This deliberate design, which is discussed in the framework of "Green by Design," demonstrates how these substances may make a substantial contribution to a cyclical and renewable society. When it comes to implementation and future reach, we have one goal in mind: to steer bio-materials development and use towards an environmentally friendly future through the smooth integration of biomimicry.

By doing this, we adopt a comprehensive viewpoint that stresses these materials' ability to function as ecological alternate resources in addition to their revolutionary potential across a range of sectors. "Nature-Inspired Bio-Materials: An Alternative Resource" is essentially a lighthouse pointing us in the direction of a future in which biomimicry, bio-materials, and nature work together to create a society that is not only inventive but also beautifully ecological. With nature's deep knowledge informing and inspiring the very elements that drive advancement, this complete comprehension opens up a world of options and creates an environmentally friendly route ahead for future generations to follow.

REFERENCES

Abedanzadeh, Sedigheh, Maryam Nourisefat, and Zainab Moosavi-Movahedi. 2021. Bioinspiration and Biomimicry in Lifestyle. In Ali Akbar Moosavi-Movahedi (Ed.), *Rationality and Scientific Lifestyle for Health*, 9. Springer.

Alagumalai, Vasudevan, Vigneshwaran Shanmugam, Navin Kumar Balasubramanian, Yoganandam Krishnamoorthy, Velmurugan Ganesan, Michael Försth, Gabriel Sas, Filippo Berto, Avishek Chanda, and Oisik Das. 2021. Impact response and damage tolerance of hybrid glass/kevlar-fibre epoxy structural composites. *Polymers,* 13(16). https://doi.org/10.3390/polym13162591

Almgren, Karin M., and E. Kristofer Gamstedt. 2010. Characterization of interfacial stress transfer ability by dynamic mechanical analysis of cellulose fiber based composite materials. *Composite Interfaces,* 17(9): 845–61.

Appleton, Ken. 2013. Elementary science teaching. In Sandra K. Abell, Ken Appleton, Deborah L. Hanuscin (Eds.), *Handbook of Research on Science Education*, 493–535. Routledge.

Ausubel, Ken. 2012. *Dreaming the Future: Reimagining Civilization in the Age of Nature*. Chelsea Green Publishing.

Barthlott, Wilhelm, M. Daud Rafiqpoor, and Walter R. Erdelen. 2016. Bionics and biodiversity–bio-inspired technical innovation for a sustainable future. In Jan Knippers, Klaus G. Nickel, Thomas Speck (Eds.), *Biomimetic Research for Architecture and Building Construction: Biological Design and Integrative Structures*, 11–55. Springer.

Biswal, Trinath, Sushant Kumar BadJena, and Debabrata Pradhan. 2020. Sustainable biomaterials and their applications: A short review. *Materials Today: Proceedings,* 30: 274–82.

Chen, Jihua. 2024 Biomimetic Manufacturing and Healthcare: Sustainable Pattern, Algorithm, and Assembly. https://doi:10.31224/3466

Coppens, Marc-Olivier, and Ayomi S. Perera. 2018. Re-designing materials for biomedical applications: From biomimicry to nature-inspired chemical engineering. *Philosophical Transactions of the Royal Society A: Mathematical, Physical and Engineering Sciences,* 377(2138): 20180268.

Cosma, Mattia Pancrazio, and Roberto Brighenti. 2023. From responsiveness in biological matter to functional materials: Analogies and inspiration towards the systematic design and synthesis of new smart materials and systems. *Applied Materials Today,* 32: 101842.

Danks, Sharon Gamson. 2010. *Asphalt to Ecosystems: Design Ideas for Schoolyard Transformation*. New Village Press.

Dinesh, Thiyagarajan, Arumugam Kadirvel, and P. Hariharan. 2020. Thermo-mechanical and wear behaviour of surface-treated pineapple woven fibre and nano-silica dispersed mahua oil toughened epoxy composite. *Silicon,* 12(12): 2911–20. https://doi.org/10.1007/s12633-020-00387-4

Ferreira, Pedro, Arlete Apolinário, and Gabriela Forman. 2023. Optimising textile biomaterial selection for sustainable product and circular design: Practical guidelines for a greener future. *Materials Circular Economy,* 5(1): 14.

Ganesan, Velmurugan, and Babu Kaliyamoorthy. 2020. Utilization of Taguchi technique to enhance the interlaminar shear strength of wood dust filled woven jute fiber reinforced polyester composites in cryogenic environment. *Journal of Natural Fibers*. https://doi.org/10.1080/15440478.2020.1789021

Ganesan, Velmurugan, Vigneshwaran Shanmugam, and Vasudevan Alagumalai. 2024. Composites part C : Open access optimisation of mechanical behaviour of calotropis gigantea and prosopis juliflora natural fibre-based hybrid composites by using Taguchi-grey relational analysis. *Composites Part C: Open Access,* 13(January): 100433. https://doi.org/10.1016/j.jcomc.2024.100433

Ganesan, Velmurugan, Vigneshwaran Shanmugam, Babu Kaliyamoorthy, Sekar Sanjeevi, Suresh Kumar Shanmugam, Vasudevan Alagumalai, Yoganandam Krishnamoorthy, et al. 2021. Optimisation of mechanical properties in saw-dust/woven-jute fibre/polyester structural composites under liquid nitrogen environment using response surface methodology. *Polymers,* 13(15). https://doi.org/10.3390/polym13152471

Iñiguez-Moreno, Maricarmen, Montserrat Calderón-Santoyo, Gabriel Ascanio, Frida Zoé Ragazzo-Calderón, Roberto Parra-Saldívar, and Juan Arturo Ragazzo-Sánchez. 2023. Harnessing emerging technologies to obtain biopolymer from agro-waste: Application into the food industry. *Biomass Conversion and Biorefinery,* 14: 29265–29282. https://doi.org/10.1007/s13399-023-04785-7

Jagadish, Maran Rajakumaran, and Amitava Ray. 2020. Investigation on mechanical properties of pineapple leaf–based short fiber–reinforced polymer composite from selected Indian (Northeastern Part) cultivars. *Journal of Thermoplastic Composite Materials,* 33(3): 324–42. https://doi.org/10.1177/0892705718805535

Jing, Linzhi, Xiang Wang, Bin Leng, Ningping Zhan, Hang Liu, Shifei Wang, Yuyun Lu, Jie Sun, and Dejian Huang. 2021. Engineered nanotopography on the microfibers of 3D-printed PCL scaffolds to modulate cellular responses and establish an in vitro tumor model. *ACS Applied Bio Materials,* 4(2): 1381–94.

John, Godfaurd, Derek Clements-Croome, and George Jeronimidis. 2005. Sustainable building solutions: A review of lessons from the natural world. *Building and Environment,* 40(3): 319–28.

Kim, Raphael. 2021. Virus as quasi-living bio-material for interaction design: Practical, ethical, and philosophical implications. In *Extended Abstracts of the 2021 CHI Conference on Human Factors in Computing Systems*, 1–7, May 8–13, 2021, Yokohama, Japan. https://doi.org/10.1145/3411763.3451770

Lakshmaiya, Natrayan, Seeniappan Kaliappan, Pravin P. Patil, Velmurugan Ganesan, Joshuva Arockia Dhanraj, Chattariya Sirisamphanwong, Tanakorn Wongwuttanasatian, et al. 2022. Influence of oil palm nano filler on interlaminar shear and dynamic mechanical properties of flax/epoxy-based hybrid nanocomposites under cryogenic condition. *Coatings,* 12(11). https://doi.org/10.3390/coatings12111675

Lakshmaiya, Natrayan, Velmurugan Ganesan, Prabhu Paramasivam, and Seshathiri Dhanasekaran. 2022. Influence of biosynthesized nanoparticles addition and fibre content on the mechanical and moisture absorption behaviour of natural fibre composite. *Applied Sciences (Switzerland),* 12(24). https://doi.org/10.3390/app122413030

Latha, Arunkumar D., A. Suresh Kumar, S. Jasgurpreet Singh, and Chohan Velmurugan. 2023. Experimental investigations of flammability, mechanical and moisture absorption properties of natural flax/nano SiO_2 based hybrid polypropylene composites. *Silicon.* https://doi.org/10.1007/s12633-023-02611-3

Mohan, Tamilselvan, Andreja Dobaj Štiglic, Marco Beaumont, Johannes Konnerth, Fazilet Gürer, Damjan Makuc, Uroš Maver, Lidija Gradišnik, Janez Plavec, and Rupert Kargl. 2020. Generic method for designing self-standing and dual porous 3D bioscaffolds from cellulosic nanomaterials for tissue engineering applications. *ACS Applied Bio Materials,* 3(2): 1197–209.

Mohanty, Amar Kumar, Manjusri Misra, and L. T. Drzal. 2002. Sustainable bio-composites from renewable resources: Opportunities and challenges in the green materials world. *Journal of Polymers and the Environment,* 10: 19–26.

Muthusamy, Anandha Raja, Senthil Muthu Kumar Thiagamani, Senthilkumar Krishnaswamy, Chandrasekar Muthukumar, Harikrishnan Pulikkalparambil, Jeyanthi Subramanian, and Naganandhan Madasamy. 2022. Erosion performance of natural fiber reinforced vinyl ester hybrid composites: Effect of layering sequences. *Materials Today: Proceedings,* 64: 200–6.

Nadar, Nandini Robin, Richelle M. Rego, Gara Dheeraj Kumar, H. Jeevan Rao, Ranjith Krishna Pai, and Mahaveer D. Kurkuri. 2023. Demystifying the influence of design parameters of nature-inspired materials for supercapacitors. *Journal of Energy Storage,* 72: 108670.

Ng, Lin Feng, Mohd Yazid Yahya, and Chandrasekar Muthukumar. 2022. Mechanical characterization and water absorption behaviors of pineapple leaf/glass fiber-reinforced polypropylene hybrid composites. *Polymer Composites,* 43(1): 203–14. https://doi.org/10.1002/pc.26367

Nichols, Eva Megan. 2017. *Hybrid Approaches to Bio-Materials and Bio-Inorganic Electrochemical Carbon Dioxide Reduction.* University of California.

Oguntona, Olusegun Aanuoluwapo, and Clinton Ohis Aigbavboa. 2017. Biomimetic materials and technologies for carbon neutral cities in South Africa: A literature review. *Procedia Engineering,* 196: 152–8.

Perera, Ayomi S., and Marc-Olivier Coppens. 2019. Re-designing materials for biomedical applications: From biomimicry to nature-inspired chemical engineering. *Philosophical Transactions of the Royal Society A,* 377(2138): 20180268.

du Plessis, Anton, and Chris Broeckhoven. 2019. Looking deep into nature: A review of micro-computed tomography in biomimicry. *Acta Biomaterialia,* 85: 27–40.

Ramesh, Velumayil, Krishnasamy Karthik, Robert Cep, and Muniyandy Elangovan. 2023. Influence of stacking sequence on mechanical properties of basalt/ramie biodegradable hybrid polymer composites. *Polymers,* 15(4). https://doi.org/10.3390/polym15040985

Sanjeevi, Sekar, Vigneshwaran Shanmugam, Suresh Kumar, Velmurugan Ganesan, Gabriel Sas, Deepak Joel Johnson, Manojkumar Shanmugam, et al. 2021. Effects of water absorption on the mechanical properties of hybrid natural fibre/phenol formaldehyde composites. *Scientific Reports,* 11(1): 1–11. https://doi.org/10.1038/s41598-021-92457-9

Sekar, Sanjivi, Shanmugam Suresh Kumar, Shanmugam Vigneshwaran, and Ganesan Velmurugan. 2020. Evaluation of mechanical and water absorption behavior of natural fiber-reinforced hybrid biocomposites. *Journal of Natural Fibers,* 00(00): 1–11. https://doi.org/10.1080/15440478.2020.1788487

Senthil Muthu Kumar, T., K. Senthilkumar, M. Chandrasekar, S. Karthikeyan, Nadir Ayrilmis, N. Rajini, and Suchart Siengchin. 2021. Mechanical, thermal, tribological, and dielectric properties of biobased composites. In Anish Khan, Sanjay M. Rangappa, Suchart Siengchin, and Abdullah M. Asiri (Eds.), Biobased *Composites: Processing, Characterization, Properties, and Applications,* 53–73. John Wiley & Sons, Inc. https://doi.org/10.1002/9781119641803.ch5

Settlage Jr, John. 1994. Conceptions of natural selection: A snapshot of the sense-making process. *Journal of Research in Science Teaching,* 31(5): 449–57.

Shukla, Ashish K, Shirsendu Mitra, Shikha Dhakar, Arnab Maiti, Sudhanshu Sharma, and Krishna K Dey. 2022. Electrochemical energy harvesting using microbial active matter. *ACS Applied Bio Materials,* 6(1): 117–25.

Silva, Luciano P. 2019. Current trends and challenges in biofabrication using biomaterials and nanomaterials: Future perspectives for 3D/4D bioprinting. In Mohammed Maniruzzaman (Ed.), *3D and 4D Printing in Biomedical Applications: Process Engineering and Additive Manufacturing*, 373–421. Wiley-VCH Verlag GmbH & Co. KGaA.

Spencer, Sarah. 2019. *Think Like a Tree: The Natural Principles Guide to Life.* Swarkestone Press.

Suresh Kumar, Shanmugam, S. Thirumalai Kumaran, Ganesan Velmurugan, A. Perumal, Sanjivi Sekar, and M. Uthayakumar. 2021. Physical and mechanical properties of various metal matrix composites: A review. *Materials Today: Proceedings,* 50(xxxx): 1022–31. https://doi.org/10.1016/j.matpr.2021.07.354

Taneja, Akriti, Ruchi Sharma, Shreya Khetrapal, Avinash Sharma, Rupak Nagraik, Baskar Venkidasamy, Manju Nath Ghate, Shavkatjon Azizov, Somesh Sharma, and Deepak Kumar. 2023. Value addition employing waste bio-materials in environmental remedies and food sector. *Metabolites,* 13(5): 624.

Velmurugan, Ganesan. 2023. Experimental investigations of physical and mechanical properties of pineapple leaf/carbon fiber–reinforced unsaturated polyester composites at liquid nitrogen environment. *Biomass Conversion and Biorefinery*, 14: 22019–34. https://doi.org/10.1007/s13399-023-04263-0

Velmurugan, Ganesan, and Kaliyamoorthy Babu. 2020. Statistical analysis of mechanical properties of wood dust filled jute fiber based hybrid composites under cryogenic atmosphere using grey-Taguchi method. *Materials Research Express,* 7(6). https://doi.org/10.1088/2053-1591/ab9ce9

Velmurugan, Ganesan, and Lakshmaiya Natrayan. 2023. Experimental investigations of moisture diffusion and mechanical properties of interply rearrangement of glass/kevlar-based hybrid composites under cryogenic environment. *Journal of Materials Research and Technology*, February. https://doi.org/10.1016/J.JMRT.2023.02.089

Velmurugan, Ganesan, Kaliyamoorthy Babu, Nagaraj, and John Presin Kumar. 2023. Investigations of flame retardancy, mechanical and thermal properties of woven hemp/PP hybrid composite for insulating material reinforced with synthetic silicon and zinc oxides. *Silicon*, no. 0123456789. https://doi.org/10.1007/s12633-023-02408-4

Velmurugan, Ganesan, Lakshmaiya Natrayan, Jasgurpreet Singh Chohan, Palani Vasanthi, Sathiyamoorthy Angalaeswari, Palani Pravin, Sankar Kaliappan, and Dhanasekaran Arunkumar. 2023. Investigation of mechanical and dynamic mechanical analysis of bamboo/olive tree leaves powder-based hybrid composites under cryogenic conditions. *Biomass Conversion and Biorefinery*, no. 0123456789. https://doi.org/10.1007/s13399-023-04591-1

Velmurugan, Ganesan, Shanmugam Suresh Kumar, Jasgurpreet Singh Chohan, A. John Presin Kumar, Thangaraj Manikandan, David Elil Raja, Kumar Saranya, Murugan Nagaraj, and Praveen Barmavatu. 2023a. Experimental investigations of mechanical and dynamic mechanical analysis of bio-synthesized CuO/ramie fiber-based hybrid biocomposite. *Fibers and Polymers*, no. 0123456789. https://doi.org/10.1007/s12221-023-00432-0

Velmurugan, Ganesan, Shanmugam Suresh Kumar, Jasgurpreet Singh Chohan, R. Sathish, S. Panneer Selvan, S. A. Muhammed Abraar, D. Elil Raja, M. Nagaraj, and Sivaprakasam Palani. 2023b. Hybrid calotropis gigantea fibre-reinforced epoxy composites with SiO_2's longer-term moisture absorbable and its impacts on mechanical and dynamic mechanical properties. *Materials Research Express,* 11(1). https://doi.org/10.1088/2053-1591/ad0bc8

Velmurugan, Ganesan, Vionthan Siva Shankar, Mohamed Kalil Rahiman, David Elil Raja, Murugan Nagaraj, and Thangavelu Jeyalakshmi Nagalakshmi. 2023. Experimental investigation of high filler loading of SiO_2 on the mechanical and dynamic mechanical analysis of natural PALF fibre-based hybrid composite. *Silicon,* no. 0123456789. https://doi.org/10.1007/s12633-023-02464-w

Vollrath, Fritz, and David Porter. 2009. Silks as ancient models for modern polymers. *Polymer,* 50(24): 5623–32.

Wang, Zhe, Xinpei Wang, Wanting Xu, Yongxiao Li, Ruizhi Lai, Xiaohui Qiu, Xu Chen, Zhidong Chen, Bobin Mi, and Meiying Wu. 2023. Translational challenges and prospective solutions in the implementation of biomimetic delivery systems. *Pharmaceutics,* 15(11): 2623.

Wilson, Ruth. 2007. *Nature and Young Children: Encouraging Creative Play and Learning in Natural Environments*. Routledge.

Zarrinbakhsh, Nima, Tao Wang, Arturo Rodriguez-Uribe, Manjusri Misra, and Amar K Mohanty. 2016. Characterization of wastes and coproducts from the coffee industry for composite material production. *BioResources,* 11(3): 7637–53.

Chapter 7

Renewable bio-composites

Properties and their applications

Zohreh Khoshraftar

7.1 INTRODUCTION

Bio-composites are made up of a biodegradable polymer matrix and cellulosic materials as the reinforcing filler. Many studies have looked into using organic fibers and flour as reinforcements in biodegradable polymer bio-composites. These reinforcing elements possess the ability to naturally decompose through microbial action, contributing significantly to the degradation of organic substances in the ecosystem [1]. Renewable bio-composites stand at the forefront of material science innovation, shaping composite materials from sustainable origins. Rooted in biodegradable polymers sourced from natural, renewable sources, these materials epitomize a fusion of sustainability and practicality. Their hallmark lies in balancing ecological mindfulness with versatile utility across various sectors. By uniting biodegradable polymers with plant-derived reinforcements like hemp, flax, or kenaf fibers, renewable bio-composites pave the way for eco-friendly manufacturing. The resulting materials boast impressive strength, adaptability, and notably, a significantly smaller environmental impact compared to conventional alternatives. Beyond providing an alternative to traditional materials, these bio-composites signify a shift toward a circular economy. They represent a commitment to responsibly leveraging nature's resources, offering a renewable and rejuvenating choice in the realm of composite materials. With technological strides, the potential applications of renewable bio-composites appear poised to transform industries seeking sustainable solutions without compromising performance or resilience.

Composite materials, classically described as the fusion of two distinct substances, blend one as a sturdy support for bearing loads while the other, known as the matrix, bears the load and binds the reinforcing fibers together. Their escalating acclaim and expansion stem from their lightweight nature, exceptional specific properties, customizable traits, malleability, and resilience to chemical erosion and environmental impact. The unique benefits of composite materials foreshadow a promising path forward, revealing

 DOI: 10.1201/9781003470342-7

numerous prospects across various sectors, including automotive, building, aerospace, military, maritime exploration, wind energy utilization, infrastructure such as piping and containers, consumer product development, and the domains of electrical and electronic progress [2,3].

Bio-composites are a type of composite material that consists of biodegradable polymers and are reinforced with components derived from natural sources [3]. What sets these composites apart is the biodegradability of both their constituents, making them an eco-friendly solution for various applications. An intriguing historical parallel lies in the use of mud bricks in ancient times. Humans, dating back to 7000 to 3300 BC in the Indus Valley, crafted these bricks from a blend of earth, moisture, and either rice husks or wheat straws for structural purposes in constructing their dwellings [4]. This ancient practice underscores the enduring nature of utilizing sustainable materials in construction. Mud bricks, made from sun-dried mud, were among the earliest man-made bio-composite materials created by mankind for construction objectives. The use of mud bricks continued into the modern ages with certain modifications [5]. In contemporary construction practices, mud bricks or cob are still used in some regions, often with enhancements and adjustments to meet modern building standards. While traditional mud brick construction may lack the durability and resistance to weathering compared to fired bricks or other modern building materials, it is still employed in certain contexts for its affordability, availability, and ecological advantages. Modern variations of mud brick construction may involve incorporating stabilizers or reinforcements to improve the strength and durability of the structures. Additionally, techniques like rammed earth construction, adobe construction, and other earth-based building methods are employed in different regions across the globe, showcasing the continued relevance of these ancient building techniques in contemporary architecture with sustainable and eco-friendly considerations.

The use of plant fibers and bioresins to create eco-friendly materials has gained significant attention as a solution to tackle ecological and economic issues associated with the manufacturing of fresh material. It is important to mention that the current rate of consumption indicates that petroleum, a finite fossil fuel, may only last for another 50–60 years. This has led to increased awareness of environmental issues and efforts to preserve petroleum resources. One notable development in this area is the adoption of plant fiber-reinforced composites (PFRCs) in various commercial applications. These materials not only help reduce greenhouse gas emissions but also take into consideration the limited availability of fossil energy resources [6,7]. By relying less on petroleum-based polymers, we can avoid depleting landfill capacities and comply with plastic waste regulations. The demand for sustainable products has driven retailers and manufacturers to invest in cost-effective and eco-friendly materials, aiming to mitigate the impacts of global warming. By utilizing plant fibers and bioresins, we can create materials with reduced environmental footprints derived from renewable resources. This transition is crucial for preserving natural resources, managing plastic waste, and meeting the growing demand for eco-conscious products. Embracing sustainable materials not only addresses environmental concerns but also presents opportunities for manufacturers and retailers to contribute positively to mitigating global warming. By adopting these materials, they can play a part in creating a more sustainable future while also meeting the needs of environmentally conscious consumers [8].

The primary goal of this extensive study is to investigate and define the characteristics, diverse applications, and promising future potential of renewable bio-composites within the automotive, construction, packaging, and renewable energy sectors. By delving into the unique attributes of these bio-composites and assessing their viability as alternatives to conventional materials, this chapter aims to provide valuable perspectives on the discussion surrounding sustainability and eco-friendly solutions across diverse industries. Through a meticulous examination of specific applications, we aim to provide a holistic understanding of how renewable bio-composites can play a pivotal role in fostering environmentally conscious practices and facilitating a transition toward more sustainable and resilient industrial practices.

7.2 PROPERTIES OF RENEWABLE BIO-COMPOSITES

Selecting materials for various applications is a crucial process, and the choice often hinges on their inherent properties. These properties are intricately linked to whether a material is isotropic or anisotropic. In many instances, the behavior of material properties in response to different physical phenomena demonstrates linearity within a specified operational range. Opting for a linear model can significantly simplify the differential constitutive equations describing these properties. Conversely, these equations are pivotal for determining material properties, providing valuable insights into the material's behavior. Understanding the original length of a material is pivotal in gauging its dimensional changes, allowing for the calculation of gains or losses. The reliable measurement of material properties is facilitated by standardized test methods, with numerous such methodologies documented and published by user communities, often through ASTM International. The realm of renewable bio-composites, as highlighted by various researchers, underscores several key properties. Mechanical properties, thermal characteristics, optical attributes, degradability, and electrical properties emerge as vital considerations. Every aspect of these facets contributes uniquely to assessing the overall appropriateness and effectiveness of renewable bio-composite materials.

7.2.1 Lightweight nature and its benefits

Renewable bio-composites are well-known for their lightweight qualities, offering numerous benefits across various industries. This characteristic arises from the inherent properties of organic and plant-based materials used in their production, providing a sustainable alternative to traditional composite materials. The lightweight nature of renewable bio-composites leads to significant weight reduction in final products like automotive parts and aerospace structures. As a result, this contributes to enhanced fuel efficiency, playing an essential role in the growth and advancement of eco-friendly transportation solutions. When considering the impact of bio-composites on end-products such as automotive components or aerospace structures, the significance of their lightweight quality cannot be overstated. The benefits of reduced weight have a direct impact on fuel efficiency, which is especially notable in certain industries. For example, in the automotive sector, incorporating lightweight bio-composite materials into vehicle components helps reduce the overall weight of vehicles. Consequently, this reduction results in lower energy usage and improved fuel efficiency. Similarly, in the aerospace industry, utilizing these composites can lead to more fuel-efficient aircraft, leading to decreased operational expenses and a reduced environmental impact.

In addition, the importance of bio-composites being lightweight cannot be overstated when it comes to environmentally friendly transportation solutions. As concerns about environmental sustainability and carbon emissions increase worldwide, it is essential to incorporate lightweight materials. Choosing renewable bio-composites allows industries to make significant progress in adopting greener practices and complying with strict environmental regulations. Moreover, the inclusion of organic and plant-based materials in bio-composites not only diminishes dependence on fossil fuels but also aids in minimizing the overall carbon footprint. Bio-composites, which are made from renewable resources, are being increasingly sought after in order to establish a sustainable supply chain. This shift reflects the global trend toward reducing our reliance on non-renewable resources and minimizing the environmental effects of traditional manufacturing methods. In essence, the lightweight and renewable nature of bio-composites not only decreases the weight of final products but also greatly contributes to enhancing fuel efficiency. These qualities, combined with their environmentally friendly composition, make bio-composites a vital component in creating eco-conscious solutions across various sectors.

7.2.2 Strength-to-weight ratio

Despite their reduced weight, bio-composites often boast a remarkable strength-to-weight ratio. This unique combination enhances their structural integrity, adapting them to be appropriate for various purposes where both strength and weight considerations are critical, such as in lightweight construction materials.

The addition of particulates to polymers can influence composite strength in two primary ways: reinforcement and weakening effects. Reinforcement occurs when the particulates impede crack propagation, enhancing overall strength. Conversely, weakening effects create stress concentration points, potentially diminishing strength. If the reinforcing effect prevails, the composite strength surpasses that of the pure polymer. However, when the weakening effect dominates, the composite strength falls below that of the neat polymer [9].

7.2.3 Environmental sustainability

Derived from sustainable sources, bio-composites are inherently eco-friendly. The use of plant-based fibers or agricultural byproducts ensures a reduced carbon footprint compared to traditional composites, contributing to the global effort to mitigate climate change.

7.2.4 Biodegradability and end-of-life management

Bio-composites offer advantages in terms of end-stage handling. Many of these materials are biodegradable, providing a sustainable solution for products with a limited lifecycle. This feature addresses concerns related to environmental impact during disposal, offering a viable alternative to conventional non-biodegradable composites.

In the present era, the widespread use of synthetic polymers derived from petroleum, particularly polyolefins in items like packaging, bottles, and molded products, has become pervasive in modern society [10]. However, the environmental consequences of this dependence are becoming increasingly evident, with almost 150 million tons of petrochemical plastics being discarded globally each year. This alarming disposal rate compounds environmental issues, posing a threat to ecosystems as the production and consumption of these materials continue to rise. A concerning aspect of petroleum-based plastics is their resistance to microbial degradation, exacerbating their environmental impact as these persistent plastic wastes accumulate in soil, rivers, and marine environments, introducing harmful pollutants and disrupting delicate ecosystems. The longevity of these plastics in the environment raises concerns about their lasting effects on both terrestrial and aquatic habitats.

In response to this ecological crisis, there is a growing emphasis on environmentally friendly biodegradable polymers as a viable alternative to their petroleum-based counterparts. The promise lies in their possibility of full breakdown back into natural ecosystems, including diverse environments. This signifies a shift from the enduring presence of traditional plastics, offering a more sustainable approach to waste management. What distinguishes these eco-friendly polymers is their ability to alter chemicals, aided by the activity of biological enzymes or microorganisms. This dynamic quality allows them to undergo a natural breakdown process, ensuring that they reintegrate harmoniously into the environment without leaving a lasting impact. The capacity of these polymers to undergo biodegradation aligns with broader initiatives aimed at minimizing the ecological consequences of human activities and promoting a more environmentally friendly coexistence with the planet. Researchers emphasize the utmost significance of biodegradability in the realm of bio-composites, ensuring that the life cycle's end phase of these samples is not only environmentally responsible but also actively contributes to the restoration of ecosystems. As scientific exploration continues into the intricacies of biodegradable polymers, the potential for mitigating the detrimental impact of plastic waste on the environment becomes increasingly promising. Essentially, the transition to eco-friendly biodegradable polymers signifies a conscientious effort to address the environmental challenges posed by petroleum-based plastics. By adopting materials that seamlessly integrate into the natural world, we pave the way for a more sustainable and harmonious coexistence with the planet we call home.

Kim et al. [1] delved into the breakdown of poly(butylene succinate) and a PBS bio-composite containing rice-husk flour (RHF) in natural and aerobic compost soils. The study assessed the weight loss percentages of HDPE, PBS, and bio-composites, as illustrated in Figure 7.1a. Interestingly, the weight loss in bio-composites exhibited a noticeable decline with escalating RHF content. This suggests that microorganisms easily targeted the cellulosic materials, augmenting the degradation capability of the composites in comparison to the PBS and HDPE matrices. Furthermore, a

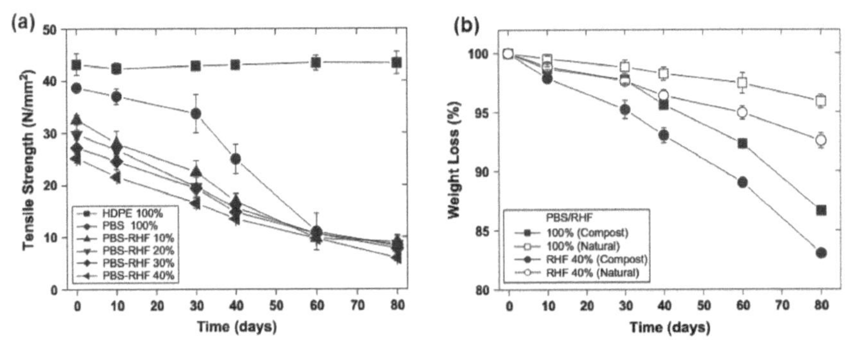

Figure 7.1 (a) Percentage weight loss of HDPE, PBS, and bio-composites for 80 days. (b) Comparison of the percentage weight loss of PBS and bio-composites at 40 wt% filler loading in natural and compost soil [1]. (Reproduced with permission from Elsevier, license number, 5694901238175.)

noteworthy observation was made regarding the weight loss percentage of bio-composites subjected to natural and compost soil environments with a 40 wt% filler loading. Intriguingly, a more rapid degradation rate was evident in the compost soil burial test compared to the natural soil burial test spanning 80 days. The accelerated biodegradation in the composting environment, as depicted in Figure 7.1b, underscores the heightened degradation capabilities of bio-composites in a chamber setting conducive to composting [11].

Yussuf et al. [12] conducted a study to assess the biodegradability differences between two types of composite materials: polylactic acid (PLA) combined with kenaf fibers (PLA-KF) and PLA combined with rice-husk flour (PLA-RHF). The investigation involved subjecting these composites to a natural soil burial test over a 90-day period. The findings revealed a slight increase in the biodegradability of both PLA-KF and PLA-RHF composites, reaching 1.2% and 0.8%, respectively, after the 90-day test period. Notably, these percentages indicate a higher level of biodegradability compared to pure PLA. This enhanced biodegradability is attributed to the presence of natural fibers, which make microorganisms more susceptible to decomposition. In essence, the study suggests that incorporating kenaf fibers or RHF into PLA composites can positively influence their biodegradation rates, potentially offering environmentally friendly alternatives to traditional plastic materials. The increased susceptibility to microbial attack in the presence of natural fibers contributes to the accelerated breakdown of these composites in natural soil environments.

Biodegradable polymers can be categorized based on their source and production methods. Agro-polymers, originating from agricultural sources such as cellulose, fall into one of these categories. Cellulose, abundant in plants, can be extracted and processed to obtain biodegradable polymers. These polymers offer advantages such as renewable availability and low environmental impact. Microbial polymers, like polyhydroxyalkanoates, are produced through the fermentation of microorganisms. These polymers have gained attention due to their ability to degrade in various environmental conditions, making them suitable for applications requiring biodegradability. Chemically synthesized biopolymers can be derived from agro-based resources or conventionally synthesized monomers. These polymers may be derived from plant-based resources or synthesized through chemical reactions using conventional monomers [13]. They offer versatility in terms of properties and can be modified to meet specific requirements. When processing biopolymers into composites, several important factors need consideration. These factors encompass the polymer's decomposition rate and temperature sensitivity, along with the material's mass per unit volume. The processing of biopolymers into composites involves considerations of degradation time, temperature, density, and reinforcement with lignocellulosic fibers to enhance properties [10].

Classification of biopolymers can be further elaborated as follows [13].

1. **Renewable Sources:**
   - **Polysaccharides:** Biopolymers originated from renewable sources such as cellulose, gums, chitosan, etc.
   - **Proteins:** Biopolymers sourced from proteins like casein, collagen, and gelatin.
   - **Lipids:** Biopolymers obtained from cross-linked glycerides.
2. **Chemically Synthesized:**
   - **Polyacids:** Biopolymers synthesized chemically, such as polyacids.
   - **Polyvinyl Alcohols:** Biopolymers produced through chemical synthesis, such as polyvinyl alcohols.
   - **Polyesters:** Biopolymers obtained by chemical synthesis using polyester monomers.
3. **Microbial Synthesis:**
   - **Poly(hydroxyalkanoates):** Biopolymers produced through microbial fermentation, such as poly(hydroxyalkanoates).
   - **Bacterial Cellulose:** Biopolymers synthesized by bacteria, specifically cellulose produced by bacterial fermentation.
   - **Xanthan, Pullulan, Curdlan:** Biopolymers synthesized by microbial processes, such as xanthan, pullulan, and curdlan.
4. **Biopolymer Composites:**
   - **Starch Blends:** Blends of biopolymers involving starch, a renewable polysaccharide, mixed with other polymers.
   - **Polyester Blends:** Blends of biopolymers including polyesters, combining their properties to achieve desired characteristics.
   - **Caseinate Blends:** Blends involving casein, a protein derived from milk, mixed with other polymers.
   - **Collagen/PVA Blends:** Blends combining collagen, a protein found in connective tissues, with polyvinyl alcohol (PVA) to create desired properties.

These categorizations offer a summary of the various kinds of biopolymers and where they are derived from. They demonstrate the diversity of biopolymers available and the potential for creating blends to enhance properties or achieve specific requirements in various applications.

7.2.5 Thermal properties

Fiber-reinforced polymer composites find extensive application in structural elements operating under extreme temperature conditions, showcasing their versatility in various industries [11]:

a. In the automotive sector, these composites are utilized for high-temperature-resistant engine components, enhancing performance and durability.

b. Aerospace and military sectors benefit from these materials due to their lightweight nature and ability to withstand both high and low temperatures, ensuring reliability in harsh environments.
c. Electronic and circuit board components benefit from the insulating properties and thermal stability of these composites, safeguarding sensitive equipment from heat-related damage.
d. Within the oil and gas industry, these composites prove invaluable in equipment exposed to extreme conditions, offering corrosion resistance and structural integrity even in high-heat environments.

Their adaptability and resilience make fiber-reinforced polymer composites a cornerstone in applications demanding robust performance under temperature extremes across diverse sectors. Yang and colleagues [14] delved into examining the influence of lignocellulosic fillers on bio-composites of thermoplastic polyester polymers. In their investigation using thermogravimetric analysis (TGA), it was noted that with an increase in filler loading, there was a slight decrease in thermal stability, coupled with an augmentation in ash content. This outcome suggested a lack of effective interaction between the filler and the polymer matrix. In an effort to enhance the interfacial adhesion, a compatible agent was introduced to ameliorate the thermal properties of the composite. Furthermore, thermal mechanical analysis (TMA) was employed to scrutinize the thermal expansion behavior of the composites. The results indicated a reduction in thermal expansion with escalating filler content, particularly when a compatibilizing agent was incorporated. This finding signifies that the lignocellulosic filler, in conjunction with the compatibilizing agent, acted to mitigate the thermal expansion of the composite materials across varying atmospheric conditions. This underscores the role of the filler in effectively restricting the thermal expansion of the composite materials.

The study conducted by Lee et al. [15] examined the polymerization of aniline on bacterial cellulose and analyzed the thermal stability of the resulting bacterial cellulose/polyaniline nanocomposite films using TGA. The TGA outcomes exhibited that the composites experienced weight loss in two distinct stages, indicating changes in cellulose structure and degradation of polyaniline. These findings offer valuable insights into the thermal behavior of bacterial cellulose/polyaniline nanocomposites.

7.3 RAW MATERIALS FOR RENEWABLE BIO-COMPOSITES

7.3.1 Various categories of organic fibers employed in bio-composites

Natural fibers (Figure 7.2), based on their composition and source, can be categorized broadly into three main types [16].

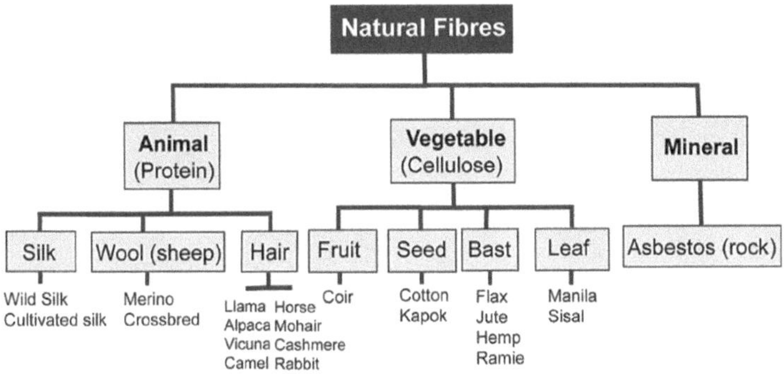

Figure 7.2 Classification of natural fibers [16]. (This content has been reproduced with permission from Elsevier, under license number, 5694891033833.)

1. **Cellulose-Based Fibers:** These fibers originate primarily from plants and are composed of cellulose, a complex carbohydrate found within the cellular structures of plants. Examples include fibers from sources like jute, cotton, hemp, flax, and bamboo. They are known for their strength, versatility, and renewable nature, making them extensively used in various industries.
2. **Protein-Based Fibers:** Derived mainly from animal sources, these fibers are primarily composed of proteins. Wool from sheep, silk from silkworms, and hair from animals like goats (mohair) and rabbits (angora) are prominent examples. These fibers often possess unique properties like warmth, softness, and resilience, suitable for applications in textiles and sometimes specialized industrial uses.
3. **Mineral-Based Fibers:** These fibers are relatively less common and are derived from inorganic sources. Examples include asbestos, basalt, and glass fibers. Though not as widely used as cellulose or protein-based fibers in natural form, these fibers are often processed and combined with other materials to enhance specific properties like strength and resistance in composite materials.

Cellulose-based fibers exhibit a further classification based on their source [17]. Natural fibers are mainly composed of cellulose, lignin, and hemicellulose as the three primary components [18,19]. Cellulose, the main component, provides natural fibers with strength and stability, while hemicellulose contributes to their structural integrity. In rural settings, these natural fibers find extensive use in non-structural applications such as crafting ropes, bags, brooms, and furniture. Additionally, they serve practical purposes in roofing and insulation. Characterized by a coarse texture, these fibers exhibit a spectrum of colors ranging from white to dark brown. This utilization of natural fibers in diverse applications underscores their versatility, from everyday tools to construction materials, reflecting a sustainable and traditional approach in rural areas [20].

The following fibers are widely utilized [16].

7.3.1.1 Bamboo

Bamboo, an expeditiously expanding grass that thrives in regions with abundant rainfall, showcases extraordinary inherent durability and resistance to extreme temperatures. When fortified with bonding agents, bamboo fibers discover utility in fashioning household furnishings, automotive cabin features, and even wave-riding boards.

7.3.1.2 Coir

Derived from the husk and outer shell of coconuts, coir fibers are categorized as seed-derived fibers. Exhibiting resistance to water and saltwater damage, coir is a versatile material. Harvesting coir from green coconuts yields a smooth, pliable white fiber, while mature coconuts produce a robust, albeit less flexible, brown fiber.

7.3.1.3 Flax

As one of the world's oldest fiber crops, flax offers ease of weaving into various fabrics, subsequently used in composite materials with diverse properties. Recognized for their eco-friendly and cost-effective nature, flax fibers, according to some studies, can rival synthetic counterparts.

7.3.1.4 Hemp

Derived from the cannabis plant family, hemp fibers find applications in textile manufacturing and the paper industry. Historical usage includes the manufacturing of ship canvas sails, rigging, nets, and caulk because of its durability and ability to withstand saltwater.

7.3.1.5 Jute

Abundantly produced in India, Bangladesh, and China, jute stands out as one of the most cost-effective fibers. Boasting a specific modulus approaching that of glass fibers, jute remains stable up to 200°C without compromising its properties.

7.3.1.6 *Bagasse*

Obtained from the remnants of squeezed sugarcane, bagasse fibers serve multiple purposes. Traditionally used as a combustible material in the sugarcane industry, it has also found applications in the paper industry. The versatility of bagasse makes it a valuable resource with eco-friendly attributes. The characteristics of certain plant fibers pertaining to their physical attributes are indicated in Table 7.1.

7.3.2 Processing techniques for obtaining suitable fibers and polymers

Raw materials are crucial in the advancement of renewable bio-composites, with particular emphasis on obtaining suitable fibers and polymers. The quest for sustainability has spurred innovative processing techniques that harness the potential of nature's offerings while minimizing environmental impact.

Table 7.1 Characteristics of certain plant fibers pertaining to their physical attributes [16]

Types of fibers	*Fibers*	*Density (g/cm^3)*	*Elongation (%)*	*Tensile strength (MPa)*	*Young modulus*
Stem fibers	Bamboo	0.6–0.91	1.4	193–600	20.6–46.0
	Flax	1.50	1.2–3.2	345–2,000	15–80
	Hemp	1.48	1.60	550–900	26–80
	Jute	1.30	1.16–1.5	393–800	13–55
	Kenaf	1.45	1.60	157–930	22.1–60
	Ramie	1.50	1.2–3.8	400–938	61.4–128
Leaf fibers	Banana	0.72–0.88	2.0–3.34	161.8–789.3	7.60–9.40
	Pineapple	1.07	2.20	126.6	4.40
	Sisal	1.50	3.0–7.0	468–700	9.40–22
Fruit fibers	Coir	1.20	17–47	175	4.0–6.0
	Oil palm	0.7–1.55	4–18	50–400	0.57–9.0
Wood fibers	Softwood Kraft (spruce)	1.50	—	1,000	18–40
	Hardwood Kraft (brich)	1.20	—	—	37.9
Synthetic fibers	E-glass	2.50	2.50	2,000–3,500	70
	S-glass	2.50	2.80	4,570	86
	Aramide	1.40	3.3–3.7	3,000–3,150	63.0–67.0

Source: This content has been reproduced with permission from Elsevier, under license number, 5694891033833.

In the realm of fiber extraction, cutting-edge methods are emerging to maintain the integrity of renewable sources. Delicate fibers, such as those derived from bamboo or flax, demand meticulous processing to maintain their inherent strength and resilience. Employing advanced mechanical separation techniques, researchers delicately harvest these fibers, ensuring a sustainable supply without compromising the ecosystems from which they originate. Moreover, the selection of polymers in bio-composite formulation is a critical facet of the process. Scientists are pioneering methods to extract polymers made from sustainable sources, steering away from traditional petroleum-based alternatives. Innovative chemical processes are being harnessed to transform these raw materials into versatile polymers, fostering the creation of eco-friendly composites. One cutting-edge approach involves utilizing enzymatic reactions to break down complex natural polymers into more manageable components. This not only streamlines the extraction process but also enhances the compatibility of the polymers with the fibers, resulting in bio-composites with superior mechanical properties. The synergy between advanced fiber extraction and polymer processing techniques signifies a harmonious collaboration between science and nature. This holistic approach ensures that the renewable bio-composites produced are not only environmentally responsible but also exhibit enhanced performance characteristics, creating a foundation for a future that is both sustainable and resilient in material science.

The manipulation of fibers and polymers involves a diverse array of methods tailored to create materials with specific characteristics for a multitude of applications. The selection of a particular methodology hinges on variables such as the nature of the polymer, the desired goal of the final product, and the specific characteristics sought in the fibers. Below are several pivotal processing techniques employed in the acquisition of appropriate fibers and polymers [21–24]:

1. **Extrusion:**
   Extrusion is a widely employed method for converting polymers into fibers. In this process, the polymer is melted, and it is pushed through a die to create a continuous fiber. The extruded fiber then undergoes a cooling and solidification process. This method is commonly used for synthetic polymers like polyethylene, polypropylene, and polyester.
2. **Spinning:**
   Spinning involves pushing a polymer solution or melt through small openings to create fibers. Various spinning methods, such as melt spinning, wet spinning, and dry spinning, cater to different polymer types and applications. For example, wet spinning is frequently used in producing fibers from polymers that dissolve in a liquid.

3. **Drawing:**
   Drawing entails elongating fibers to improve their tensile strength and orientation. It is often applied to partially oriented fibers obtained through processes like extrusion or spinning. Drawing can be done in a uniaxial or biaxial manner, depending on the desired properties of the final fiber.
4. **Electrospinning:**
   Electrospinning is an advanced technique that uses an electric field to extract polymer fibers from a solution or melt. This method can produce ultrafine fibers with nanometer-scale diameters. This property makes it beneficial for applications like tissue engineering and filtration, thanks to the increased surface area of the fibers produced.
5. **Melt-Blowing:**
   Melt-blowing stands as a manufacturing method wherein molten polymer is forced through fine nozzles, creating polymer fibers that undergo swift cooling via hot air. Widely utilized in the production of nonwoven materials, this process finds application in crafting filters and disposable fabrics.
6. **Solution Spinning:**
   Solution spinning involves dissolving the polymer in a solvent to produce a spinning solution, which is then extruded through a spinneret to generate fibers. This technique is suitable for polymers that pose challenges in processing through alternative methods.
7. **Polymer Blending:**
   Polymer blending is a method in which two or more polymers are mixed to achieve specific properties in the resulting fibers. By combining different polymers, it is possible to enhance characteristics such as strength, flexibility, and chemical resistance.

7.4 MANUFACTURING TECHNIQUES OF BIO-COMPOSITES

There are two traditional methods for manufacturing: out-of-autoclave (OoA) and autoclave techniques. These techniques offer a significant advantage over conventional manufacturing methods by allowing the fabrication of more complex geometrical components without the need for molds. This means that these advanced techniques eliminate the traditional requirement of molds, opening up new possibilities in manufacturing [25].

Various traditional methods have been extensively utilized in the manufacturing of bio-composites [26–28]. Over time, considerable progress has been made in refining these methods, with researchers dedicated to enhancing bio-composite quality and cost-effectiveness through modifications to existing techniques. An emerging trend in composite production involves the utilization of additive manufacturing methods. These innovative methods offer the advantage of creating complex and lightweight composite structures without requiring extra tools. Moreover, additive manufacturing enables the creation of multi-material composites, a feat beyond the capabilities of traditional manufacturing approaches. The selection of fabrication techniques is contingent on specific application requirements, such as desired fiber dispersion, aspect ratio, and reinforcement orientation [29]. When deciding on a manufacturing process and technique, several factors come into play. These factors include production costs, the desired shape and the ultimate blueprint, characteristics of the primary materials, sizing factors, and restrictions imposed by the manufacturing process. The ultimate choice is influenced by a comprehensive evaluation of these elements, ensuring an optimal balance between performance and practical considerations.

7.4.1 Extrusion injection molding

Extrusion stands as a pivotal manufacturing technique, involving the melting of thermoplastic polymer pellets. In this process, these molten pellets are blended with fibers through the utilization of either one or two rotating screws, with the latter being the more prevalent choice due to its superior capability to disperse fibers within the matrix. The amalgamated material is then forcefully pushed out of the chamber via a die. Twin screws are favored for their ability to provide effective dispersion of both fiber and matrix components. The resulting product from the extrusion process serves as a precursor for subsequent stages, particularly the injection molding process, as outlined by Pickering et al. [30]. Numerous studies have delved into the realm of producing poly lactic acid/cellulose fiber composites through the extrusion method, succeeded by injection molding. Notable contributions in this domain include the works of Chaitanya and Singh [31], Jaszkiewicz et al. [32], and Komal et al. [33]. To visualize this process, refer to the illustration of extrusion injection molding presented in Figure 7.3a. In the initial stage of the manufacturing process, polymer pellets and short fibers are introduced into the hopper of a single or twin-screw extruder. As this combination progresses through the extruder's barrel, the application of heat causes the resin to melt, while the screw's motion facilitates a thorough blending of the polymer melt with the fibers. The resulting mixture is then extruded through a die. Following this, the composite undergoes a cooling process before being shaped into pellets using a pelletizer. These pellets made of composite materials are used as feedstock for the injection molding machine, where they are transformed into the desired dimensions. An inherent challenge with PLA composites lies in the suboptimal interfacial bonding between cellulose fibers and the PLA matrix. To address this limitation and enhance the adhesion between the matrix and fibers, a two-step manufacturing process is employed. This involves an initial extrusion stage, followed by injection molding. This meticulous process ensures an improved and more robust composite material, ready for application in various industries [34].

7.4.2 Extrusion compression molding

Extrusion compression molding stands as another crucial method for producing environmentally friendly PLA-based green composites. This process involves two steps, similar to extrusion injection molding, as illustrated in Figure 7.3b. To commence, the initial step involves feeding PLA pellets and cellulose fibers into a single or twin-screw extruder. Within this machine, the materials undergo a transformation into a molten mixture of PLA and fibers, forming what is known as PLA/fiber melt. Subsequently, this PLA/fiber melt is subjected to compression molding. During this stage, the molten composite is shaped and compressed within a mold under high pressure and temperature. The pressure assists in ensuring the uniform distribution of fibers within the PLA matrix, contributing to the desired structural integrity and mechanical properties of the final green composite product. This method enables the creation of composite materials that leverage the biodegradability and sustainable nature of PLA while incorporating the reinforcing benefits of cellulose fibers, resulting in eco-friendly and durable composite materials suitable for various applications. The procedure involves transferring this molten composite to a preheated mold set at the desired temperature. Subsequently, the mold is positioned within a compression molding machine and subjected to hot pressing to achieve the final composite product. An intriguing study conducted by a research group utilized this method to fabricate composites comprising PLA and banana fibers. This involved compounding PLA pellets and fibers in an extruder, amalgamating them into a molten form collected in a preheated mold. The subsequent step involved hot pressing the mold at 170°C under a pressure of 40 bar. The composite was taken out of the mold once it reached a temperature of 50°C. To facilitate easy removal, polyester sheets were employed. This process showcases a unique approach to developing composite materials, harnessing the combined properties of PLA and banana fibers. By precisely controlling temperature and pressure and employing specialized materials like polyester sheets, the research group successfully crafted durable composites suitable for diverse applications [33,34].

(a)

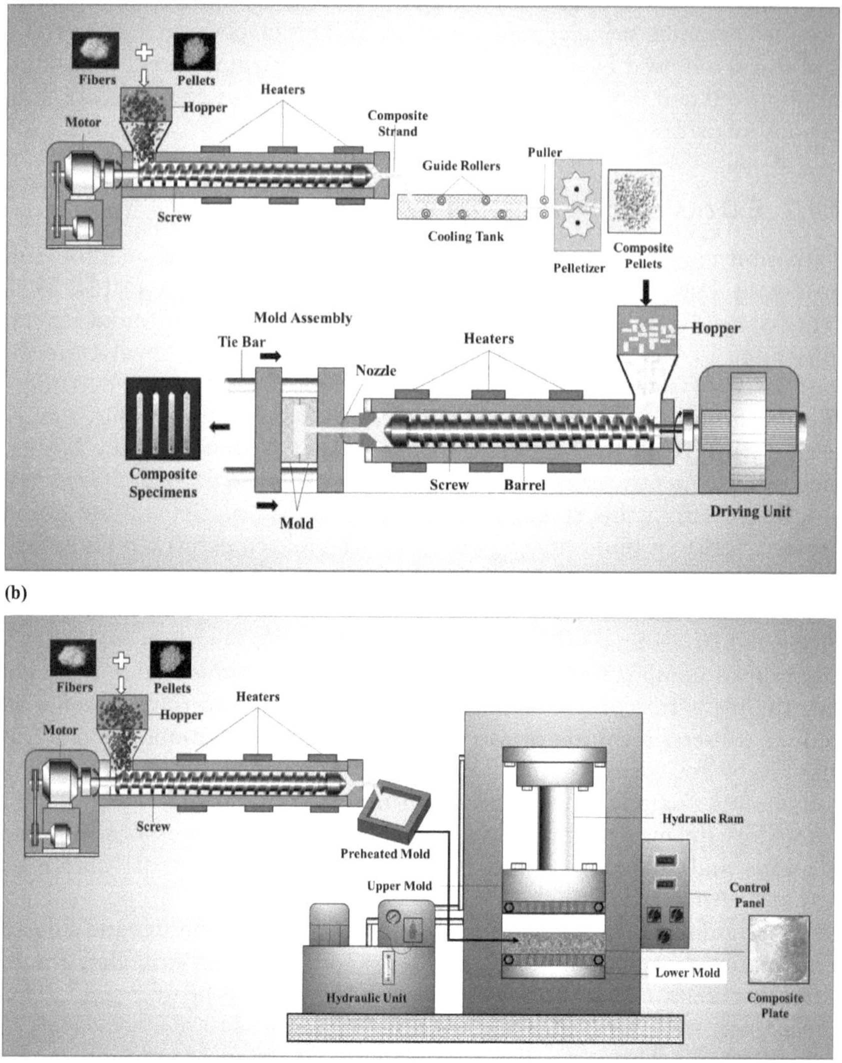

(b)

Figure 7.3 (a) The extrusion injection molding process layout. (Reproduced with permission from Elsevier, license number, 5694880814135) [30]. (b) The procedure for extrusion compression molding. (Reproduced with permission from Elsevier, license number, 5694880814135) [30].

7.5 APPLICATIONS IN AUTOMOTIVE INDUSTRY

A burgeoning trend in the automotive industry involves a shift toward employing bio-composite materials as the predominant substance in the manufacturing of new and upcoming vehicles. This is evident in the exploration of innovative materials such as kenaf and hemp fiber reinforcements combined with polyurethane foam in a conceptual car. This approach exemplifies the potential for developing lightweight and sustainable materials tailored for load-bearing applications. This progressive utilization of bio-composites underscores the industry's commitment to exploring eco-friendly alternatives that balance both weight considerations and structural integrity. The incorporation of kenaf and hemp fibers into polyurethane foam showcases a promising avenue for achieving materials that not only minimize the overall weight of vehicles but also enhance their sustainability profile. It is important to note that while not every conceptual idea materializes into a market-ready product, these exploratory concepts play a crucial role in forecasting the feasibility of eco-friendly materials specifically designed for lightweight yet load-bearing applications in the automotive sector. These experiments and prototypes contribute valuable insights into the potential for advancing the industry's commitment to environmentally conscious practices. In essence, the increasing focus on bio-composite materials in automotive manufacturing reflects a commitment to pushing the boundaries of sustainable innovation. The utilization of kenaf and hemp fibers in conjunction with polyurethane foam represents a forward-thinking approach to meeting the industry's dual objectives of reducing vehicle weight and embracing environmentally friendly solutions for load-bearing components [28].

During the 1930s, Henry Ford, the individual who established Ford Motor Company, envisioned utilizing bio-based materials for vehicle parts as a means to support farmers economically. His inspiration stemmed from their struggles, prompting him to explore the feasibility of using soil-derived components for car manufacturing. This initiative aligned with Ford's pursuit of environmentally friendly lightweight vehicles, offering the potential to comply with weight and emissions regulations. Motivated by these factors, Ford Motor Company embarked on an ambitious research endeavor focused on integrating natural fibers into automobiles. This initiative culminated in the introduction of the T concept car in 1941, a groundbreaking innovation featuring a body predominantly constructed from soy resin-reinforced materials. Notably, the vehicle's weight was significantly reduced, amounting to two-thirds that of a regular car. Moreover, the car was designed to run on hemp oil fuel, taking advantage of the abundant oil resources available at the time. However, despite the promising advancements, the outbreak of the Second World War led to a shift in priorities for the automotive industry toward highly mechanized vehicles for wartime needs. Consequently, this shift affected the progress of Ford Motor Company's research into bio-based materials for automobiles, slowing down its development [35].

In 1957, a significant milestone occurred in the realm of bio-composites within the automotive industry. East Germany unveiled the Trabant car, featuring a groundbreaking monocoque construction. This innovative design incorporated various components such as the roof, bootlid, bonnet, wings, and doors, all crafted from a thermosetting phenolic resin reinforced with cotton [24]. Fast forward to 1982, FIAT contributed to the exploration of bio-based materials in automobiles by introducing wood floor-filled door panels in the FIAT Punto [36]. However, it wasn't until the late 20th century that the practical application of natural fibers and bio-composites gained momentum. The increasing awareness of fossil resource depletion, a driving force behind the industry's growth, coupled with mounting environmental concerns, propelled stakeholders to consider more sustainable alternatives. The quest for fuel efficiency, environmental friendliness, lighter vehicles, and renewable materials gained prominence [35,37,38]. In response to these challenges, Daimler–Benz initiated intensive research in 1991 to replace traditional glass fiber reinforcements with natural fibers in certain auto parts. A subsidiary of Daimler–Benz in Sao Paulo, Brazil, embarked on the ambitious "Beleem project," utilizing coconut fibers in the production of commercial vehicle parts [39]. This marked the beginning of a broader industry trend, with several key players joining the movement toward incorporating eco-friendly and renewable materials in vehicle manufacturing. Over the years, substantial progress has been achieved both in research laboratories and production facilities, as explored in greater detail in the subsequent sections. The commitment of industry leaders to these advancements underscores a collective effort to address environmental concerns and usher in a new era of sustainable automotive technologies.

In 2008, Ford Motor Company joined forces with Lear Corporation to pioneer the development of soy-based seat foam, a groundbreaking initiative that found its way into the Ford Mustang. The subsequent introduction of bio-based cushions marked a significant stride, impacting over one million vehicles under the Ford, Lincoln, and Mercury brands by 2010. This innovative approach extended to various other Ford models, including the Expedition, F-150, Focus, Escape, Mercury Mariner, Lincoln Navigator, and Lincoln MKS [40]. Furthering its commitment to sustainable materials, Ford embraced wheat straws, recognizing their commendable mechanical properties and durability. As a result, these straws were incorporated into the injection-molded storage bins of the 2010 Ford Flex crossover. The utilization of hemp-polypropylene composites led to the formation of the front grill of the Ford Montageträger [35,41], showcasing the versatility of eco-friendly materials. Additionally, a noteworthy environmental milestone was achieved through the integration of 100% castor bean oil-derived nylon-11 in the manufacturing of fuel tank tubes. This strategic move resulted in an impressive reduction of close to 1 million pounds of CO_2 emissions annually (Figure 7.4). By exploring and implementing sustainable alternatives in various components across their vehicle lineup, Ford not only demonstrated a commitment to environmental responsibility but also

Figure 7.4 A depiction of the incorporation of natural fibers in the assembly of the Ford Flex quarter trim panel, storage bin, and cover line, utilizing polypropylene reinforced with 20% wheat straw.

contributed significantly to reducing the automotive industry's ecological footprint [35]. Table 7.2 outlines various initiatives aimed at advancing the improvement of bio-composites specifically tailored for use in automotive purposes.

7.6 CONSTRUCTION AND BUILDING MATERIALS

The utilization of composite materials can be traced back to earlier civilizations, where they were employed for various household applications. One unique example of this dates back 3,000 years ago when the Egyptians ingeniously utilized straw-reinforced clay to construct walls. This early utilization of natural fibers for reinforcement in composite materials showcased the innovative thinking and resourcefulness of ancient civilizations.

In the past few decades, wood has seen widespread application in bolstering cement composites, finding its way into various products like non-structural building materials. Contemporary instances of wood-cement hybrids encompass slate roofing, cladding, architectural features, and alternatives to traditional lumber, such as fascia, trim soffit, and corner boards. Wood fiber emerged as a substitute for asbestos in cement-based construction materials, leveraging its superior qualities, such as enhanced durability, impressive fire resistance, and resilience against moisture. Moreover, this innovation has undergone treatments to fortify it against decay and

Table 7.2 Examples of applications of bio-composites in the automobile industry [28]

Manufacture	*Model*	*Bio-composite materials*	*Applications*
Ford	2012 Ford focus 2013 Ford F-250 2014 F-series pickups. 2015 Ford Mustang sports	Coir fiber blended with polypropylene (PP) Coconut shells and rubber from shredded tires. Rice hulls and PP Coconut powder and thermoplastic polyolefin	Onboard battery housing, structural guards, lightweight armrests, door cladding and applique, brackets
BMW	BMW i3 BMW iFE.20 BMW 5 & 7 series	Hemp composites Flax bio-composites	Headliners, dashboard, and interior panels. Interior and dashboard design. Hood, trunk, and rear wing
Mercedes Benz	EvoBus, Setra, and Travego	Flax/polyester composites	Engine encapsulations, door lining, seat backrests, rear trunk covers, and rear panel shelves. Natural fiber composite bumpers
Volkswagen	Audi A4, A8, Roadster Porsche Cayman 718 GT4 CS MR	Wood fiber and PP bio-composite Bcomp bio-composite material	Boot lining, spare tire lining, door panes Full bodywork kit and interior panels
General Motors	Cadillac DeVille GMC	Wood and kenaf/PP bio-composites Flax/polyester bio-composites	Seatbacks, door panels, cargo floor Roof lining, package trays, trim
Toyota	2003 Raum	Hemp fibers bio-composites and potato-derived PLA composites reinforced with sugarcane bagasse	Interior panels. Springboard
Tesla	2017 Model S P100DL	Flax laminates	Conventional body panels, interior panels

safeguard it from insect damage, ensuring a long-lasting and resilient solution for the construction industry. Wood-based composites play a pivotal role in producing composite panels, a significant cornerstone in the realm of building materials. These panels, crafted from wood-fiber-reinforced plastic composites (WPC), hold appeal owing to the ready accessibility of wood fiber found abundantly in sawmill byproducts, coupled with its remarkable strength and lightweight nature [42–44].

Manufacturers of construction materials are increasingly turning to composite materials for building structures like dams, bridge decks, roads, sidewalks, and pavements, placing a strong emphasis on enhancing durability. To bolster this durability, various additives, including fly ash, slag, and silica fume, are employed to mitigate chemical interactions between the matrix and fibers. Fibers play a pivotal role, serving as primary reinforcements to amplify composites that can enhance the strength and durability of materials used in applications such as pavements and slabs, helping to prevent cracking caused by environmental stressors [43,44]. In essence, fibers function as vital connectors that span across cracks, imparting post-cracking ductility to the structure. Their presence acts as a deterrent to crack propagation, absorbing and dissipating energy when integrated into the matrix materials. This integration culminates in an overall improvement in the strength and resilience of the composite, showcasing the multifaceted role of fibers in enhancing the performance and longevity of construction materials.

In the 1920s, the utilization of agasse, a fiber derived from sugar cane processing, marked the inception of manufacturing composite panels in America [8]. Presently, residential structures in the Philippines, Jamaica, and Ghana benefit from the incorporation of building panels and roofing sheets constructed with bagasse fiber-reinforced phenolic composites [42]. Unlike conventional roof tiles, predominantly composed of materials like clay, cement, and steel, these innovative panels address the drawbacks of excessive weight, leading to elevated design (dead weight) and installation loads (live weight). Furthermore, conventional materials exhibit unfavorable environmental consequences, including high embodied energy, material waste, and pollution from manufacturing processes affecting air, land, and water.

Novel advancements in sheet materials have introduced eco-friendly alternatives to conventional options, notably through the integration of recycled paper reinforced with cellulose fibers. In a study by Dweib et al. [45], bio-based materials were employed to fabricate an innovative NFC (natural fiber composite) sandwich roof structure along with structural beams. The incorporation of recycled paper into composite sheets and structural unit beams demonstrated strength and stiffness levels that rivaled those of the conventional materials commonly employed in roof construction. This pioneering approach signifies a sustainable shift in material utilization, emphasizing both environmental consciousness and structural integrity.

The prospect of natural fiber composites replacing petroleum-based materials in structural applications signifies a shift toward environmentally

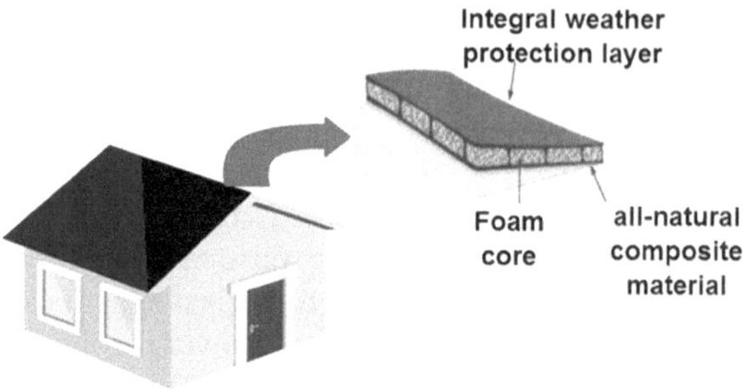

Figure 7.5 The diagram illustrates a monolithic house roof constructed using natural composites, which provides protection against hurricanes.

friendly alternatives. In the aftermath of Hurricane Andrew, the ACRES group in South Carolina, USA, pioneered the development of a sustainable and hurricane-resistant housing structure (Figure 7.5). This innovative dwelling, designed for high performance at a low cost, was crafted from unconventional yet eco-friendly materials, including chicken feathers, recycled paper, straws, and a resin derived from soybean oil, resulting in a fully bio-based composite. The construction of the housing structure employed advanced techniques, such as vacuum infusion molding, to create a roof composite. This composite, in the form of a sandwich panel, was not only resilient but also exhibited thermal insulation properties. The integration of these green materials not only demonstrated the feasibility of sustainable construction but also showcased the potential of bio-based composites to serve as both a structural and environmentally conscious solution [43].

Renewable materials have found diverse applications in the construction industry, extending beyond the basic structural elements to encompass a wide range of building components. From flooring and furniture to roofing sheets, bridges, and pavement pathways, these sustainable alternatives are gradually supplanting conventional petroleum-based systems. In the realm of interior spaces, the evolution of green kitchens stands out as an exemplar of bio-based innovation. Cupboards are now fashioned from bioproducts like wheat or rice straw fibers reinforced with polyurethane resin, replacing traditional formaldehyde-laden counterparts. Moreover, countertops are crafted from hemp fibers due to their exceptional strength, demonstrating a conscious choice to employ renewable resources in high-stress applications. The revolution in bridge construction is exemplified by the advent of stay-in-place bridge forms (SIP), constructed from bio-composites rather than traditional steel forms. The porous and breathable nature of these bio-composite SIP forms facilitates water evaporation through the structure, mitigating corrosion risks. Beyond corrosion resistance, these bio-composite SIP forms offer a distinct advantage in terms of weight—they are lighter compared to steel forms, enabling faster and more cost-effective installations. This multifaceted approach to incorporating renewable materials not only addresses environmental concerns but also emphasizes the practical benefits of sustainability in modern construction practices [43].

Additional instances comprise utilizing cereal straw fiber/PLA for architectural cladding screens through cold press molding. Furthermore, a pedestrian bridge constructed from PLA reinforced with hemp and flax fibers serves as another illustration. Considered the "most sustainable building in Africa," the hemp house in South Africa incorporates hempcrete, hemp insulation, and hemp plaster in its construction [43].

The utilization of agricultural residues, as well as other renewable and sustainable biomass, is experiencing a rapid growth in green building applications. This trend is reflected in the estimated market size of green building materials, which is projected to reach approximately USD 19 billion in 2021. Lignocellulosic agricultural residues offer numerous advantages for green building applications. These residues, which are available in abundance at low cost, serve as renewable and sustainable sources. What sets them apart is that many of these agricultural residues and biomasses possess exceptional and distinct properties. These properties make them highly desirable for building applications, as they exhibit excellent thermal insulation, noise reduction, and other key characteristics. Thus, utilizing these unique agricultural residues presents a viable and sustainable solution in the production of green building materials for both outdoor and interior applications, such as insulation, framing, siding, roofing, and more [46]. Bio-based materials have been employed to enhance the properties of various construction materials. They have been utilized as reinforcement in gypsum plasters, as binders for cement-based structural elements, and even as substitutes for cement itself. By incorporating bio-based materials, building materials can provide improved insulation against sound and heat, as well as reduced density. These materials have also been successfully used to reinforce concrete, offering unique advantages in terms of sustainability and performance [46].

7.6.1 Durability of bio-composites

Hayes et al. [47] highlight the myriad advantages of fiber-reinforced polymer composites compared to conventional materials. These benefits encompass resistance to moisture, ultraviolet radiation, chemical attacks, dynamic loadings, freeze-thaw cycles, and the detrimental effects of physical aging on material properties. Goldstein [48] further underscores that the incorporation of fiber-reinforced polymer composites substantially enhances the durability of construction projects. Consequently, the utilization of these composites in construction sites emerges as a highly advantageous strategy, promising to significantly extend the lifespan and resilience of projects [49].

7.6.2 Long-term durability and performance of bio-composites in construction

When choosing construction materials, it is crucial for contractors and builders to consider green requirements and practical features. These features include stiffness, affordability, durability, versatility, and ease of use. While conventional manufacturing materials may possess one or more of these features, none of them typically provide all the desired characteristics. Manufacturers strive to create materials that not only fulfill the fundamental criteria of construction components but also demonstrate a commitment to environmental sustainability. However, these environmentally friendly materials often come with higher costs. For instance, the production of materials such as glass, carbon, and aramid requires a significant amount of energy, and in the case of glass, it relies on non-renewable resources [49].

On the other hand, resins can be considered environmentally friendly when compared to certain materials like metal. Resins are often derived from renewable resources and have relatively lower environmental impacts. They offer a viable alternative for reducing the environmental footprint of construction materials. In conclusion, manufacturers strive to create construction materials that possess the desired practical features while also being environmentally sustainable. Though some materials may seem environmentally friendly, they can be associated with higher costs and energy consumption. It is a continuous challenge for the industry to find a balance between meeting construction requirements and minimizing environmental impact. Composite materials possess an exceptional ability to merge multiple basic components together, resulting in a bulk material that surpasses the individual properties of its constituents. This characteristic grants composites enhanced traits such as strength, stiffness, and durability, making them highly desirable in various applications. In order to integrate the use of new types of fiber composites in the construction industry, it is crucial to prioritize the development of sustainable methods, systems, and standards. This involves implementing guidelines that promote the utilization of eco-friendly materials and ensuring responsible sourcing of raw materials. Additionally, minimizing waste and emissions during the production process and encouraging environmentally conscious manufacturing practices are integral for sustainable composite usage. By adopting sustainable practices, the construction industry can contribute to environmental preservation and alleviate the negative impacts associated with conventional construction materials. Incorporating sustainable fiber composites not only improves the overall sustainability of construction projects but also allows for enhanced performance while reducing resource consumption and environmental harm. Moreover, focusing on sustainable practices and standards specific to composite materials ignites opportunities for innovation and the exploration of eco-friendly alternatives. This constant research and development effort drives the discovery of novel materials and manufacturing techniques that further enhance the sustainability and performance of composites in construction. To conclude, the exceptional ability of composite materials to combine basic components and generate bulk materials with superior properties paves the way for the construction industry to embrace sustainability. By prioritizing sustainable practices and fostering the advancement of new fiber composites, the industry can spearhead positive change and contribute to the creation of an environmentally friendly and efficient built environment.

7.7 PACKAGING AND CONSUMER GOODS

7.7.1 Biodegradable and eco-friendly packaging materials

Packaging plastics are frequently disposed of after a single use, contributing significantly to an ongoing waste issue. Traditionally, efforts to tackle this problem have focused on incorporating natural fibers into non-biodegradable plastics or developing biodegradable polymers. However, advancements in understanding these biomaterials have opened doors to their combined effective utilization. The significance of biodegradable polymers in promoting sustainability cannot be overstated. By utilizing biodegradable plastics, we can establish a continuous, renewable cycle from sourcing raw materials, manufacturing, utilization, collection, and composting. Composting these biopolymers holds the potential to generate carbon-rich soil, which in turn aids in reducing reliance on chemical fertilizers. This integrated approach not only minimizes waste but also contributes to the enhancement of soil quality, aligning with the principles of a more sustainable future [50–52].

The most recent national waste data report from 1997 indicated that packaging accounted for approximately 10%±14% of the overall landfill waste [52]. It is important to note that this percentage encompasses various types of packaging materials, not solely plastics. The figure provided represents a range of 10%–14% of total landfill waste attributed to packaging materials. This includes a diverse array of materials used for packaging purposes, such as paper, cardboard, glass, metals, and plastics. The report's data underscores the significant contribution of packaging to the overall waste generated, emphasizing the necessity for sustainable and eco-friendly packaging solutions across different material types to reduce landfill waste. The evolution of natural-fiber-biopolymer composites marks a pivotal stride toward delivering eco-friendly consumer goods that align with specific usage requirements while minimizing environmental impact. Despite the existence of a substantial body of research, the commercial availability of such products, especially in the packaging sector, remains relatively limited. Nevertheless, an increasing trend toward adoption is propelled by advancements in the affordability and accessibility of biopolymers. Additionally, governmental regulations addressing landfill practices and recycling further contribute to the growing acceptance of these sustainable materials in the market [52].

Challenging the dominance of wood and cardboard in secondary and tertiary packaging poses a formidable task, considering their inherent qualities as natural-fiber composites that are highly recyclable and predominantly biodegradable. Take corrugated cardboard, for instance, with density values ranging between 30 and 90 kg/m^3 in its packaging form and 180 and 300 kg/m^3 when compacted. Similarly, wood, spanning from balsa wood to pine, exhibits densities between 170 and 500 kg/m^3. To compete effectively, natural-fiber composites must offer additional value while maintaining extremely low densities. One strategy to achieve this is through plastic foaming, enhancing material properties. Additionally, addressing water absorption concerns can contribute to establishing a competitive edge.

Tang and colleagues undertook an extensive examination focusing on the application of biopolymers and nanocomposites derived from biopolymers in the realm of food packaging [53]. In their work, they classified biopolymers for packaging into six distinct categories, namely starch-based polysaccharides, cellulose-based polysaccharides, other polysaccharide-based materials, protein-based polymers, polyhydroxyalkanoate (PHA)-based polymers, and PLA-based polymers. Additionally, the review explores various biopolymer blends, encompassing six additional types. In the realm of bionanocomposites, the study identifies seven distinct material types: starch-based, cellulose-based, pectin-based, gluten-based, gelatin-based, soy protein-based, and polymers derived from PLA and polyhydroxybutyrate (PHB). The exploration of these diverse material categories underscores the versatility and potential of biopolymers and their nanocomposites for the development of innovative and sustainable food packaging solutions. This categorization provides a nuanced understanding of the rich landscape of materials available for creating packaging that meets the complex requirements of the food industry while minimizing environmental impact [54].

Youssef [55] advocates for the utilization of nanocomposites endowed with biofunctional properties in the realm of food packaging. Among the pivotal attributes of biodegradable packaging materials, their capacity to minimize environmental impact stands out prominently. In an evaluation by Davis and Song, the environmental implications of waste management were scrutinized concerning biodegradable packaging derived from crop-based raw materials, juxtaposed with petroleum-based polymer packaging materials. The study's findings underscored the suitability of biodegradable packaging for single-use disposable applications, particularly in scenarios where postconsumer packaging could undergo local composting for effective material recycling. Nonetheless, for the widespread adoption of these biodegradable materials to attain environmental success, the establishment of efficient collection, transportation, and treatment technologies is deemed imperative [56]. In a complementary exploration, Guan et al. [57] introduced acetylated starch/PLA as a material for loose-fill packaging. The researchers delved into diverse compositions and preparation methodologies aimed at achieving the desired packaging properties. This innovative approach not only enhances the potential for eco-friendly loose-fill packaging but also highlights the significance of tailoring material characteristics to meet specific application requirements [54].

Exploring the integration of natural fibers as reinforcing agents in biodegradable foamed packaging materials represents a distinctive and intriguing research avenue. An article authored by Carr et al. [58] sheds light on this intriguing possibility, focusing on the impact of incorporating short cassava fibers or wheat fibers (at concentrations of 1%, 2%, and 3%) on the mechanical properties of cassava starch foams. The study's findings reveal a nuanced relationship between fiber content and key mechanical properties. Specifically, the cassava starch foam enriched with 1% fibers exhibits the highest compression strength. Interestingly, as additional fibers are introduced, a subsequent decline in compression strength is observed. This suggests that an optimal balance exists, and exceeding a certain fiber concentration may compromise the foam's compressive properties. Similarly, the study investigates the effect on flexibility and identifies a parallel trend. The foam's flexibility initially benefits from the incorporation of fibers, reaching an optimum at 2%. Beyond this point, however, the introduction of more than 2% fibers results in a diminishing flexibility of the foam. This intricate interplay between fiber content and mechanical attributes underscores the need for a nuanced approach in designing biodegradable foamed packaging materials with tailored and optimized properties for specific applications [51].

7.8 RENEWABLE ENERGY COMPONENTS

7.8.1 Bio-composites utilized in the construction of wind turbine blades

Wind turbine blades are made of composite materials consisting of fibers and polymers that are reinforced together. By varying the types of fibers and polymers used, different combinations with distinct properties can be achieved. Currently, carbon and glass fibers are commonly used along with thermoset polymers due to their strength advantages. However, recycling these materials poses challenges. To address these issues, researchers have been investigating the potential applications of advanced materials such as natural composites. These materials have properties such as high specific strength, low cost, lightweight, recyclability, and ease of manufacturing, making them a preferable option compared to synthetic composites.

Additionally, composites with thermoplastic polymers are being introduced to create eco-friendly and biodegradable materials. However, processing these materials is more complex compared to thermoset polymer composites. Furthermore, hybrid composites, which incorporate multiple fiber materials in one composite, offer additional advanced properties compared to single-material fiber composites. The bonding parameters between these different fibers and resins play a noteworthy role in defining the overall combination's performance. Additionally, adding nanomaterials to composites increases their overall performance and characteristics [59].

In contemporary contexts, the paradigm of global energy evolution has witnessed a shift toward prioritizing renewable sources, marking a superior choice for both exploitation and consumption. Notably, among the spectrum of renewable energy sources, wind energy emerges as a standout contender. This prominence can be attributed to its distinctive characteristics and contributions to the ongoing transformation of the global energy landscape. Wind energy, in particular, has garnered significant attention and acclaim due to its capacity to harness the kinetic energy found in air mass movement. This form of renewable energy stands out for its sustainability, environmental friendliness, and efficiency in power generation. By harnessing the power of the wind through advanced technologies such as wind turbines, societies can tap into a clean and replenishable energy source, mitigating the environmental impact associated with traditional energy extraction methods. Furthermore, the significance of wind energy lies in its potential to reduce dependence on finite and environmentally detrimental resources. As nations strive to meet their energy demands while addressing climate change concerns, the adoption of wind energy plays a pivotal role in achieving a more sustainable and resilient energy infrastructure.

Wind turbine blades, along with other components, have the highest development and installation costs. They are typically made from fiber-reinforced composites, which are susceptible to damage like any other parts made from these materials. Engineers are searching for innovative designs that utilize new materials with commendable properties. Currently used materials for blade development are being challenged by studies exploring the possibility of replacing them with materials that are cheaper, easier to produce, lighter, more reliable, biodegradable, and eco-friendly. Blade designers must prioritize materials that are cost-effective, lightweight, easy to process, recyclable, high performing, and long-lasting. Furthermore, as wind turbine technology advances and expands into offshore environments, the environmental compatibility of the materials becomes crucial. For example, in humid or offshore environments, water absorption and corrosion characteristics are important factors to consider [59].

Swolfs (2018) characterizes fiber-hybrid composites as materials that incorporate multiple fiber types within a matrix, offering increased flexibility in design [60]. Despite considerable advancements in recent years, there is still a need for a thorough understanding of these hybrid fiber composites. Research has predominantly centered on pseudo-ductility in such composites, particularly to overcome toughness challenges in applications like wind turbine blades. While many composites display impressive strength and stiffness, their lack of ductility remains a significant obstacle. The utilization of fiber hybridization serves as a remedy to these issues, providing a means to introduce ductility, a critical quality in engineering applications. Ductility plays a crucial role in predicting failure patterns in components, systems, or structures, enabling early detection and prevention of damage. Consequently, the incorporation of diverse fibers in composites not only addresses existing limitations but also contributes to the creation of materials with distinctive properties essential for robust engineering applications. In a study conducted by Akil et al. [61], a combination of glass fiber and high modulus jute was utilized, resulting in a composite with significantly enhanced mechanical properties, both in terms of tensile and flexural strength. The notable improvement in these properties can be attributed primarily to the inclusion of glass fiber in the composite. Additionally, the presence of glass fiber played a crucial role in reducing the water absorption properties of the composite. This innovative hybridization approach not only bolstered mechanical performance but also addressed concerns related to water absorption, showcasing the potential of synergistic combinations in composite material engineering. Additionally, the fusion of two natural fibers to create a composite material holds promise for enhancing material properties while concurrently lowering costs. This hybridization approach sparks particular interest in the realm of biodegradable composites, where both the fiber (acting as reinforcement) and the polymers comprising the matrix are biodegradable. This innovative synergy not only improves material characteristics but also aligns with the growing emphasis on sustainable and environmentally friendly materials [62].

Certainly, the evident advantages of synthetic-natural hybrid composites underscore their superior performance, making them promising candidates for the advancement of wind turbine blade materials. However, the cost-effectiveness inherent in natural-natural fiber hybrid composites, hybrid nanocomposites, and their distinct characteristics should serve as a catalyst for extensive research aimed at enhancing the performance of these natural hybrid composites. This pursuit not only introduces an environmentally friendly material but also aims to create a substance capable of rivaling purely synthetic composites or synthetic-natural hybrid composites in mechanical properties. The endeavor holds the potential to yield a unique and competitively robust material for various applications [62].

The durability of wind turbines heavily depends on how they are manufactured, as this affects the quality of composite materials and their lifespan. An important concern arises from inconsistencies in resin content and fiber shape during production. If the resin's viscosity is low before solidifying, it can move unevenly due to gravity or slight differences in pressure and tension, causing variations in resin amounts across the composite. This imbalance weakens some areas while overloading others, leading to deformities, structural weaknesses, and cracks. Therefore, choosing the right manufacturing method is crucial to reduce these disparities, ensuring a more reliable and robust composite that minimizes the risk of damage and failure during the turbine's lifetime [63].

The hand layup or wet layup method is currently the preferred and efficient technique for manufacturing composite wind turbine blades. This manual process involves skilled technicians carefully layering composite materials onto a mold, allowing for better control and precision in positioning reinforcing fibers. The method offers flexibility and adaptability, making it suitable for producing blades of different shapes and sizes. Additionally, hand layup is a cost-effective solution compared to automated techniques, as it requires fewer resources and equipment. Given the growing importance of wind energy, the hand layup method provides a reliable and customizable solution for the production of wind turbine blades [63].

7.8.2 Energy storage applications

Spent coffee grounds (SCGs) represent a significant biomass waste stream, often overlooked and discarded, contributing to environmental concerns. In a study, Hu et al. [64] proposed an innovative approach for the high-value utilization of SCGs in the development of composite phase change materials (CPCMs). By incorporating 1–5 wt% of reduced graphene oxide (RGO), SCGs were transformed into a stable supporting material, addressing issues such as leakage and low light-to-heat conversion efficiency. The resulting RGO-modified SCGs, termed SCGs@RGO, exhibit a superior three-dimensional porous structure compared to pristine SCGs. This unique morphology enables the absorption of 60.3 wt% polyethylene glycol (PEG), ensuring excellent leakage-proof performance. Additionally, the introduction of RGO imparts slight enhancements in thermal conductivity, further improving the overall performance of the CPCMs. SCGs@RGO demonstrates strong absorption across the full UV–Visible–Near Infrared (NIR) spectrum, indicating exceptional light-to-heat conversion capabilities. The material efficiently captures and stores solar thermal energy within a remarkably short time frame (245 s) of exposure to light. This property holds significant promise for applications requiring rapid and efficient solar energy utilization. Furthermore, PEG/SCGs@RGO CPCMs exhibit excellent reusability, chemical stability, crystalline stability, and thermal stability in phase change tests over 100 cycles. The robust performance observed in these tests underscores the potential of the developed composite material for practical and sustainable energy storage applications. In summary, this work not only addresses the environmental threat posed by SCGs but also demonstrates their transformation into a valuable resource through the creation of advanced CPCMs. The incorporation of RGO enhances structural stability, light absorption, and overall performance, paving the way for sustainable and efficient solar thermal energy storage systems. This research contributes to the growing field of biomass valorization for environmentally friendly and economically viable energy solutions.

In the rapidly evolving field of materials science, the quest for sustainable, cost-effective, and high-performance materials has never been more critical. Polymer composites have emerged as one of the frontrunners in this arena, with a unique blend of versatility and utility. However, the environmental implications and economic constraints associated with synthetic reinforcements have prompted researchers to explore natural alternatives. Enter the world of PVAc/AR bio-composites. By harnessing the inherent properties of poly (vinyl acetate) (PVAc) and the natural attributes of Asparagus Racemosus (AR), researchers have pioneered a green, sustainable material with promising applications [65]. The preparation of these bio-composites via in situ emulsion polymerization is not just a methodological choice but a testament to sustainable manufacturing practices. This eco-friendly approach minimizes energy consumption and waste generation, aligning perfectly with the global push toward greener technologies.

7.8.2.1 Characterization: beyond the surface

The meticulous characterization of the PVAc/AR composite films unveils a harmonious blend of the polymer matrix with the natural filler.

1. **FT-IR Analysis:** The distinct AR band observed at 3,300 cm^{-1} in the FT-IR spectra stands as a hallmark of successful composite formation, reaffirming the integration of AR within the PVAc matrix.
2. **Structural Insights:** X-ray diffraction studies offer intriguing insights into the structural dynamics. The observed shift in the amorphous peak of PVAc/AR composites relative to pure PVAc signifies the alteration in the material's crystalline structure, potentially enhancing its mechanical properties.
3. **Morphological Homogeneity:** Advanced atomic force microscopy techniques (AFM) and FE-SEM paint a vivid picture of the composite's microstructure. The uniform dispersion of AR within the polymer matrix is not merely aesthetically pleasing but pivotal for ensuring consistent material properties.
4. **Thermal and Mechanical Robustness:** DSC and TGA analyses provide compelling evidence of enhanced thermal stability and an elevated glass transition temperature in the presence of AR. Such enhancements can be attributed to the reinforcing effect of AR, fortifying the polymer matrix against thermal degradation.

5. **Mechanical and Electrical Attributes:** The observed improvements in tensile strength, hardness, dielectric constant, and AC conductivity with boehmite inclusion signify the potential of PVAc/AR composites in demanding applications, ranging from energy storage to flexible electronics.

The development of environmentally friendly PVAc/AR composites not only underscores the feasibility of integrating natural fillers in advanced materials but also paves the way for sustainable innovations. As the global emphasis on green technologies intensifies, these bio-composites stand poised to revolutionize sectors reliant on polymers, offering a harmonious blend of performance and sustainability. In conclusion, the synthesis and characterization of PVAc/AR bio-composites herald a new era in material science, where nature-inspired solutions hold the key to unlocking unparalleled advancements [65].

7.9 CASE STUDY

7.9.1 Airbus: aim of developing a fully recyclable aircraft cabin interior

https://renewable-carbon.eu/news/airbus-to-research-use-of-natural-fibres-for-aircraft-components/

The South African Council for Scientific and Industrial Research (CSIR) has partnered with Airbus to research the use of natural fiber-based polymer composites for environmentally friendly aircraft manufacturing. This collaborative research aims to examine how materials like flax, hemp, and kenaf can be incorporated into the manufacturing process of different interior aircraft components. The key objective of this research project is to find sustainable solutions to improve fuel efficiency, reduce aircraft weight, and minimize the environmental impact of aircraft. Additionally, the project aims to develop a cabin interior that is fully recyclable. The Airbus Research and Technology team is particularly interested in finding a natural fiber material that meets specific requirements for cabin interiors. These requirements include the strength, fire-retardant properties, flexibility, lightweight, and biodegradability of the material. Two natural fibers, kenaf and sisal, which are widely grown in South Africa, have caught the attention of the Airbus team due to their low cost, biodegradability, low density, nonabrasiveness, and recyclability. The team is also exploring the potential use of nanotechnology and bioresins to enhance the properties of these natural fibers. The CSIR team has expressed their belief that natural fibers have already been successfully used in the automobile sector, and they are excited to extend this research to aircraft interiors for the first time. This project constitutes CSIR's initial exploration of natural fiber applications in the aircraft industry. In summary, this research project aims to find environmentally friendly and sustainable alternatives for aircraft cabin interiors. The focus is on identifying a suitable natural fiber material that meets the required properties and can be enhanced using nanotechnology and bioresins [66].

7.10 CONCLUSION

This chapter has offered a thorough examination of the diverse facets surrounding renewable bio-composites, shedding light on their properties, applications, and future potential. By delving into the eco-friendly nature of these materials and their associated benefits, we have gained a profound understanding of their significance in the quest for sustainable alternatives to traditional materials across various industries. Throughout the chapters, we have navigated the historical evolution of renewable bio-composites, recognizing their lightweight nature, impressive strength-to-weight ratio, biodegradability, and remarkable thermal and acoustic insulation properties. The customization capabilities of these bio-composites have also been highlighted, showcasing their adaptability to meet specific industry requirements. This book has gone beyond the surface, exploring the raw materials that form the foundation of renewable bio-composites and delving into the manufacturing techniques employed. Furthermore, we have touched upon the emerging advancements in the field, providing a glimpse into the exciting developments that hold promise for the future. With a focus on real-world applications, this resource has examined the pivotal role of renewable bio-composites in key sectors such as automotive, construction, packaging, and renewable energy. By doing so, it has demonstrated the tangible impact these materials can have on driving sustainability and ushering in a new era of environmentally conscious practices. As we conclude this journey into the realm of renewable bio-composites, it is evident that this resource serves as a valuable tool for researchers, professionals, and policymakers alike. It is our hope that the insights garnered from these pages will inspire further exploration, innovation, and collaboration in harnessing the complete capability of renewable bio-composites to create a future that is both sustainable and resilient.

REFERENCES

1. Kim, H.-S., Kim, H.-J., Lee, J.-W. & Choi, I.-G. Biodegradability of bio-flour filled biodegradable poly (butylene succinate) bio-composites in natural and compost soil. *Polym. Degrad. Stab.* 91, 1117–1127 (2006).
2. Guna, V., Ilangovan, M., Ananthaprasad, M. G. & Reddy, N. Hybrid bio-composites. *Polym. Compos.* 39, E30–E54 (2018).

3. Shaker, K., Nawab, Y. & Jabbar, M. Bio-composites: eco-friendly substitute of glass fiber composites. In Kharissova. O. V., Torres-Martínez, L. M., & Kharisov, B. I. (Eds.), *Handbook of Nanomaterials and Nanocomposites for Energy and Environmental Applications,* 1–25 (2020).
4. Khan, A. & Lemmen, C. Bricks and urbanism in the Indus Valley rise and decline. *arXiv Prepr. arXiv1303.1426* (2013).
5. Goodhew, S. & Griffiths, R. Sustainable earth walls to meet the building regulations. *Energy Build.* 37, 451–459 (2005).
6. Mohanty, A. K., Misra, M. & Drzal, L. T. *Natural Fibers, Biopolymers, and Biocomposites.* (CRC press, 2005).
7. Ramesh, M., Palanikumar, K. & Reddy, K. H. Mechanical property evaluation of sisal--jute--glass fiber reinforced polyester composites. *Compos. Part B Eng.* 48, 1–9 (2013).
8. Ramesh, M., Palanikumar, K. & Reddy, K. H. Plant fibre based bio-composites: sustainable and renewable green materials. *Renew. Sustain. Energy Rev.* 79, 558–584 (2017).
9. Das, O., Kim, N. K. & Bhattacharyya, D. The mechanics of biocomposites. In Ambrosio, L. (Ed.), *Biomedical Composites,* 375–411 (Elsevier, 2017).
10. Khoshraftar, Z. & Ghaemi, A. Adsorption of organic pollutants from pesticides using polymeric adsorbents. In Ghaemi, A., Norouzbeigi, R., Masoumi, H. (Eds.), *Polymeric Adsorbents,* 461–512 (Elsevier, 2024).
11. Reddy, T. R. K., Kim, H.-J. & Park, J.-W. Renewable biocomposite properties and their applications. *Compos. Renew. Sustain. Mater.* 177–197, (2016).
12. Yussuf, A. A., Massoumi, I. & Hassan, A. Comparison of polylactic acid/kenaf and polylactic acid/rise husk composites: the influence of the natural fibers on the mechanical, thermal and biodegradability properties. *J. Polym. Environ.* 18, 422–429 (2010).
13. Mhatre, A. M., Raja, A. S. M., Saxena, S. & Patil, P. G. Environmentally benign and sustainable green composites: current developments and challenges. In Muthu, S. S. (Ed.), *Green Composites: Sustainable Raw Materials,* 53–90 (Springer, 2019).
14. Yang, H.-S., Kim, H.-J., Park, H.-J., Lee, B.-J. & Hwang, T.-S. Effect of compatibilizing agents on rice-husk flour reinforced polypropylene composites. *Compos. Struct.* 77, 45–55 (2007).
15. Lee, B.-H., Kim, H.-J. & Yang, H.-S. Polymerization of aniline on bacterial cellulose and characterization of bacterial cellulose/polyaniline nanocomposite films. *Curr. Appl. Phys.* 12, 75–80 (2012).
16. Shekar, H. S. S. & Ramachandra, M. Green composites: a review. *Mater. Today Proc.* 5, 2518–2526 (2018).
17. Mwaikambo, L. Review of the history, properties and application of plant fibres. *African J. Sci. Technol.* 7, 121 (2006).
18. Khoshraftar, Z. & Ghaemi, A. Evaluation of pistachio shells as solid wastes to produce activated carbon for CO_2 capture: isotherm, response surface methodology (RSM) and artificial neural network (ANN) modeling. *Curr. Res. Green Sustain. Chem.* 5, 100342.
19. Khoshraftar, Z. & Ghaemi, A. Enhanced carbon dioxide adsorption using lignin-derived and nitrogen-doped porous carbons: a machine learning approaches, RSM and isotherm modeling. *Case Stud. Chem. Environ. Eng.* 100668 (2024).

20. Ticoalu, A., Aravinthan, T. & Cardona, F. A review of current development in natural fiber composites for structural and infrastructure applications. *Proceedings of the Southern Region Engineering Conference (SREC 2010)*, 11–12 November 2010, Toowoomba, Australia (2010).
21. Subramanian, M. N. *Polymer Blends and Composites: Chemistry and Technology*. (John Wiley & Sons, 2017).
22. Jiang, S. et al. Electrospun nanofiber reinforced composites: a review. *Polym. Chem.* 9, 2685–2720 (2018).
23. Islam, M. S., Ang, B. C., Andriyana, A. & Afifi, A. M. A review on fabrication of nanofibers via electrospinning and their applications. *SN Appl. Sci.* 1, 1–16 (2019).
24. Fang, J., Wang, X. & Lin, T. Functional applications of electrospun nanofibers. *Nanofibers Prod. Prop. Funct. Appl.* 14, 287–302 (2011).
25. Preethikaharshini, J. et al. Review of advanced techniques for manufacturing biocomposites: non-destructive evaluation and artificial intelligence-assisted modeling. *J. Mater. Sci.* 57, 16091–16146 (2022).
26. Balla, V. K., Kate, K. H., Satyavolu, J., Singh, P. & Tadimeti, J. G. D. Additive manufacturing of natural fiber reinforced polymer composites: processing and prospects. *Compos. Part B Eng.* 174, 106956 (2019).
27. Bajpai, P. K., Ahmad, F. & Chaudhary, V. Processing and characterization of bio-composites. In Martínez, L. M. T. (Ed.), *Handbook of Ecomaterials*, 1–18 (Springer, 2017).
28. Phiri, R., Rangappa, S. M., Siengchin, S., Oladijo, O. P. & Dhakal, H. N. Development of sustainable biopolymer-based composites for lightweight applications from agricultural waste biomass: a review. *Adv. Ind. Eng. Polym. Res.* 6, 436–450 (2023).
29. Gholampour, A. & Ozbakkaloglu, T. A review of natural fiber composites: properties, modification and processing techniques, characterization, applications. *J. Mater. Sci.* 55, 829–892 (2020).
30. Pickering, K. L., Efendy, M. G. A. & Le, T. M. A review of recent developments in natural fibre composites and their mechanical performance. *Compos. Part A Appl. Sci. Manuf.* 83, 98–112 (2016).
31. Chaitanya, S. & Singh, I. Processing of PLA/sisal fiber biocomposites using direct-and extrusion-injection molding. *Mater. Manuf. Process.* 32, 468–474 (2017).
32. Jaszkiewicz, A., Bledzki, A. K. & Franciszczak, P. Improving the mechanical performance of PLA composites with natural, man-made cellulose and glass fibers—a comparison to PP counterparts. *Polimery* 58, 435–442 (2013).
33. Komal, U. K., Lila, M. K. & Singh, I. PLA/banana fiber based sustainable biocomposites: a manufacturing perspective. *Compos. Part B Eng.* 180, 107535 (2020).
34. Rajeshkumar, G. et al. Environment friendly, renewable and sustainable poly lactic acid (PLA) based natural fiber reinforced composites--a comprehensive review. *J. Clean. Prod.* 310, 127483 (2021).
35. Akampumuza, O., Wambua, P. M., Ahmed, A., Li, W. & Qin, X.-H. Review of the applications of biocomposites in the automotive industry. *Polym. Compos.* 38, 2553–2569 (2017).

36. Mann, D., Van Den Bos, J. C. & Way, A. Markets. In *Automotive Plastics and Composites* (Second Edition), 175–210 (Elsevier, 1999). https://www.sciencedirect.com/science/article/abs/pii/B9781856173490500118
37. Schuh, T. G. Natural Fibre Performance Forum. Copenhagen, Denmark (1999).
38. Park, H. S., Dang, X. P., Roderburg, A. & Nau, B. Development of plastic front side panels for green cars. *CIRP J. Manuf. Sci. Technol.* 6, 44–52 (2013).
39. John, M. J. & Thomas, S. Biofibres and biocomposites. *Carbohydr. Polym.* 71, 343–364 (2008).
40. Andresen, C., Demuth, C., Lange, A., Stoick, P. & Pruszko, R. Biobased automobile parts investigation. *A Rep. Dev. USDA Off. Energy Policy New Uses* (2012).
41. Holbery, J. & Houston, D. Natural-fiber-reinforced polymer composites in automotive applications. *JOM* 58, 80–86 (2006).
42. Staiger, M. P., and Nick Tucker. Natural-fibre composites in structural applications. In *Properties and Performance of Natural-Fibre* Composites, 269–300 (Woodhead Publishing, 2008). https://doi.org/10.1533/9781845694593.2.269.
43. Musikwa, M. The development of natural fibre reinforced composites roof sheet. (2018).
44. Ahmad, H., Chhipi-Shrestha, G., Hewage, K., & Sadiq, R. A comprehensive review on construction applications and life cycle sustainability of natural fiber biocomposites. *Sustainability* 14(23), 15095 (2022). https://doi.org/10.3390/su142315905
45. Dweib, M. A., Hu, B., Shenton Iii, H. W. & Wool, R. P. Bio-based composite roof structure: manufacturing and processing issues. *Compos. Struct.* 74, 379–388 (2006).
46. Sheng, D. D. C. V., Ramegowda, N. S., Guna, V. & Reddy, N. Groundnut shell and coir reinforced hybrid bio composites as alternative to gypsum ceiling tiles. *J. Build. Eng.* 57, 104892 (2022).
47. Hayes, M. D. et al. Laboratory and field testing of composite bridge superstructure. *J. Compos. Constr.* 4, 120–128 (2000).
48. Goldstein, H. Catching up on composites. *Civ. Eng.* 66, 47 (1996).
49. Yazdi, M. F. A. et al. Bio-composite materials potential in enhancing sustainable construction. *Desalin. Water Treat.* 52, 3631–3636 (2014).
50. Jayaraman, K. & Bhattacharyya, D. Mechanical performance of woodfibre--waste plastic composite materials. *Resour. Conserv. Recycl.* 41, 307–319 (2004).
51. Duhovic, M., Peterson, S. & Jayaraman, K. Natural-fibre--biodegradable polymer composites for packaging. In Pickering, K. L. (Ed.), *Properties and Performance of Natural-Fibre Composites*, 301–329 (Elsevier, 2008).
52. Averous, L. Biodegradable multiphase systems based on plasticized starch: a review. *J. Macromol. Sci. Part C Polym. Rev.* 44, 231–274 (2004).
53. Tang, X. Z., Kumar, P., Alavi, S. & Sandeep, K. P. Recent advances in biopolymers and biopolymer-based nanocomposites for food packaging materials. *Crit. Rev. Food Sci. Nutr.* 52, 426–442 (2012).

54. Georgios, K., Silva, A. & Furtado, S. Applications of green composite materials. *Biodegrad. Green Compos.* 16, 312–337 (2016).
55. Youssef, A. M. Polymer nanocomposites as a new trend for packaging applications. *Polym. Plast. Technol. Eng.* 52, 635–660 (2013).
56. Davis, G. & Song, J. H. Biodegradable packaging based on raw materials from crops and their impact on waste management. *Ind. Crops Prod.* 23, 147–161 (2006).
57. Guan, J., Eskridge, K. M. & Hanna, M. A. Acetylated starch-polylactic acid loose-fill packaging materials. *Ind. Crops Prod.* 22, 109–123 (2005).
58. Carr, L. G., Parra, D. F., Ponce, P., Lugão, A. B. & Buchler, P. M. Influence of fibers on the mechanical properties of cassava starch foams. *J. Polym. Environ.* 14, 179–183 (2006).
59. Thomas, L. & Ramachandra, M. Advanced materials for wind turbine blade-a review. *Mater. Today Proc.* 5, 2635–2640 (2018).
60. Swolfs, Y., Verpoest, I. & Gorbatikh, L. Recent advances in fibre-hybrid composites: materials selection, opportunities and applications. *Int. Mater. Rev.* 64, 181–215 (2019).
61. Akil, H. M., Santulli, C., Sarasini, F., Tirillò, J. & Valente, T. Environmental effects on the mechanical behaviour of pultruded jute/glass fibre-reinforced polyester hybrid composites. *Compos. Sci. Technol.* 94, 62–70 (2014).
62. Samuel, B., Sumaila, M. & Dan-Asabe, B. Potential natural fiber polymer composites for wind turbine blade development: a short review. *J. Mek.* 45, 1–14 (2022).
63. Desai, N., Bhatt, P. & Solanki, M. Application of fully green bio-composites in manufacturing of wind turbine blades: a strategic review. *J. Environ. Friendly Mater.* 5, 77–87 (2021).
64. Hu, X., Huang, H., Hu, Y., Lu, X. & Qin, Y. Novel bio-based composite phase change materials with reduced graphene oxide-functionalized spent coffee grounds for efficient solar-to-thermal energy storage. *Sol. Energy Mater. Sol. Cells* 219, 110790 (2021).
65. Ramesan, M. T., Sameela, T. P., Meera, K., Bahuleyan, B. K. & Verma, M. In situ emulsion polymerization of poly (vinyl acetate) and asparagus racemosus biopolymer composites for flexible energy storage applications. *Polym. Compos.* 44, 4168–4177 (2023).
66. Arockiam, N. J., Jawaid, M. & Saba, N. Sustainable bio composites for aircraft components. In Jawaid, M., Thariq, M. (Eds.), *Sustainable Composites for Aerospace Applications,* 109–123 (Elsevier, 2018).

Chapter 8

Sustainable and environment-friendly manufacturing

Yuvaraj Subramaniyan, Elango Krishnan Soundararajan, Vivek Deivasigamani, Saravanakumar Ramasamy, and Fathimathul Fuhaima Puthiya Purayil

8.1 INTRODUCTION

The growing focus on environmental concerns, driven by factors such as climate change, excessive waste generation, and diminishing natural resources is prompting governments worldwide to implement initiatives aimed at mitigating the adverse effects of these issues on both current and future generations. Thus, it becomes apparent that the manufacturing industry must swiftly transition toward environmentally friendly practices to mitigate its detrimental effects on the planet's resources. Simultaneously, environmental consciousness should be coupled with economic growth, which can contribute to achieving social objectives by ensuring satisfactory working conditions for all employees. In alignment with this perspective, the concept of sustainability, encompassing environmental, economic, and social dimensions, has garnered considerable attention in recent times. Numerous production and manufacturing firms are adapting their operational methods to align with the demanding standards set by global governance and to advocate for sustainable development [1]. As a result, the literature is experiencing a surge in papers addressing these topics. Given the growing emphasis on sustainable practices within the manufacturing sector, this paper seeks to perform an analytical review to categorize the insights gleaned from research papers on the subject.

8.2 GREEN MANUFACTURING

Green manufacturing revolves around altering corporate procedures, manufacturing techniques, and stakeholder perspectives to mitigate the repercussions caused by industries toward environmental hurdles such as climate change. Embracing sustainable approaches empowers companies to enhance their success, competitiveness, and profitability [2]. It transcends

 DOI: 10.1201/9781003470342-8

mere ethical considerations and extends to strategic business advantages. It involves manufacturing methods aimed at mitigating the impacts of environmental disasters. Also referred to as green production, it seeks to modernize outdated production methods and embrace sustainable manufacturing solutions that minimize waste generation. It requires individuals and businesses aspiring to adopt green practices to explore, develop, or utilize advanced technologies to identify optimal approaches for reducing their environmental footprint. Green manufacturing encompasses manufacturing techniques geared toward alleviating the effects of environmental disasters. Alternatively called as green production, it endeavors to update obsolete production methods and adopt sustainable manufacturing solutions which reduce the generation of waste. Individuals and businesses aspiring to embrace green practices are tasked with exploring, developing, or employing cutting-edge technologies to pinpoint the most effective strategies for reducing their environmental impact. Green computing focuses on lessening the ecological footprint of technology by diminishing energy usage, decreasing waste, and advocating for sustainable growth. Its objective is to alleviate the carbon emissions produced by the IT sector and its associated industries.

8.2.1 Need for green manufacturing

Manufacturing plays a vital role in various avenues of economic advancement. Transitioning away from practices like casting and forging toward sustainable development is underway. Addressing economic challenges [3] within the manufacturing sector, such as low-profit margins coupled with high capital costs and market volatility, is imperative. Small-scale producers need to improve their manufacturing processes and gain training and tools despite encountering tight profit margins. Prioritizing eco-friendly efforts and sustainability yields advantages in terms of efficiency. The time for change is undoubtedly upon us. Green manufacturing confronts environmental challenges at a systemic level, representing a burgeoning industrial trend with promising long-term sustainability. Various tools available on the market can aid in monitoring inventory and production, facilitating the swift adoption of green manufacturing practices by companies. Technologies enable real-time tracking of inventory movements and identification of potential dead stock, allowing for prompt removal. Additionally, these tools can analyze material usage patterns, establish reorder points to prevent stock-outs, and develop strategies to optimize resource utilization. There are a range of features and strategies associated with green manufacturing practices. These approaches encompass various intelligent and energy-efficient measures designed to facilitate the transition of manufacturing operations toward environmentally sustainable practices that matches with the sustainability targets of the industry. Key aspects are minimizing energy consumption, decreasing water usage, lowering waste generation throughout processes, and reducing emissions. Additionally, actions such as recovery, redesign, recycling, re-manufacturing, and reprocessing provide further advantages to these approaches, enhancing the conceptualization of sustainable manufacturing operations. Change in climate is indicative of a broader issue of modern over-consumption, necessitating the extraction of natural resources and the production and provision of commodities for our contemporary civilization. Contributions to greenhouse gas emissions arise from different sectors such as transportation and construction to industrial production and agriculture [4]. At the heart of these industries, which significantly impact pollution and climate, are individuals who demand automobiles, aircraft, energy, large-scale enterprises, residential structures, telecommunications, technology, entertainment, and more. It is paramount to thoroughly evaluate every input and adopt sustainable business practices to achieve sustainable production. Authentic sustainable production methodologies transcend mere environmental assessments, encompassing the welfare and livelihoods of individuals dedicated to manufacturing consumer goods. Although technology has improved efficiency and manufacturing methods, numerous products are made from untreated materials that are challenging to repair, leading to accumulations of plastic and electronic waste that pollute the environment. 3D printing, which builds 3D parts layer by layer using composite materials, is an emerging technology known for its cost-effectiveness and efficiency in producing intricate and lightweight items [5].

8.2.2 Applications and future scope of green manufacturing

Green manufacturing encompasses the overhaul of manufacturing procedures and adoption of eco-conscious methods within the industrial realm [6]. Fundamentally, the shift toward greener manufacturing involves workers reducing their consumption of natural resources, cutting down on pollution and waste, re-purposing materials through recycling and reuse, and curbing emissions in their processes. Companies committed to green manufacturing are proactively exploring, innovating, or embracing technologies and methodologies geared toward lessening their ecological footprint. Here are some examples showcasing the diverse applications of green manufacturing for fostering sustainability.

8.2.2.1 Consumption of materials

The production of goods has direct and indirect impacts on the environment. Green manufacturing plays a crucial role in reducing material consumption, including emissions generated during material transport and product disposal. As urbanization continues to grow and demand for new goods increases, production significantly influences the well-being of communities. In addition to the above, green manufacturing offers benefits and profitability, thereby disrupting traditional production methods with sustainable approaches. Embracing sustainability requires a shift in mindset, focusing on addressing the root causes of waste throughout the production process rather than solely addressing end-of-life cycle issues [7]. It is imperative to establish a comprehensive strategy for managing materials from the outset, rather than resorting to disposal at the project's conclusion, to maximize efficiency and minimize environmental impact.

8.2.2.2 Products reuse

Establishment of circular return loops for repair of products along with the accumulation, refurbishment and dispersion of components as well as the reclamation, reconditioning, and re-purposing of materials. Achieving this relies on manufacturing processes operating in a circular manner, where materials are withdrawn from the life cycle and reintroduced strategically. Automation can contribute to this efficiency, but it demands a reassessment of the overall system and an adjustment of the business model to ensure economic viability. Utilizing the full potential of the supply chain and the consumer base of production facilities constitutes a pragmatic approach to advancing sustainable practices. The Industrial Internet of Things (IIoT) introduces unprecedented opportunities to engineer innovative processes for the development of sustainable materials, decarbonization of energy consumption, exploitation of digital advancements for augmented productivity, and extension of product life cycles, all within the "zero waste to waste" paradigm of Industry 4.0. Significant possibility exists to substitute materials with lower carbon intensity, while green manufacturing techniques can enhance material processing systems. Furthermore, electronic businesses now have the capacity to adopt greener packaging and utilize environmentally friendly new bio-based materials [8].

8.2.2.3 Reduction of cost

In the pursuit of economic considerations, the circular economy internalizes costs to prioritize enhanced service and performance while mitigating risks. Unlike the linear approach, where responsibility ends at the point of sale, the circular economy extends a company's responsibility beyond this point, considering the entire life cycle of products. While the linear model generates waste streams that burden communities and society, the circular model transforms these into value streams. Linear systems prioritize uniformity for increased productivity and ease of use. Despite the significant benefits humanity has derived from the Industrial Revolution, such as increased productivity, there have been long-term repercussions, including pollution and resource depletion. Today, both companies and consumers recognize the need to evaluate activities for their sustainability. This realization has sparked a revolution in the industry sector, driven by a collective awareness of past excesses. Green manufacturing encompasses an array of cutting-edge technologies, such as lean waste reduction systems, renewable energy sources, and comprehensive product life cycle management frameworks [9].

8.2.2.4 Elimination of waste

Implementing a process aimed at minimizing waste helps decrease occurrences of errors during its production. Sustainable manufacturing encourages a cooperative effort among all engineering departments, promoting greater transparency and competence. This aims to cultivate a future personnel who will be capable of utilizing the latest tools more efficiently, collaboratively, and safely. Achieving sustainable production necessitates breaking down silos within organizations. Manufacturing engineers often have to troubleshoot issues on the production floor without disrupting operations. The advent of industrialization and manufacturing has disrupted the delicate balance of various ecosystems. Although automation has provided the economic groundwork for societal advancement and growth, it has faced difficulties in fostering a symbiotic connection with nature. Lean production embodies an innovative facet of industrial and socioeconomic frameworks. The notion of the 6R's underscores essential resource management principles as complementary methods to attain sustainable manufacturing [10].

8.2.2.5 Sustainable development

Increasing demands are driving manufacturers to focus on sustainable development, influencing practices in the manufacturing sector. Sustainable manufacturing involves adopting methods that mitigate adverse environmental impacts, prioritize energy conservation, minimize waste, and enhance community resilience. Its significance extends not only to global welfare but also to the well-being of producers and consumers alike [11]. Advancements in technology are poised to revolutionize manufacturers' production processes and service delivery while providing deeper insights into product life cycles. GPS satellite technology, for instance, enables manufacturers to optimize performance in real-time and swiftly identify and address any issues.

8.2.2.6 Saving in energy

Adopting a green manufacturing approach lies in the significant energy savings it offers, aligning both environmental and commercial interests. With the ongoing decarbonization of energy, a promising future is anticipated. Furthermore, substantial reductions in decarbonization costs can be achieved by curbing the rise of high-carbon transportation. The arduous conditions faced by employees as they convert raw materials into usable textiles or components constitute one of the lesser-known facets of industrial manufacturing [12]. An exponential growth in population has led to a corresponding increase in demand for enhanced food production. It is crucial that all parties involved, from local to global levels, encompassing both public and private sectors in agriculture and industry, embrace coordinated actions to drive transformation. Sustainability has garnered significant attention lately, fueled by growing awareness of the environmental impacts of consumer choices.

8.2.2.7 Process innovation

Innovative digital processes have the capacity to promote more effective and eco-conscious manufacturing methods, even within sectors that heavily rely on energy. Utilizing the IIoT, developers can emphasize resource efficiency endeavors alongside aims for increased productivity by incorporating intelligent interconnected devices. These linked assets provide insights into both past and current energy usage throughout a facility, empowering industrial enterprises to optimize energy and resource management from the ground level [13]. Given that electricity typically constitutes a major expense for businesses, this holds considerable importance. Moreover, sustainability endeavors can stimulate innovation. For instance, prompting engineers and machine operators to reduce waste or enhance recycling during manufacturing often results in additional operational enhancements. These efforts can significantly enhance the organization's profitability while also reducing its carbon footprint and mitigating pollutants emitted into the atmosphere. The resulting benefits, including improved air and water quality, reduced waste disposal, and increased adoption of renewable resources, ultimately contribute to a better future for generations to come.

8.2.2.8 Minimize carbon emissions

As technology advances, green manufacturing practices are predictable to become more widespread. And as factories worldwide seek strategies to minimize carbon emissions, they must prioritize increasing their energy efficiency. Industrial companies continuously face enduring challenges due to the persistent issues of natural resource scarcity, environmental degradation, and climate change. Consequently, these companies have been diligently endeavoring to devise sustainable solutions that facilitate economic advancement without exacerbating environmental degradation [14]. The escalating culture of consumerism is propelling carbon emissions associated with consumption. It is also important to recognize that carbon emissions are often attributed solely to industries, overlooking the role of consumers in influencing production through their consumption habits. This lapse signifies a lost chance for consumers to advocate for sustainably produced goods and services, encouraging companies to reevaluate their manufacturing processes and lower their carbon emissions. In today's society, characterized by over-consumption and excessive manufacturing, there is a proliferation of convenient, disposable plastic products. This phenomenon results in the release of greenhouse gases and pollutants both during manufacturing and disposal.

8.2.2.9 Enhanced production flexibility

Nanotechnology and additive manufacturing stand ready to transform production and logistics, offering heightened operational flexibility and resource conservation. Progress in additive manufacturing suggests that producing individual parts on-site will soon be viable, eradicating the necessity for ordering and awaiting delivery. Furthermore, the Internet of Things (IoT) [15] paradigm implies that a sophisticated automated monitoring system could identify worn components on devices autonomously. The prevailing tendency in industrial automation is gravitating toward intelligent manufacturing and robotic automation. The notion of a "smart producer" is anticipated to bolster overall efficacy, furnishing augmented worker safety and entrepreneur profitability by amalgamating IoT-connected devices, data analytics, robotics, and artificial intelligence [16]. One of the principal advantages of intelligent manufacturing and automated robotics lies in their capacity to gradually cultivate eco-friendly facilities and nurture a healthier ecosystem. Technologies such as blockchain are tailored to fulfill the requirements of both consumer goods and manufacturing enterprises. Contemporary value networks are intricate, spanning across international boundaries and linking numerous markets and firms.

8.2.2.10 Resources and material management

Effective management of resources and materials has become imperative due to the increasing cost and scarcity of global raw resources. There is a pressing need for long-term regenerative solutions that provide alternatives to natural systems. Inventories and sources of materials undergo changes influenced by various factors such as natural occurrences, societal shifts, and governmental policies [17]. As resources dwindle, consumer behaviors evolve, and nationalistic interests intensify, the demand for nature's contributions increases significantly, driven by dynamic and cyclical demand patterns. Transitioning to environmentally friendly production methods can help firms reduce offset electricity costs. Implementing energy-efficient light bulbs, adopting energy-saving modes and smart waste management techniques [18] are among the activities that can immediately contribute to expense reduction. While the financial impacts of these changes may be evident in the short term, it may take longer to realize additional benefits associated with transitioning to sustainable production techniques.

8.2.2.11 Environmentally friendly factories

In addition to achieving energy savings, manufacturers operating environmentally friendly facilities may prioritize sustainability as a key component of their environmental strategy. Recognizing the importance of maintaining consumer satisfaction for long-term success, companies understand the significance of addressing ecological concerns to meet the expectations of their customer base. Environmental preservation is a priority for many consumers, leading businesses to adopt products and practices aimed at protecting the environment. Companies at every level of the value chain encounter challenges related to water, electricity, and chemical consumption, highlighting the need for assistance in managing these resources. In modern society, sustainability stands as a fundamental developmental objective for humanity. The social dimension of sustainability carries significant societal implications, affecting human health and well-being, including considerations related to labor costs and environmental damage. To address social, environmental, and economic challenges within the manufacturing industry, it is essential to explore the interconnectedness between these three perspectives [19]. By pursuing balanced development that accounts for societal, environmental, and economic factors, the manufacturing sector can be guided toward sustainability.

8.2.2.12 Reduction of pollution

Sustainable production involves techniques for manufacturing products that minimize opposing impacts on the surrounding environment while conserving energy and resources. The situation also takes into account the consequences of products for employees, communities, and consumers. Achieving this goal entails advancing product and life cycle management, leading to diminished environmental and societal footprints. Lately, eco-friendly production has surfaced as a pivotal tactic for automotive firms, who are increasingly integrating it to distinguish themselves in the market. Auto manufacturers utilize diverse eco-friendly methods to cut expenses and foster an eco-aware brand identity, thereby bolstering their competitive edge. These methods predominantly revolve around conserving energy and water, minimizing waste, and reducing the emission of CO_2. Embracing sustainable manufacturing strategies enables enterprises spanning various industries, from consumer goods to manufacturing, to trim costs, capitalize on groundbreaking innovation, and enhance their competitive position.

The art of green manufacturing has an enlightening forthcoming scope. Preserving our resources, both natural and human, is essential for safeguarding that future. As population growth continues to strain ecosystems, the rise of automation and artificial intelligence further burdens employees. The increasing desire for openness across the complete supply chain, primarily motivated by concerns for sustainability and ethics, reflects a shift in consumer attitudes, with more people unwilling to endorse deceit, exploitation, or environmental damage. Businesses are increasingly adopting automation and IoT/IIoT-driven manufacturing [20], allowing products and processes to communicate and self-regulate, thus reducing reliance on human assembly workers. Concurrently, manufacturers are employing virtual reality to boost effectiveness and involve both employees and customers in eco-friendly manufacturing practices. The challenges and complexities of eco-conscious manufacturing are expected to endure, with sustainability considerations encompassing the entire product lifecycle, extending far beyond production facilities. Sophisticated systems will be necessary to monitor products from inception to disposal, placing greater emphasis on producers to enhance productivity. To achieve sustainable production, companies must reevaluate and enhance their waste reduction and recovery methods. The future production landscape, supporting rapid technological advancements, is anticipated to undergo significant transformation compared to its current state.

8.3 4D PRINTING AS A SUSTAINABLE CIVIL ENGINEERING SOLUTION

Skylar Tibbits introduced the notion of 4-dimensional printing (4DP) that was unveiled during a TED talk presentation. Tibbits described 4DP as the capability to program materials to change their shape, properties, and even process information over time. Later its inception, scientists and engineers globally have dedicated themselves to advancing and refining this expertise, exploring its significant uses in domains such as medical, architecture, and aerospace. With the emergence of new technologies, civil engineering evolves, aiming to discover more effective approaches for designing, building, and sustaining the infrastructures in our environment. Among these advancements, 4DP stands out as a potentially transformative innovation for civil engineering [14]. In this context, we will delve into the concept of 4DP and its prospective impacts on the realm of civil engineering. 4DP involves the fabrication of detailed 3-dimensional objects by depositing the material in different layers. While 3D printing has been extensively used in civil engineering for various purposes, including architectural models and structural components, 4DP takes this technology to a higher level. The key feature of 4DP involves the utilization of materials capable of altering their behaviors over time in reaction to external factors. The various advantages the 4DP has in civil engineering include its adaptive designs and its ability to create complex geometric designs. Though the process is highly inventive and promising, it has its own setbacks, which are illustrated in Figure 8.1. The scalability of the projects that can be carried out using this technology is limited. Though it can easily create small-scale designs and infrastructure, it is difficult to scale up the project to large structures that require high detailing. Concerns about numerous factors such as compatibility of the materials used, cost-effectiveness of the project, durability of the structure [21], and availability of smart materials also affect the implementation of 4DP technology on a large scale. However, this opens up for further research in the area which can create results in the near future.

The implementation of 4DP demands a substantial shift in the workers. Professionals have to necessitate reskilling and acquiring fresh proficiencies to understand and successfully employ 4DP technologies and its materials. It entails a coordinated endeavor among academic institutions, professionals, and industry experts to ensure that upcoming experts are adequately prepared with the information and capabilities to connect the possibility of 4DP. To conclude, the assimilation of 4DP into civil engineering presents promising prospects for shaping future infrastructure.

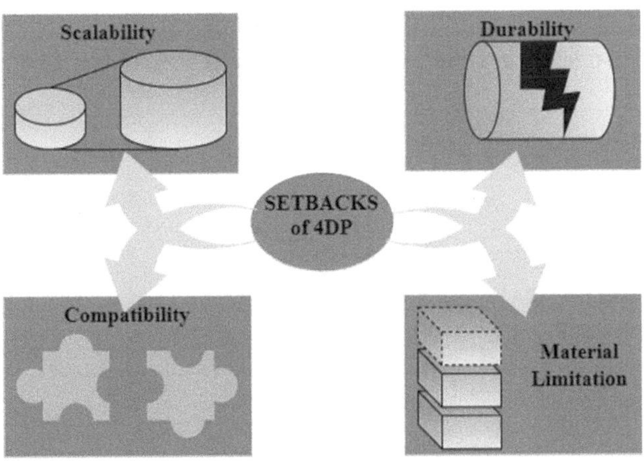

Figure 8.1 Setbacks of 4DP.

8.3.1 Scale of 4DP in civil engineering

An essential advantage of integrating 4DP technology into civil engineering is its ability to manufacture complex and refined structures across a range of dimensions. This innovation facilitates the creation of structures spanning from detailed small elements to extensive constructions and infrastructure endeavors. Figure 8.2 demonstrates these benefits in the domain of civil engineering. These constructions have uses across various domains, encompassing biomedical engineering, environmental surveillance, and regulation. On a broader spectrum, 4DP can be utilized for the construction of full-size structures. For instance, scientists at the Singapore Center for 3D Printing have formulated a 4DP concrete technique, enabling the development of buildings and structures that self-assemble. In this, the structures are designed to modify to the shifting atmospheric circumstances like earthquakes and can be implemented in various situations from disaster response to urban planning and design. Besides buildings and infrastructure projects, 4DP technology is useful in bridges, tunnels, roads, and different transportation infrastructures. In addition, 4DP technology can be used to create materials and structures for renewable energy sectors like wind turbines and solar panels. This technology has the potential to revolutionize various aspects of the built environment by enabling the development of more efficient, sustainable, and adaptable structures. 4DP technique could significantly enhance the creation of a more resilient and improved built environment for future generations. It is important to recognize that the size and complexity of the structure being printed will influence both the time and cost of the printing process. Moreover, 4DP technology has more contribution to the development of the environment for future generations [11]. It is crucial to acknowledge that the intricacy and size of the printed structure will impact both the duration and expense of the printing process. Many steps will be involved for printing structures with more complexity and also took more time for completion. Consequently, the process of 4DP of materials is slow and expensive and it may not be suitable for

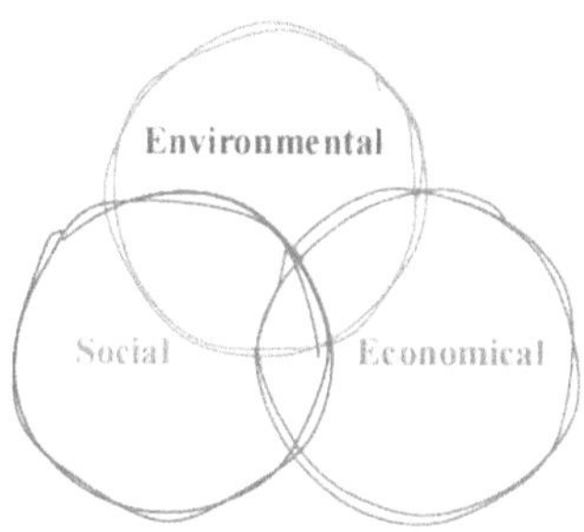

Figure 8.2 Triple Bottom Line of Sustainability.

many applications. Nevertheless, advancement in 4DP technology leads to cost-effectiveness and it can be used in many civil engineering applications. Research and development in 4DP lead to the finding of new materials and printing techniques that expand the possibilities of this technology. Also, 4DP offers substantial potential in civil engineering by enabling the creation of adaptive, sustainable, and innovative structures of various scales. From small-scale sensors and actuators to large-scale buildings and infrastructure projects, 4DP has the capacity to revolutionize the way we design, construct, and manage our built environment [5].

8.3.2 Practical applications of 4DP in expansive civil constructions

4DP technology utilization in large-scale projects is currently in its early stages, and considerable research and development efforts have been undertaken to translate this concept into reality. One notable example of such endeavors involves the searching of 4DP for fabricating components used in bridges. Moreover, various environmental stimuli are adopted in various research. Through the application of 4DP technology, this prototype bridge incorporates components that can alter their shape in response to changing load conditions. This adaptive capability enhances the structural integrity and resilience of the bridge. In another context, the potential of 4DP technology is being investigated for creating self-assembling components for tunnels. These components are engineered to autonomously assemble into the complete tunnel structure under specific environmental conditions, thereby significantly reducing manual labor requirements and improving the speed and efficiency of tunnel construction. Although these applications are currently in their initial stages and focused on research, they demonstrate the potential of 4DP technology in the construction of large civil structures. With further evolution and refinement over time, these applications may ultimately lead to the widespread adoption of 4DP in the construction of dams, bridges, tunnels, and beyond. Although the application of 4DP technology in major civil structures is currently in its early phases, considerable research and development endeavors have been undertaken to actualize this concept. One notable illustration of such endeavors is the exploration of 4DP for fabricating components utilized in bridges. For instance, a group of researchers has commenced the development of a 4D-printed bridge capable of adjusting to various environmental stimuli [3]. By employing 4DP technology, this prototype bridge was engineered with components capable of altering their shape in response to fluctuating load conditions. This adaptive feature enhances the bridge's structural integrity and resilience. In another scenario, the potential of 4DP technology is being investigated to fabricate self-assembling components for tunnels. These components are designed to autonomously assemble into the complete tunnel structure under specific environmental conditions, thereby significantly reducing the manual labor required and improving the speed and efficiency of tunnel construction. While these applications are currently in their early stages and primarily in experimental stage, they underscore the potential of 4DP technology in the construction of large civil structures. The growth of these applications would lead to the implementation of 4DP in the construction of dams, bridges, tunnels, and various other infrastructure projects.

8.3.3 Construction materials used in 4DP

The application of intelligent materials emerges as a crucial element in the success of 4DP, where these materials showcase sensitivity to external triggers like electric fields and temperature. This flexibility empowers 4D-printed items to react to their environment, presenting numerous potential uses, particularly in civil engineering. Shape memory alloys (SMAs) and shape memory polymers (SMPs) [22] are classified as intelligent materials, which can change their form in response to heat or light. Likewise, SMAs can return to their original shape when exposed to heat. These materials hold promise for constructing civil structures capable of adjusting to environmental shifts, such as bridges that expand or contract with temperature changes. Another promising avenue in exploring 4DP materials for civil engineering involves the advancement of smart concrete [23]. Unlike conventional concrete, widely used in construction for its strength and durability but lacking inherent responsiveness, researchers are investigating ways to infuse concrete with "smart" attributes. A notable instance is self-healing concrete, containing bacteria that generate limestone upon contact with water, effectively repairing any cracks that may appear over time. Additionally, an emerging technology is thermochromic concrete, which alters color in response to temperature fluctuations, potentially serving as a visual indicator of structural issues. Hydrogels represent another category of intelligent materials applicable in 4DP. These substances exhibit significant expansion or contraction in response to environmental changes, such as temperature or pH variations. This trait holds potential for civil engineering applications, allowing the creation of structures capable of adapting to environmental changes, potentially improving their resilience and lifespan.

Despite exciting progress in 4DP materials, significant challenges remain. The high expense and limited accessibility of intelligent materials may hinder their widespread use. Nevertheless, the potential benefits of 4DP in civil engineering, including improved resilience, adaptability, and durability of structures, underscore its importance as an area of exploration. The selection of materials holds considerable significance in integrating 4DP technology into civil engineering. It not only dictates the functionality and effectiveness of the resulting 4D printed structures but also impacts their sustainability and environmental impact. Currently, the market is predominantly dominated by petroleum-based substrates, valued for their cost-effectiveness and mechanical properties compared to biological alternatives. However, these petroleum-derived materials carry significant environmental consequences, stemming from their non-renewable nature and adverse effects on ecosystems. With the growing demand for sustainable solutions, there is increasing interest in exploring eco-friendly options. Among these alternatives, natural fibers and other bio-based materials stand out. These materials, including cellulose, hemp, and flax, offer promising avenues for reducing the environmental footprint of 4DP practices in civil engineering while enabling the creation of intricately functional printed structures [24].

Despite their advantages, incorporating natural fibers into 4DP presents certain challenges [25]. One significant obstacle is their inherent lack of stimulus-responsive properties, a defining feature of 4D-printed structures. Recent research endeavors have aimed to tackle this limitation by introducing intelligent materials capable of responding to various environmental stimuli into natural fiber compositions. Despite the hurdles, the potential benefits presented by these bio-derived materials—their renewable nature, biodegradability, and ability to produce intricately functional structures—establish them as promising candidates for advancing 4DP in civil engineering. In summary, 4DP in civil engineering offers an intriguing and hopeful field of exploration. As our comprehension of intelligent materials advances, the opportunities for 4DP applications in construction will also expand [26].

8.3.4 Applications of 4DP

The potential uses of 4DP in construction are extensive, covering both architectural and structural aspects. 4D-printed building elements have the ability to adopt to the surrounding conditions and self-repair that leads to sustainable construction. 4DP is employed to develop structures that respond to external stimuli like light and temperature. Moreover, opening or closing to control indoor temperatures was adjusted by smart facades by minimizing the need for artificial heating or cooling and improving energy efficiency. In terms of structure, 4DP can facilitate the production of self-assembling components, streamlining construction processes by minimizing manual labor and enabling the creation of more intricate and accurate structures. Investigations into 4D-printed concrete have indicated its potential for promoting more sustainable construction practices. This entails incorporating 4D properties into concrete to enable it to adapt and respond to its surroundings over time. Such advancements could lead to structures capable of self-repairing, addressing cracks and faults automatically, thereby enhancing safety and longevity while cutting maintenance expenses. Despite its potential, implementing 4DP in construction poses notable challenges, including the intricacies of the design process, the necessity for novel construction methodologies, the high expenses associated with smart materials, and technological limitations related to scalability and speed. Nevertheless, ongoing research and technological advancements hold promise for the transformation of the construction sector through 4DP.

8.4 SUSTAINABILITY IN THE CONSTRUCTION INDUSTRY

The global economy heavily relies on the construction sector, yet it also contributes significantly to energy consumption and carbon emissions, unlike manufacturing which has a much smaller environmental impact. Consequently, sustainability in construction is a matter of great concern for policymakers, industry practitioners, and scholars. Achieving sustainability in construction involves not only addressing ecological and environmental issues but also considering economic, social, and technical factors such as construction costs, time, and societal impacts. A primary goal of sustainable business plans is to generate profits for shareholders while also positively impacting the environment and society. With increasing awareness of global challenges, corporate leaders are recognizing the importance of sustainable business models for long-term profitability. Understanding and implementing sustainability strategies are vital for achieving success in the construction industry. The triple bottom line concept, depicting the interconnectedness of economic, social, and environmental aspects of sustainable development is illustrated in Figure 8.2. After concentrating solely on environmental issues in the previous decade, the research community in the construction industry has shifted toward a more holistic consideration of sustainability, encompassing the triple bottom line. Additionally, it is pointed out that the building industry has predominantly emphasized environmental factors exclusively [23] and the buildings are intricately connected to regional culture and sustainability, playing a crucial role in the cultural and geographic diversity of a region. The subjective nature of sustainability, laden with values, leads individuals to interpret the term differently based on their perceptions of society and its acceptance by others. Common terms in construction discussions about sustainability, such as "green" and "sustainable" building, are subject to diverse interpretations by stakeholders involved in construction projects, influenced by factors like education, age, and cultural background [8]. Nevertheless, international research on building sustainability faces challenges in effectively aligning with local contexts due to additional barriers (Table 8.1).

Table 8.1 Various effects on construction works over the environment

Sl. no.	*Effects on environment*	*Description*
1	Natural resources	The utilization of non-renewable materials poses a significant challenge in the construction sector, contributing to waste generation. According to the World Watch Institute, the industry accounts for 40% of global consumption of raw stone, gravel, and sand, as well as 25% of annual virgin wood usage. Despite the abundance of natural resources, it is imperative to recognize their finite nature, which implies that their depletion is inevitable over time. Some companies are gradually adopting new technologies such as 3D printers and biodegradable materials to reduce resource consumption. However, the construction industry remains one of the least digitized sectors, potentially delaying the effectiveness of such shifts [5].
2	Biodiversity	Biodiversity stands as the second most vital consideration. Contemplate the impact of buildings on wildlife. Instantly, thoughts of nighttime construction sites or noisy machinery may come to mind [4]. Concerning animals, the disturbance of their usual circadian rhythms, especially for bats, badgers, and birds, can be particularly harmful. Consequently, this aspect represents just a fraction of a broader concern. Over time, wildlife is inevitably affected by ongoing construction and development.
3	Atmospheric conditions	Activities associated with construction, transportation, and the manufacturing of building materials collectively make substantial contributions to this elevated statistic. Additionally, dust generated from construction sites emerges as a notable source of air pollution. Particulate Matter 10 (PM10), composed of concrete, wood, or stone particles, can be easily observed using a basic magnifying glass. If this dust persists in the air for extended periods, it has the potential to pose significant health risks to both humans and animals.
4	Waste disposal	Waste is ubiquitous, with the construction sector alone responsible for 59% of total waste production. This staggering statistic underscores the magnitude of the issue. The construction industry generates a significant amount of waste due to its reliance on short-term, cost-effective solutions that frequently necessitate replacement. Valuable resources are squandered because recycling is not commonly practiced on construction sites.

8.4.1 Various Barriers to sustainability in the construction industry

Based on existing literature, the subsequent factors are identified as obstacles to achieving sustainability within the construction sector. There are numerous factors that are considered barriers to attaining sustainability in the construction arena. The major contributing factor that acts as a barrier to attain sustainability is the inattention of clients and other stakeholders. Most of the time due to neglect and ignorance, and due to unawareness of the fact, many stakeholders tend to look over the causes of poor levels of sustainability in a workplace. Also, there is a trend of reluctance among the workers to go through with the practice of sustainability which ultimately leads to various undesired issues. The law and policy makers also create a barrier to sustainability due to their negligence to make the stakeholders of any construction project abide and thrive by the rules of law and the issue of corruption and other discrepancies pave the way to a barrier being created.

8.4.2 Proposed measures for achieving sustainability in construction

Merely designing sustainable structures is insufficient; construction companies must also consider the following factors: Ensuring the long-term viability of a project is among the most thought-provoking aspects of this profession, particularly concerning sustainability. Building construction relies on the utilization of natural resources, which are often discarded once they have served their purpose. Renovation projects may require the removal of previously installed elements to accommodate necessary alterations, rapidly depleting natural resources and causing significant environmental impacts. The use of sustainable building materials does not have to be a binary choice. Construction activities themselves can have a substantial environmental footprint, considering the energy required to establish construction sites, operate equipment, and machinery. The sourcing and utilization of energy can significantly impact emissions. Moreover, the use of fossil fuel-powered heavy machinery exacerbates the environmental impact of projects. An alternative approach to mitigate environmental harm is the adoption of lean production methods within closely monitored facilities. Indoor factories, equipped to minimize waste and recycle resources, can manufacture products under controlled conditions. Subsequently, these prefabricated components are transported to the construction site and assembled, resulting in a reduced environmental footprint.

8.5 CONCLUSION

In conclusion, the transition to green manufacturing practices is imperative in addressing global environmental concerns while promoting economic growth and social objectives. This shift toward sustainable development involves adopting green initiatives and leveraging advanced technologies to minimize waste and reduce carbon footprints. The introduction of 4DP in civil engineering presents a transformative approach to infrastructure design, albeit with challenges such as scalability and material compatibility that require further research [27]. Embracing 4DP holds promise for creating adaptive, sustainable, and innovative built environments, facilitated by the utilization of smart materials and ongoing technological advancements. Despite obstacles, the potential applications of 4DP in construction are extensive, offering hope for transformative change toward a more sustainable future. Collaboration among industry stakeholders and educational institutions is crucial in preparing the workforce and overcoming challenges to realize the full potential of 4DP in construction [28]. Overall, the integration of green manufacturing practices and 4DP technologies underscores the importance of sustainability in shaping the future of manufacturing and infrastructure development.

REFERENCES

1. C.W. Isaac, F. Duddeck, Current trends in additively manufactured (3D printed) energy absorbing structures for crashworthiness application–a review, *VirtualPhys. Prototyp.* 17 (4) (2022) 1058–1101, https://doi.org/10.1080/17452759.2022.2074698
2. F.E. Azhar, A 4D Printing Communication Framework for Designers and Engineers (Doctoral Dissertation), Brunel University, London, 2022.
3. U.K. Vates, S. Mishra, N.J. Kanu, Biomimetic 4D printed materials: a state-of-the-art review on concepts, opportunities, and challenges, *Mater. Today: Proc.* 47 (2021) 3313–3319, https://doi.org/10.1016/j.matpr.2021.07.148
4. C. Maraveas, I.S. Bayer, T. Bartzanas, 4D printing: perspectives for the production of sustainable plastics for agriculture, *Biotechnol. Adv.* 54 (2022), 107785, https://doi.org/10.1016/j.biotechadv.2021.107785.
5. J.L. Alves, L. Santana, B. Rangel, *4D Printing and Construction: Reality, Future, Or Science Fiction in: 3D Printing for Construction with Alternative Materials*. Springer International Publishing, Cham, 2023, pp. 155–175, https://doi.org/10.1007/978-3-031-09319-7_7.
6. M. Mehrpouya, A. Dehghanghadikolaei, B. Fotovvati, A. Vosooghnia, S.S. Emamian, A. Gisario, The potential of additive manufacturing in the smart factory industrial 4.0: a review, *Appl. Sci.* 9 (18) (2019) 3865, https://doi.org/10.3390/app9183865.

7. H. Yi, Y. Kim, Prototyping of 4D-printed self-shaping building skin in architecture: design, fabrication, and investigation of a two-way shape memory composite (TWSMC) façade panel, *J. Build. Eng.* 43 (2021), 103076, https://doi.org/10.1016/j.jobe.2021.103076.
8. X. Liu, M. Wei, Q. Wang, Y. Tian, J. Han, H. Gu, Z. Gu, Capillary-force-driven self-assembly of 4D-printed microstructures, *Adv. Mater.* 33 (22) (2021), 2100332, https://doi.org/10.1002/adma.202100332.
9. M.I. Farid, W. Wu, L. Guiwei, Z. Yu, Research on imminent enlargements of smart materials and structures towards novel 4D printing (4DP: SMs-SSs), *Int. J. Adv. Des. Manuf. Technol.* 126 (7–8) (2023) 2803–2823, https://doi.org/10.1007/s00170-023-11180-z.
10. C.K. Khoo, F. Salim, J. Burry, Designing architectural morphing skins with elastic modular systems, *Int. J. Architect. Comput.* 9 (4) (2011) 397–419, https://doi.org/10.1260/1478-0771.9.4.397.
11. M.Y. Khalid, Z.U. Arif, W. Ahmed, 4D printing: technological and manufacturing renaissance, *Macromol. Mater. Eng.* 307 (8) (2022), 2200003, https://doi.org/10.1002/mame.202200003.
12. H.J. Chen, C.F. Peng, C.W. Tang, Y.T. Chen, Self-healing concrete by biological substrate, *Materials* 12 (24) (2019) 4099, https://doi.org/10.3390/ma12244099.
13. C. Yuan, F. Wang, Q. Ge, Multimaterial direct 4D printing of high stiffness structures with large bending curvature, *Extreme Mech. Lett.* 42 (2021), 101122, https://doi.org/10.1016/j.eml.2020.101122.
14. L. Joharji, R.B. Mishra, F. Alam, S. Tytov, F. Al-Modaf, N. El-Atab, 4D printing: a detailed review of materials, techniques, and applications, *Microelectron. Eng.* (2022), 111874, https://doi.org/10.1016/j.mee.2022.111874.
15. N. Makul, Advanced smart concrete-a review of current progress, benefits and challenges, *J. Clean. Prod.* 274 (2020), 122899, https://doi.org/10.1016/j.jclepro.2020.122899.
16. C. Debrah, A.P. Chan, A. Darko, Artificial intelligence in green building, *Autom. ConStruct.* 137 (2022), 104192, https://doi.org/10.1016/j.autcon.2022.104192.
17. L. Santo, F. Quadrini, D. Bellisario, L. Iorio, Applications of shape-memory polymers, and their blends and composites, In: J. Parameswaranpillai, S. Siengchin, J. George, S. Jose (eds) *Shape Memory Polymers, Blends and Composites. Advanced Structured Materials*, vol 115. Springer, Singapore *Adv* (2020) 311–329, https://doi.org/10.1007/978-981-13-8574-2_13.
18. K.L. Ameta, V.S. Solanki, S. Haque, V. Singh, A.P. Devi, R.S. Chundawat, Critical appraisal and systematic review of 3D & 4D printing in sustainable and environment-friendly smart manufacturing technologies, *Sustain Mater Technol.* (2022), e00481, https://doi.org/10.1016/j.susmat.2022.e00481.
19. B. Jian, F. Demoly, Y. Zhang, H.J. Qi, J.C. André, S. Gomes, Origami-based design for 4D printing of 3D support-free hollow structures, *Eng. Times* 12 (2022) 70–82, https://doi.org/10.1016/j.eng.2021.06.028.
20. H. Vahabi, F. Laoutid, M. Mehrpouya, M.R. Saeb, P. Dubois, Flame retardant polymer materials: an update and the future for 3D printing developments, *Mater. Sci. Eng. R Rep.* 144 (2021), 100604, https://doi.org/10.1016/j.mser.2020.100604.

21. Z. Zhang, K.G. Demir, G.X. Gu, Developments in 4D-printing: a review on current smart materials, technologies, and applications, *Int. J. Soc. Netw. Min.* 10 (3) (2019) 205–224, https://doi.org/10.1080/19475411.2019.1591541.
22. A. Subash, B. Kandasubramanian, 4D printing of shape memory polymers, *Eur. Polym. J.* 134 (2020), 109771, https://doi.org/10.1016/j.eurpolymj.2020.109771.
23. E. Pei, J. Shen, J. Watling, Direct 3D printing of polymers onto textiles: experimental studies and applications, *Rapid Prototyp. J.* 21 (5) (2015) 556–571, https://doi.org/10.1108/RPJ-09-2014-0126.
24. S. Ram Kishore, A.P. Sridharan, U. Chadha, D. Narayanan, M. Mishra, S.K. Selvaraj, A.E. Patterson, Natural fiber biocomposites via 4D printing technologies: a review of possibilities for agricultural bio-mulching and related sustainable applications, *Prog. Addit. Manuf.* (2023) 1–31, https://doi.org/10.1007/s40964-023-00433-8.
25. Q. Ge, H.J. Qi, M.L. Dunn, Active materials by four-dimension printing, *Appl. Phys. Lett.* 103 (13) (2013), https://doi.org/10.1063/1.4819837.
26. M. Quanjin, M.R.M. Rejab, M.S. Idris, N.M. Kumar, M.H. Abdullah, G.R. Reddy, Recent 3D and 4D intelligent printing technologies: a comparative review and future perspective, *Proc. Comput. Sci.* 167 (2020) 1210–1219, https://doi.org/10.1016/j.procs.2020.03.434.
27. S. Tibbits, 4D printing: multi-material shape change, *Architect. Des* 84 (1) (2014) 116–121, https://doi.org/10.1002/ad.1710.
28. F. Momeni, X. Liu, J. Ni, A review of 4D printing, *Mater. Des.* 122 (2017) 42–79, https://doi.org/10.1016/j.matdes.2017.02.068

Chapter 9

Optimizing sustainable manufacturing

A comprehensive analysis of innovative technologies and strategic interventions

Ramaganesh Marimuthu and Bathrinath Sankaranarayanan

9.1 INTRODUCTION

Sustainable manufacturing signifies a transformative change in industrial practices, driven by a global recognition of the pressing need for ecologically answerable and communally conscious production processes. The background of sustainable manufacturing is rooted in the acknowledgment of the detrimental impacts traditional manufacturing has had on the environment, resource depletion, and social equity. Historically, industrial activities have contributed significantly to pollution, waste generation, and the depletion of finite resources. Consequently, the quest for sustainable manufacturing has gained momentum, aiming to reconcile economic growth with ecological preservation and social well-being (Ribeiro et al., 2014). This shift underscores the importance of adopting cleaner and more competent skills, adopting the tenets of circular economy principles, and fostering responsible supply chain management. As a multidimensional concept, sustainable manufacturing seeks to balance eco-friendly stewardship, fiscal feasibility, and social responsibility, flagging the method for a more resilient and harmonious industrial landscape. In this context, understanding and promoting innovative technologies play a pivotal part in attaining the objectives of sustainable manufacturing, marking a crucial intersection between technological advancements and responsible industrial practices (Sarkis et al., 2011).

The significance of innovative technologies in the realm of sustainable manufacturing cannot be overstated, representing a cornerstone in the ongoing quest for added eco-friendly and communally conscious industrial methods. As industries face the imperative to reduce their ecological footprint and tackle urgent global issues such as climate change and dwindling resources, the role of innovation becomes paramount. Innovative technologies offer transformative solutions by optimizing production processes, enhancing energy efficiency, and minimizing waste. Beyond environmental considerations, these technologies can also stimulate economic growth, improve

DOI: 10.1201/9781003470342-9

competitiveness, and open up new avenues for industries to flourish in a fast-changing market (Blecker and Abdelkafi, 2015). Furthermore, the combination of innovative technologies in sustainable manufacturing fosters a positive feedback loop, encouraging ongoing research and development, while simultaneously addressing societal expectations for ethically and sustainably produced goods. As such, grasping the elements that shape the embrace and successful implementation of these technologies is pivotal for steering industrial practices toward a more supportable and robust future. This research aims to unravel the complexities surrounding these factors, shedding light on the dynamics that shape the adoption of innovative technologies within the context of sustainable manufacturing (Zhang et al., 2017).

With an increasing global awareness of the adverse impacts of conventional manufacturing practices, the imperative to transition toward sustainability is more pressing than ever. Innovative technologies emerge as linchpins in this transition, holding the potential to revolutionize manufacturing processes, minimize environmental footprints, and bolster economic resilience. The multifaceted nature of sustainable manufacturing demands a comprehensive understanding of the features that either enable or hinder the implementation of these technologies (Wuest et al., 2015). By delving into this research, we seek to contribute valuable understandings that can update industry practitioners, representatives, and scholars alike. The outcomes of this study are anticipated to not only enhance our understanding of the intricate dynamics governing the integration of innovative technologies but also provide practical recommendations to propel sustainable manufacturing practices forward, aligning industrial progress with environmental stewardship and societal well-being. This study is important in addressing contemporary challenges at the intersection of industry, environment, and society. This study encourages an examination of the subsequent questions.

1. What are the key technological innovations currently adopted in sustainable manufacturing, and how do they contribute to environmental and social sustainability?
2. What are the environmental impacts and benefits associated with the adoption of innovative technologies in sustainable manufacturing?
3. What barriers, both internal and external, impede the widespread adoption of innovative technologies in the context of sustainable manufacturing?
4. How can the identified problems be interconnected to facilitate the effective integration of sustainability initiatives?

This study seeks to demonstrate the intricate interconnections among diverse factors influencing innovative technologies in sustainable manufacturing through the application of the multi-criteria decision-making (MCDM) method. Specifically, the decision-making trial and evaluation laboratory (DEMATEL) practice in the fuzzy framework has been employed for this investigation. The preference for the fuzzy DEMATEL technique instead of other MCDM methods is justified by its capacity to emphasize interconnections and identify the significance of specific factors. Furthermore, the fuzzy DEMATEL approach classifies problems into distinct cause-and-effect relationships, offering nuanced insights. The outcomes of this analysis are poised to shed light on critical factors necessitating immediate attention. By providing a clear delineation of cause-and-effect relationships, the findings from fuzzy DEMATEL have the potential to assist industry leaders and professionals in formulating targeted approaches for achieving success in the integration of sustainability initiatives.

9.2 LITERATURE REVIEW

This section outlines the methodology employed for a systematic literature review (SLR) designed to discern the issues influencing advanced technologies for sustainable manufacturing. The search strategy involved the use of targeted keywords applied across major databases, with the integration of Boolean operators to enhance the precision of the search criteria. Following a thorough analysis, the study successfully pinpointed 10 key factors associated with innovative technologies in the context of sustainable manufacturing.

9.2.1 Challenges influencing innovative technologies for sustainable manufacturing

Fu et al. (2018) explored the fragmented landscape of sustainable technology implementation by analyzing 34 selected articles from databases like Elsevier and Web of Science. Factors influencing adoption, including environmental regulations and firm characteristics, are examined, with a focus on CO_2 reduction, material efficiency, and recycling technologies. They identified coercion, market pressure, technology capability, internal support, and cooperation as key influencers. Firm characteristics and technology types are acknowledged as variables affecting outcomes, laying the foundation for future research directions. Garetti and Taisch (2012) examined the trends and research obstacles in sustainable manufacturing. Adapting to sustainable manufacturing requires innovative products, operations, and organizational models. The dynamic nature of sustainability, influenced by factors like new technology and environmental laws, means that what's considered sustainable can change over time. The European Commission emphasizes the global significance of research in manufacturing to address challenges like globalization, service orientation, knowledge intensity, and environmental concerns, calling for collaborative solutions from the scientific community worldwide. Ngibe and Lekhanya (2020) highlighted the critical role of manufacturing SMEs in South Africa's economic growth and job creation. It emphasizes the urgent need for attention to environmental factors, both internal (inadequate support and capital) and external (social factors and technological changes), to ensure the sustainable growth and success of these enterprises, with recommendations provided for their enhancement. Kuo et al. (2022) used DEMATEL to identify key factors influencing sustainable manufacturing, highlighting eco-design, laws, regulations, and waste management as crucial criteria. The recommendation prioritizes issues like eco-design and waste management for effective resource allocation, emphasizing compliance with regulations and international standards to achieve sustainable development. Suresh and Dharunanand (2023) utilized the total interpretive structural modeling (TISM) method for identifying and analyzing 12 pivotal elements affecting sustainable maintenance practices within manufacturing industries. Through literature review, expert opinions, and questionnaire-based interviews, factors such as availability rate, adoption of governmental regulations, and investment in training and educational programs were identified as pivotal in sustainable maintenance. The proposed model provides practical insights for experts and maintenance managers to enact effective sustainable maintenance protocols within manufacturing facilities. Shahzad et al. (2022) developed and validated a green innovation adoption (GIA) framework grounded in the unified theory of acceptance and use of technology (UTAUT) to understand and improve the adoption of green innovation technology in Pakistani manufacturing industries. Results from structural equation modeling (SEM) and artificial neural network (ANN) investigations indicate that factors like result expectation, effort expectation, and organizational size significantly affect green behavioral intention (GBI), emphasizing the importance of these factors in accelerating sustainable development by embracing eco-friendly technologies. Khurana et al. (2021) examined critical factors (CFs) for implementing Sustainable Oriented Innovation (SOI) in micro, small, and medium enterprises (MSMEs) in the Delhi NCR region of India. Utilizing the analytical hierarchy process (AHP), the study identifies "Top management support," "Government initiatives," and "Financial resources" as the top crucial issues influencing the successful implementation of SOI practices in MSMEs, providing valuable insights for companies, academics, and governmental professionals to enhance sustainable development in this sector. Niederprüm (2018) explored factors influencing the implementation of environmentally sustainable process technologies within manufacturing sectors to achieve UN SDG goals 12 and 13. It reveals that state regulation positively influences adoption, while financial and technological capabilities are necessary conditions. Firm size acts as a precondition, and financial costs, coupled with regulatory pressures, can impact the adoption of sustainable technologies. Jasiulewicz-Kaczmarek (2018) emphasized the need for a holistic approach to sustainable maintenance management, covering strategic, tactical, and operational levels.

The result highlights the importance of considering economic, ecological, and social effects in assessing maintenance processes for effective implementation in sustainable production. Yin et al. (2020) examined the implementation challenges of green technology innovation (GTI) in collaborative efforts among multiple agents to promote environmentally friendly manufacturing practices. It proposes a comprehensive evaluation system, highlighting key factors such as green R&D expenditure, new green product share, and regulatory incentives to enhance GTI capability and competitiveness. The approach combines subjective and objective methods for a nuanced assessment of GTI capability in manufacturing enterprises. Agrawal (2019) used TISM to analyze factors influencing sustainable Additive Manufacturing (AM), identifying "Flexibility in manufacturing," "Time to develop new product," and "Local availability of technology" as dominant factors. The research contributes an original structural model that can assist AM experts in systematically understanding and addressing factors affecting sustainable AM. Chen (2016) introduced a theoretical framework for the sustainable innovation capacity of enterprises, encompassing knowledge, production, and market novelty capabilities. Through factor analyses and a practical case simulation using system dynamics, the study validates the framework, offering a foundation for future research and practical guidance for firms seeking to enhance their sustainable innovation. Dwivedi et al. (2023) investigated the role of technological revolution in sustainability within manufacturing organizations, particularly in an emerging economy with a circular economy perspective. The findings emphasize the significance of entrepreneurship and market orientation as key factors influencing the role of technological innovation in advancing sustainability objectives, suggesting the importance of circular economy practices, sustainable education, and stakeholder engagement for competitiveness in developing countries. Jawahir et al. (2013) highlighted the importance of the cleaning operation in automotive remanufacturing and identified seven key factors contributing to high cleaning costs. These factors are categorized under technical aspects related to product and process cleaning and business-related considerations of the remanufacturer. Bhanot et al. (2017) introduced a holistic sustainability framework for manufacturing industries by finding influential factors through the maximum mean de-entropy algorithm and decision-making trial and evaluation laboratory (DEMATEL). The research integrates results with interpretive structural modeling, statistically validating enablers and factors through SEM, aiming to bridge opinion differences among researchers and industry professionals for effective sustainability implementation. Mazzanti and Zoboli (2006) examined the factors of environmental novelty at the firm level utilizing information from manufacturing firms in the Emilia Romagna region, Northern Italy. The study finds that environmental innovation is influenced by a combination of exogenous factors (e.g., environmental policy pressure) and endogenous elements such as environmental R&D, networking activities, and innovative-oriented industrial relations, shedding light on the nuanced dynamics of environmental innovation within a specific industrial context. Ullah et al. (2022) identified key factors of environmentally sustainable innovation within the manufacturing sector, utilizing a hybrid methodology including fuzzy Delphi method, interpretive structural modeling (ISM), and MICMAC analysis. The results highlight "cost reduction" and "government support" as the top crucial factors, emphasizing the importance of these factors in motivating the incorporation of eco-friendly innovation within the manufacturing industry in Pakistan. Table 9.1 represents factors influencing innovative technologies in sustainable manufacturing.

Table 9.1 Factors influencing innovative technologies in sustainable manufacturing

S.no	*Sustainable factor*	*Explanation*	*References*
1.	Energy efficiency (F1)	Adopt technologies that enhance energy efficiency in manufacturing processes, minimizing overall energy consumption, and minimizing ecological effect	Garetti et al. (2012), Menghi et al. (2019), Chen et al. (2020b)
2.	Renewable energy integration (F2)	Explore and incorporate renewable energy sources like solar and wind, or hydropower to energize manufacturing facilities, reducing dependence on non-renewable resources	Zhang et al. (2018), Gupta et al. (2021)
3.	Material efficiency and recycling (F3)	Implement technologies that promote the efficient use of materials and incorporate recycling processes to reduce waste generation and support a closed-loop economy	Ghisellini et al. (2016), Kim et al. (2017)
4.	Smart manufacturing and IoT (F4)	Utilize Internet of Things (IoT) devices and smart manufacturing technologies to optimize production processes, monitor resource usage in real-time, and identify areas for improvement	Mourtzis et al. (2018), Tao et al. (2018)
5.	Advanced robotics and automation (F5)	Integrate robotics and automation to boost efficiency, minimize mistakes, and increase overall productivity, leading to resource and energy savings	Dornfeld and Min (2011), Kyratsis et al. (2018), Al-Jumaily et al. (2019)
6.	Water conservation and management (F6)	Implement technologies that focus on reducing water usage in manufacturing processes and improving water treatment and recycling systems to minimize water waste	Hoekstra et al. (2011), Suresh et al. (2023), Kargari and Tabatabai (2018)

(Continued)

Table 9.1 (Continued) Factors influencing innovative technologies in sustainable manufacturing

S.no	*Sustainable factor*	*Explanation*	*References*
7.	Supply chain transparency (F7)	Leverage technologies like blockchain and data analytics to enhance supply chain visibility, traceability, and accountability, ensuring sustainability during the complete product life span	Kim and Kim (2016), Wang et al. (2019)
8.	Life cycle assessment (LCA) (F8)	Incorporate life cycle assessment tools to assess the ecological effect of products from raw material extraction to disposal at the end of their life cycle, helping in the selection of eco-friendly alternatives	Song et al. (2019) Notter et al. (2010)
9.	Collaborative and open innovation (F9)	Foster collaboration and open innovation by leveraging digital platforms and technologies that facilitate knowledge-sharing and partnerships, accelerating the progress of sustainable solutions	Chesbrough (2011), Huizingh (2011)
10.	Regulatory compliance and standards (F10)	Stay informed about and adhere to evolving environmental regulations and standards. Ensure that adopted technologies meet or exceed compliance requirements to avoid negative environmental impacts and legal issues	Dornfeld and Min (2011), Zhang et al. (2017)

9.3 RESEARCH METHODOLOGY

Figure 9.1 shows the framework of the projected work, delineated into two separate stages. During the initial stage, expert insights are utilized to address the issues influencing innovative technologies in sustainable manufacturing. The second phase employs fuzzy DEMATEL to unveil the weights and interconnections among these issues.

- Phase 1: Confirmation of factors

  To confirm the identified factors, we collected expert insights via a questionnaire distributed to a group of ten individuals, each highly experienced in implementing innovative technologies in sustainable manufacturing. The questionnaire addressed the ten factors recognized in our research, and its trustworthiness was evaluated using Cronbach's alpha, yielding a score of 0.802, which exceeds the acceptable threshold of 0.70 and reinforces the credibility of the survey. The validity of the questionnaire was assessed using the Kaiser–Meyer–Olkin (KMO) measure and Bartlett's test of sphericity.

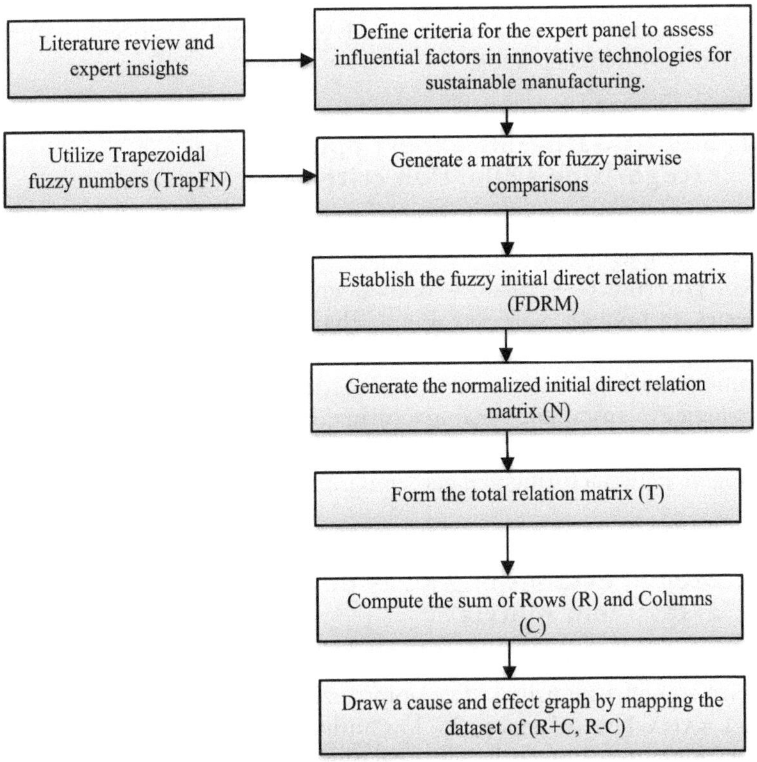

Figure 9.1 Structure of the projected work.

The results indicated a significant value of 0 for Bartlett's test and a KMO value of 0.716, supporting the suitability for further analysis. The questionnaire was designed in a YES/NO format to gather expert opinions, asking respondents to indicate YES if they agreed with the listed factors and NO if they disagreed. Experts were encouraged to review the factors, suggest modifications, and make revisions as necessary. Regular reminders were sent to prompt feedback. Ultimately, consensus among the experts confirmed the validation of the factors identified in the study.

- Phase 2: Fuzzy DEMATEL method

  The DEMATEL method, pioneered by Gabus et al. (1972), proves invaluable in scrutinizing intricate causal relationships among factors. This structural modeling technique employs a digraph to vividly illustrate the causal links and interconnections between various factors (Marimuthu et al., 2021). In adopting this method, factors are methodically classified into categories of causes and effects, enhancing the discernment of the causal relations among them. The elucidation of the fuzzy DEMATEL method procedure unfolds in the subsequent steps.

9.3.1 Step 1: Gathering expert opinions and categorizing evaluation criteria

A diverse board was assembled to provide valuable insights on pertinent issues, encompassing industry executives and leaders, researchers and academics, government regulatory bodies, environmental and sustainability advocates, technology providers, supply chain partners, employees and labor unions, investors and financial institutions, consumers, and educational institutions, each possessing extensive experience in implementing innovative technologies in sustainable manufacturing. Utilizing insights from previous research, we identified crucial factors impacting innovative technologies in sustainable manufacturing and gathered input from industry experts.

9.3.2 Step 2: Establishing fuzzy direct assessment matrix

Following the identification of crucial factors, a crucial step is to conduct a pairwise comparison using a five-point fuzzy linguistic scale (0 = no influence, 1 = very low influence, 2 = low influence, 3 = strong influence, and 4 = very high influence) to delineate relationships among factors. Specialists utilize this scale to assign ratings, leading to the formation of an assessment matrix. The insights provided by specialists pay to fuzzy evaluations, influencing linguistic elements and producing the optimistic TrapFN shown in Table A1. TrapFN is denoted by a quadruplet, i.e., $(l_1ij, m_1ij, n_1ij, o_1ij)$, where $l_1 \leq m_1 \leq n_1 \leq o_1$. Let $A_1\frac{k}{ij} = \left(l_1\frac{k}{ij}, m_1\frac{k}{ij}, n_1\frac{k}{ij}, o_1\frac{k}{ij}\right)$, where $1 \leq k \leq k$, $1 \leq i \leq n$; $1 \leq j \leq n$ is the fuzzy sorting representing the rating provided by the kth decision maker regarding the influence of factor i on factor j.

9.3.3 Step 3: Creating fuzzy initial direct relation matrix (FDRM)

Defuzzification involves the conversion of fuzzy records into their precise values, resulting in the acquisition of the fuzzy direct assessment matrix (FDRM) using Eq. (9.1). Subsequently, the average k is computed from the $n \times n$ FDRM across all decision makers, where k denotes the total count of decision makers:

$$D_t = \frac{1}{4}(l_1 + m_1 + n_1 + o_1) \tag{9.1}$$

9.3.4 Step 4: Construct the normalized, initial, direct relation matrix (*N*) using Eqs. (9.2) and (9.3)

$$n = \min\left[\frac{1}{\max\sum_{i=1}^{n}|aij|}, \frac{1}{\max\sum_{j=1}^{n}|aij|}\right] \tag{9.2}$$

$$\mathrm{N} = nA \tag{9.3}$$

9.3.5 Step 5: Prepare the total relation matrix (*T*) using Eq. (9.4) as follows:

$$T = (I - \mathrm{N})^{-1} \tag{9.4}$$

Where I is the identity matrix and $T = \left[t_{ij}\right]_{n \times n}$.
i and j signify the criteria in the pairwise comparison matrix.

9.3.6 Step 6: Calculate the sum of rows (*R*) and columns (*C*) using Eqs. (9.5) and (9.6)

$$R = \left[\sum_{j=1}^{n} t_{ij}\right]_{n \times 1} \tag{9.5}$$

$$\mathrm{C} = \left[\sum_{i=1}^{n} t_{ij}\right]_{1 \times n} \tag{9.6}$$

R indicates the influence of criterion i on criterion j, and C indicates the influence of criterion j on criterion i.

9.3.7 Step 7: Construct a cause and effect graph by plotting the dataset of (*R* + *C*, *R* − *C*)

The summation (R+C) denotes the consequence, referred to as "Prominence," of pivotal factors in sustainable manufacturing. Furthermore, (R – C) signifies the "Influence" in the context of innovative technologies in sustainable manufacturing. A positive value in (R – C) classifies the factor as part of the cause group, whereas a negative value places it in the effect group. Figure 9.1 visually denotes the workflow diagram detailing the stages in the fuzzy DEMATEL method.

9.4 IMPLEMENTATION OF THE SUGGESTED APPROACH

The suggested workflow to examine the factors that contribute to the effective adoption of innovative technologies in sustainable manufacturing. The initial identification of factors includes a literature review and is informed by expert opinions. In the case study, experts are provided with a questionnaire featuring the identified factors, and their responses are evaluated using fuzzy DEMATEL. The implementation of innovative methods in sustainable manufacturing is crucial for environmental impact reduction, cost savings, and global competitiveness in India. The implementation of the suggested approach involves several steps.

Step 1*:* Initiate the process by constructing a pairwise comparison matrix with the fuzzy linguistic scale shown in Appendix A, Table A3.

Step 2*:* Moving on to step 3 of the fuzzy DEMATEL method, derive the Initial Direct Relation Matrix (FDRM) by defuzzifying fuzzy numbers into exact values. Appendix B, Table B1, presents the fuzzy decision-making matrix (FDRM) for the identified factors influencing in sustainable manufacturing.

Step 3*:* Then, calculate the Normalized Direct Relation Matrix (N) using Eqs. (9.2) and (9.3). The fuzzy normalized direct relation matrix (Fuzzy NDRM) for the identified factors can be found in Table B2 of Appendix B.

Step 4*:* Calculate the total relation matrix (TRM) with Eq. (9.4). The fuzzy total relation matrix (Fuzzy TRM) for the factors is provided in Table B3 of Appendix B.

Step 5*:* Obtain the sum of rows (R) and columns (C) through Eqs. (9.5) and (9.6), as detailed in Table 9.2.

Table 9.2 Determination of ($R + C$) and ($R - C$) datasets of factors influencing innovative technologies in sustainable manufacturing

Factors	*R*	*C*	*R + C*	*Rank*	*R − C*	*Cause/effect*
F1	3.8070	3.2413	7.0483	3	0.5657	Cause
F2	3.0497	3.5068	6.5565	9	−0.4571	Effect
F3	3.4663	3.7773	7.2436	1	−0.3111	Effect
F4	4.0740	3.0232	7.0972	2	1.0508	Cause
F5	3.5689	3.3981	6.9670	4	0.1708	Cause
F6	3.2786	3.3714	6.6500	8	−0.0927	Effect
F7	3.5131	3.1514	6.6645	7	0.3616	Cause
F8	3.2688	3.4536	6.7224	6	−0.1848	Effect
F9	3.5147	2.8820	6.3967	10	0.6327	Cause
F10	2.5651	4.3010	6.8661	5	−1.7359	Effect

Subsequently, the categorization of factors in the integration of innovative technologies into sustainable manufacturing is established using the datasets $(R + C)$ and $(R - C)$. The assignment to cause or effect groups is determined by the results extracted from the $(R + C)$ and $(R - C)$ datasets, outlined in Table 9.2. This classification is then visually represented through a causal diagram, illustrated in Figure 9.2.

The $(R + C)$ dataset value indicates the "Prominence" level of the factors, whereas $(R - C)$ differentiates the factors into cause-and-effect categories. Leveraging these dataset values, a causal diagram is generated for the factors. This diagram serves as an important resource for policymakers and industry practitioners, highlighting areas that require improvement.

9.5 FINDINGS AND DISCUSSION

In this division, we delve into the outcomes of the fuzzy DEMATEL analysis concerning the 10 factors that impact innovative technologies in sustainable manufacturing. Illustrated in Figure 9.2 are the identified cause-and-effect groups, which serve as pivotal results. Subsequently, we engage in a comprehensive discussion focusing on the prominence, influence, and consequences associated with these factors.

9.5.1 Eminence factors

The sequence of importance of the 10 major factors that influence innovative technologies in sustainable manufacturing using a significance threshold $(R + C)$ is noted as F3>F4>F1>F5>F10>F8>F7>F6>F2>F9. Material efficiency

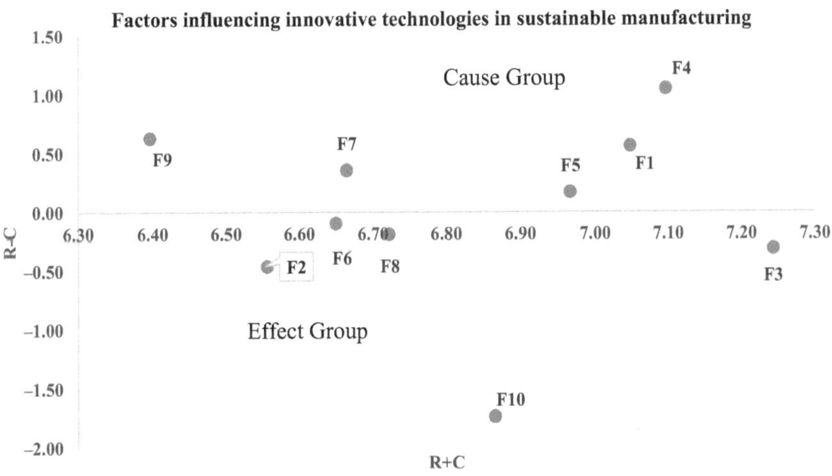

Figure 9.2 Cause and effect group of sustainable factors.

and recycling (F3) and smart manufacturing and IoT (F4) are the most important factors compared with other factors with values of 7.2436 and 7.0972, respectively. On the other end, collaborative and open innovation (F9) and renewable energy integration (F2) are the minimum important factors related with other factors with values of 6.3967 and 6.5565, respectively. Material efficiency and recycling (F3) emerge as a key factor, indicating its critical influence on the overall innovative processes in sustainable manufacturing. This factor underscores the importance of optimizing material usage and incorporating effective recycling mechanisms to enhance the sustainability of manufacturing practices. The positive correlation with other factors suggests that improvements in material efficiency can lead to positive cascading effects across the system (Ellen MacArthur, 2018). The integration of material efficiency practices not only optimizes resource utilization but also fosters a culture of sustainability within manufacturing processes. By minimizing waste and enhancing the circularity of materials, innovative technologies can be developed with a focus on longevity and eco-friendliness. Furthermore, the emphasis on recycling as a key factor ensures that end-of-life products and waste materials are reintegrated into the production cycle, reducing the environmental footprint (Kirchherr et al., 2016). These practices not only aid in resource conservation but also promote the advancement of innovative technologies that adhere to sustainability principles. The integration of material efficiency and recycling, therefore, stands as a cornerstone in the pursuit of innovative solutions within the sustainable manufacturing paradigm, fostering a holistic approach that benefits both industry and the environment (Melnyk et al., 2019). Smart manufacturing and IoT (F4) surfaces as another vital factor, highlighting the increasing significance of technology integration in manufacturing processes. The positive relationships identified suggest that advancements in smart manufacturing and the implementation of IoT technologies contribute significantly to the overall sustainability of manufacturing practices. This aligns with the current trend toward Industry 4.0 and emphasizes the role of technology in driving sustainable innovation (World Economic Forum, 2017). In the continuously changing field of sustainable manufacturing, Smart Manufacturing and the integration of the Internet of Things (IoT) stand out as transformative factors driving innovation. The conjugal of Smart manufacturing and IoT technologies introduces a paradigm shift in manufacturing processes, empowering instantaneous data gathering, evaluation, and swift decision implementation. The interconnectedness of devices, machinery, and systems facilitates efficiency improvements, predictive maintenance, and resource optimization. This interconnected intelligence not only enhances the overall performance and productivity of manufacturing operations but also plays a pivotal role in achieving sustainability goals (Al-Fuqaha et al., 2018). The capacity to monitor, manage, and optimize processes in real-time is in perfect harmony with the principles of sustainable manufacturing, allowing for precise resource utilization and energy efficiency. As Smart manufacturing and IoT continue to evolve, they emerge as cornerstones of innovation, ushering in a new era where technology and sustainability converge to shape the future of manufacturing (Monostori et al., 2017). The negative correlations identified with other factors indicate that changes in collaborative and open innovation (F9) may not necessarily lead to significant cascading effects within the system. This prompts a critical examination of the role of collaboration and open novelty in the specific perspective of sustainable manufacturing practices considered in this study. The negative relationships identified with other factors suggest that variations in renewable energy integration (F2) may not strongly impact the overall dynamics of sustainable manufacturing technologies in the studied system. As we interpret these results, it becomes apparent that while collaborative and open innovation (F9) and renewable energy integration (F2) may not have an important position in the innovation dynamics of the current study, their relevance may vary in different contexts. Future research endeavors could explore these factors in diverse manufacturing settings, shedding light on the contextual factors that influence their significance in shaping innovative technologies for sustainable manufacturing.

9.5.2 Influencing factors

The emphasis in our analysis is placed on addressing the factors confronted by the cause group, which have been prioritized based on their ($R - C$) values. These factors serve as a benchmark for devising strategies to tackle other influencing elements. Notably, Smart manufacturing and IoT (F4) emerge as the most impactful factor, boasting a substantial value of 1.0508 in comparison to all other factors influencing the cause. This underscores the pivotal role of smart manufacturing and IoT as a focal point for interventions and highlights its paramount influence in shaping the overall impact on the cause. Smart manufacturing and IoT (F4) emerge as pivotal forces with a transformative impact on factors influencing innovative technologies in sustainable manufacturing. The integration of these technologies significantly enhances operational efficiency by enabling continuous surveillance and management of manufacturing processes. This leads to a reduction in resource waste and an overall improvement in sustainability practices. Additionally, the data-driven decision-making facilitated by IoT contributes to a more informed approach, identifying areas for optimization and innovation within the manufacturing ecosystem (Monostori et al., 2017). Predictive maintenance capabilities, made possible by IoT sensors, not only minimize downtime but also extend the lifespan of equipment, aligning with sustainable practices. Furthermore, smart manufacturing and IoT take on a pivotal role in optimizing supply chain processes, minimizing energy consumption, and enhancing sustainability across the entire value chain (Tao et al., 2014). Despite the evident positive impacts, it is imperative to address potential challenges and ensure the maximization of benefits from smart manufacturing and IoT. Robust cybersecurity measures should be implemented to safeguard IoT infrastructure against potential threats. Establishing and adhering to industry-wide compatibility standards will facilitate smooth integration, fostering compatibility and scalability of these technologies (Lee et al., 2015). Investing in workforce skill development is crucial for fully leveraging of smart manufacturing and IoT, minimizing operational risks. Conducting a life cycle assessment is vital to understand and mitigate the environmental impact, ensuring that the implementation aligns with sustainable practices. Continuous monitoring and optimization of IoT devices and smart manufacturing systems are paramount for identifying opportunities for ongoing efficiency gains and sustainability improvements. Through these remedial actions, organizations can navigate the transformative landscape of smart manufacturing and IoT while ensuring sustainable and innovative manufacturing practices (Sarkis et al., 2013). With a value of 0.6327, collaborative and open innovation (F9) ranks as the second most influential causal factor in the $(R - C)$ column. This suggests its significant influence on the behavior of the analyzed system. The emphasis on collaboration and openness fosters a dynamic ecosystem where diverse stakeholders contribute ideas and expertise. This approach often results in a more deeper understanding of sustainable manufacturing challenges and facilitates the development of innovative solutions (Zeng et al., 2021). Open innovation practices, such as partnerships and knowledge-sharing, contribute to a broader pool of resources and ideas, accelerating the pace of technological advancements. The impact extends beyond individual organizations, creating a collaborative network that collectively drives sustainability in manufacturing by addressing complex challenges through shared expertise and resources (Zhang et al., 2021). To optimize the impact of collaborative and open innovation (F9) in factors influencing innovative technologies in sustainable manufacturing, several remedial actions can be considered. Firstly, organizations should establish dedicated platforms for collaboration, providing a space for stakeholders to exchange ideas and collaborate on sustainable initiatives. Emphasizing a culture of knowledge-sharing and fostering partnerships with research institutions, industry peers, and other relevant entities can enhance the effectiveness of collaborative efforts (Ritala and Huotari, 2021). Additionally, organizations should implement mechanisms to incentivize and recognize collaborative contributions, encouraging active participation from diverse stakeholders. Continuous surveillance and assessment of collaborative initiatives are essential to pinpoint areas for enhancement and guarantee conformity with sustainability goals. By implementing these remedial actions, organizations can harness the full potential of collaborative and open innovation, fostering a collaborative environment that propels sustainable manufacturing innovations (Llopis et al., 2020).

9.5.3 Resulting factors

Various obstacles affect the factors influencing innovative technologies in sustainable manufacturing within the effect group. These factors, ranked by severity, are F6, F8, F3, F2, and F10, with their respective scores indicating their impact as F6>F8>F3>F2>F10. Additionally, these five factors are shaped by causal factors related to innovative technologies in sustainable manufacturing. Among the effect factors, water conservation and management (F6) exhibit the least severe impact with an $(R - C)$ score of −0.0927. The other effect factors include life cycle assessment (LCA) (F8), material efficiency and recycling (F3), renewable energy integration (F2), and regulatory compliance and standards (F10). Notably, regulatory compliance and standards (F10) stand out with the lowest $(R - C)$ score of −1.7359, emphasizing its significant and high-impact role within the system. To address the identified obstacles and enhance the factors influencing innovative technologies in sustainable manufacturing, a comprehensive set of remedial actions is recommended. Firstly, for the factor of water conservation and management (F6), organizations should implement advanced water-saving technologies and adopt efficient water management practices (Yazdanpanah et al., 2014). Life cycle assessment (LCA) (F8) can be improved by integrating streamlined assessment methodologies and tools into the manufacturing process (Kralisch et al., 2015). Material efficiency and recycling (F3) can be enhanced through the adoption of principles of a circular economy, promoting recycling programs, and adopting environmentally conscious materials (Panghal et al., 2023). For renewable energy integration (F2), businesses should prioritize investments in renewable energy sources, such as solar and wind power, to reduce reliance on traditional energy (Pouresmaieli et al., 2023). Finally, to address regulatory compliance and standards (F10), organizations should prioritize staying updated on environmental regulations, invest in compliance monitoring systems, and actively engage in industry standards development to ensure alignment with sustainable manufacturing practices (El-Gayar and Fritz, 2006). Implementing these remedial actions collectively can contribute to a manufacturing ecosystem that is more robust and environmentally sustainable. Further, material efficiency and recycling (F3) is underscored by its substantial $(R+C)$ value of 7.2436, affirming its utmost significance in the context of sustainable manufacturing. The high $(R+C)$ value signifies its critical importance in sustainable manufacturing. The likely cause of this heightened relevance could stem from a growing awareness of environmental concerns, regulatory pressures, and an increasing emphasis on responsible resource management. To address and further enhance material efficiency and recycling in manufacturing, organizations can implement comprehensive recycling programs, adopt circular economy principles, invest in innovative recycling technologies, and collaborate with suppliers to source and use eco-friendly materials (Chen et al., 2020a). Additionally, fostering a sustainable ethos throughout the organization, providing employee training, and actively engaging in industry initiatives can contribute to a more robust and effective approach to material efficiency and recycling.

9.6 CONCLUSIONS

As societies worldwide confront escalating environmental concerns and the imperative for sustainable development, manufacturing industries show a pivotal part in shaping a more resilient and eco-friendly future. Innovative technologies hold the key to ushering in transformative changes, enabling manufacturing processes to be more resource-efficient, environmentally conscious, and socially responsible. By understanding and mitigating the elements impacting the integration of these technologies, industries can unlock substantial benefits, including enhanced operational efficiency, reduced environmental impact, and improved product life cycles. Moreover, the urgency to address these factors is underscored by the worldwide dedication to addressing climate change and attaining goals for sustainable development. Adopting innovative technologies in manufacturing is essential for achieving a harmonious balance between economic growth and environmental health, ensuring a sustainable and prosperous future for generations to come. Addressing the factors influencing innovative technologies in sustainable manufacturing is of paramount importance and necessity in the contemporary industrial landscape. The fuzzy DEMATEL technique was employed in this article to assess the top 10 factors influencing innovative technologies in sustainable manufacturing. The study's results indicate that material efficiency and recycling, alongside smart manufacturing and the IoT, stand out as the foremost factors exerting influence on innovative technologies within the realm of sustainable manufacturing. Given the pivotal role of material efficiency and recycling as a top factor influencing innovative technologies in sustainable manufacturing, managers should strategically prioritize investments in advanced recycling technologies and processes. Establishing a comprehensive material efficiency strategy that emphasizes the use of recycled materials, waste reduction, and circular economy principles can lead to both environmental benefits and cost savings. Additionally, fostering collaboration with suppliers to source eco-friendly materials and implementing employee training programs to enhance awareness and skills related to material efficiency practices should be integral components of the managerial approach. By integrating material efficiency and recycling into the core business strategy, managers can not only enhance sustainability but also establish a good edge in the active realm of sustainable manufacturing. Recognizing the prominence of smart manufacturing and the IoT as crucial factors influencing innovative technologies in sustainable manufacturing, managers should adopt a forward-thinking approach to leverage these technologies effectively. It is imperative for organizations to invest in IoT-enabled devices and smart manufacturing systems that enhance real-time data collection, analysis, and communication throughout the manufacturing processes. Additionally, fostering a culture of innovation and digital literacy among the workforce is essential. Managers should prioritize interdisciplinary collaboration, facilitating the integration of IoT technologies seamlessly into existing manufacturing operations. This enhances operational efficiency while also paving the way for predictive maintenance, resource optimization, and reduced environmental impact. Embracing smart manufacturing and IoT technologies aligns with sustainability objectives, enhances competitiveness, and positions the organization as a leader in the changing environment of sustainable manufacturing.

While acknowledging certain limitations, it is essential to highlight that this study is specifically conducted within the Indian context. Nevertheless, the procedural stages outlined in this research can serve as a valuable framework that may be useful to other emerging nation situations. By adopting the projected procedural approach, researchers in diverse emerging countries can either corroborate the factors identified in this study or uncover novel insights specific to their own unique circumstances. This adaptability underscores the potential for widespread applicability, allowing for a comprehensive exploration of the factors influencing innovative technologies in sustainable manufacturing across a range of developing countries. In the future, there is an opportunity to utilize several multi-criteria decision-making (MCDM) tools, such as fuzzy cognitive maps (FCM), analytic network process (ANP), analytic hierarchy process (AHP), ISM, and the technique for order preference similarity to the ideal solution (TOPSIS), to evaluate the factors affecting innovative technologies in sustainable manufacturing. The utilization of these alternative MCDM tools holds promise for addressing the limitations mentioned in this study and may yield robust results. Employing a diverse set of MCDM methodologies in future research endeavors can enhance the comprehensiveness and accuracy of evaluations, offering key insights into the complex dynamics that shape the landscape of innovative technologies in sustainable manufacturing.

REFERENCES

Agrawal, R. "Application of Total Interpretive Structural Modelling (TISM) for Analysis of Factors Influencing Sustainable Additive Manufacturing: A Case Study." *Rapid Prototyping Journal* 25, no. 7 (2019): 1198–1223.

Al-Fuqaha, A., Guizani, M., Mohammadi, M., Aledhari, M., & Ayyash, M. "Internet of Things in Industries: A Survey." *IEEE Transactions on Industrial Informatics* 14, no. 4 (2018): 2239–2246.

Al-Jumaily, A., Al-Jumaily, H., & Abdelnour-Nocera, J. "Automation and Robotics in Sustainable Food Processing: A Review." *Food and Bioprocess Technology* 12, no. 8 (2019): 1353–1372.

Bhanot, N., Venkateswara Rao, P., & Deshmukh, S. G. "An Integrated Approach for Analysing the Enablers and Barriers of Sustainable Manufacturing." *Journal of Cleaner Production* 142 (2017): 4412–4439.

Blecker, T., & Abdelkafi, N. "Innovations for Sustainable Manufacturing and Supply Chains: Connecting Collaborative Networks with Green Production and Consumption." Sustainability 7, no. 5 (2015): 6127–6140.

Chen, L., Wang, H., & Li, Q. "Integrating Renewable Energy Sources for Sustainable Manufacturing: A Case Study Approach." *International Journal of Green Energy* 17, no. 1 (2020a): 48–58.

Chen, S-H. "The Influencing Factors of Enterprise Sustainable Innovation: An Empirical Study." *Sustainability* 8, no. 5 (2016): 425.

Chen, T. L., Kim, H., Pan, S. Y., Tseng, P. C., Lin, Y. P., & Chiang, P. C. "Implementation of Green Chemistry Principles in Circular Economy System Towards Sustainable Development Goals: Challenges and Perspectives." *Science of the Total Environment* 716 (2020b): 136998.

Chesbrough, H. "Open Innovation: The Next Decade." Research Policy 40, no. 8 (2011): 117–119.

Dornfeld, D., & Min, S. "Sustainable Manufacturing: Trends and Research Challenges." *Proceedings of the Institution of Mechanical Engineers, Part B: Journal of Engineering Manufacture* 225 no. 10 (2011): 1939–1949.

Dwivedi, A., Sassanelli, C., Agrawal, D., Gonzalez, E. S., & D'Adamo, I. "Technological Innovation Toward Sustainability in Manufacturing Organizations: A Circular Economy Perspective." *Sustainable Chemistry and Pharmacy* 35 (2023): 101211.

El-Gayar, O., & Fritz, B. D. "Environmental Management Information Systems (EMIS) for Sustainable Development: A Conceptual Overview." *Communications of the Association for Information Systems* 17, no. 1 (2006): 34.

Ellen MacArthur Foundation. "Circular Economy: An Integrated Evaluation of Resource Efficiency and Economic Sustainability in Manufacturing." *Journal of Industrial Ecology* 22, no. 3 (2018): 400–412.

Fu, Y., Kok, R. A. W., Dankbaar, B., Ligthart, P. E. M., & van Riel, A. C. R. "Factors Affecting Sustainable Process Technology Adoption: A Systematic Literature Review." *Journal of Cleaner Production* 205 (2018): 226–251.

Gabus, A., & Fontela, E. J. B. G. R. C. "World problems, an invitation to further thought within the framework of DEMATEL." *Battelle Geneva Research Center*, Geneva, Switzerland 1 no. 8 (1972): 12–14.

Garetti, M., & Taisch, M. "Sustainable Manufacturing: Trends and Research Challenges." *Production Planning & Control* 23, no. 2–3 (2012): 83–104.

Ghisellini, P., Cialani, C., & Ulgiati, S. "Circular Economy in Manufacturing: A Comprehensive Review." *Resources, Conservation and Recycling* 114, (2016): 11–32.

Gupta, M., Kumar, A., & Tiwari, M. "A Framework for Assessing the Factors Influencing Renewable Energy Adoption in Manufacturing Industries." Energy 214, (2021): 118944.

Hoekstra, A. Y., Mekonnen, M. M., & Chapagain, A. K. "Water Efficiency in Manufacturing: Best Practices and Case Studies." Water Efficiency in Manufacturing 25, no. 3 (2011): 275–290.

Huizingh, E. K. R. E. "Open innovation: State of the art and future perspectives. *Technovation*, 31, no. 1 (2011): 2–9.

Jasiulewicz-Kaczmarek, M. "Identification of maintenance factors influencing the development of sustainable production processes–a pilot study." In *IOP Conference Series: Materials Science and Engineering*, vol. 400, p. 062014. IOP Publishing, 2018.

Jawahir, I. S., Badurdeen, F., & Rouch, K. E. "Innovation in sustainable manufacturing education." *GCSM* (2013): 9–16.

Kargari, A., & Tabatabai, S. A. "Water Reuse and Recycling in Industrial Wastewater Treatment: A Review." *Journal of Environmental Management 217,* (2018): 341–354.

Khurana, S., Haleem, A., Luthra, S., & Mannan, B. "Evaluating Critical Factors to Implement Sustainable Oriented Innovation Practices: An Analysis of Micro, Small, and Medium Manufacturing Enterprises." *Journal of Cleaner Production* 285 (2021): 125377.

Kim, S., Lee, J., & Choi, Y. "Advances in Material Efficiency and Recycling Technologies for Sustainable Manufacturing." Resources, Conservation and Recycling 119, (2017): 78–85.

Kim, Y., & Davis, D. F. "The Impact of Supply Chain Transparency on Consumer Awareness and Business Performance." *International Journal of Production Economics 180,* (2016): 525–534.

Kirchherr, J., Reike, D., & Hekkert, M. "Circular Economy in Manufacturing: A Review." *Journal of Cleaner Production* 115 (2016): 11–32.

Kralisch, D., Ott, D., & Gericke, D. "Rules and Benefits of Life Cycle Assessment in Green Chemical Process and Synthesis Design: A Tutorial Review." *Green Chemistry* 17, no. 1 (2015): 123–145.

Kuo, Y.-C., Wu, Y.-M., & Liu, Y.-X. "Identifying Key Factors for Sustainable Manufacturing and Development." *Review of Integrative Business and Economics Research* 11, no. 1 (2022): 30–50.

Kyratsis, P., Chryssolouris, G., & Mavrikios, D. "Advanced Robotics in Manufacturing: A Review." *Journal of Manufacturing Systems 48,* (2018): 52–60.

Lee, J., Bagheri, B., & Kao, H. A. "Cyber-physical Systems and Manufacturing Revitalization: A Review." *Journal of Manufacturing Systems* 37 (2015): 517–527.

Llopis, O., Bernabeu, R., & Peris-Ortiz, M. "Knowledge Sharing and Sustainable Innovation: A Systematic Literature Review." *Sustainability* 12, no. 20 (2020).

Marimuthu, R., Sankaranarayanan, B., Ali, S. M., de Sousa Jabbour, A. B. L., & Karuppiah, K. "Assessment of Key Socio-Economic and Environmental Challenges in the Mining Industry: Implications for Resource Policies in Emerging Economies." *Sustainable Production and Consumption* 27 (2021): 814–830.

Mazzanti, M., & Zoboli, R. "Examining the Factors Influencing Environmental Innovations. *Ecological Economics* 57, no. 4 (2006): 731–746.

Melnyk, S. A., Closs, D. J., & Calantone, R. "Closing the Loop on the Circular Economy: Barriers to Adoption of End-of-Life Strategies in Manufacturing Firms." *Production and Operations Management* 28, no. 7 (2019): 1735–1752.

Menghi, R., Papetti, A., Germani, M., & Marconi, M. "Energy Efficiency of Manufacturing Systems: A Review of Energy Assessment Methods and Tools." *Journal of Cleaner Production* 240 (2019): 118276.

Monostori, L., Kádár, B., Bauernhansl, T., Kondoh, S., & Kumara, S. "Smart Manufacturing: Characteristics, Technologies and Enabling Factors." *Procedia CIRP* 63 (2017): 217–222.

Mourtzis, D., Doukas, M., & Bernidaki, M. "Internet of Things (IoT) in Manufacturing: A Comprehensive Review." Procedia CIRP, 72 (2018): 442–447.

Ngibe, M., & Lekhanya, L. M. "Environmental Factors Affecting Innovative Leadership Towards Sustainable Growth of Manufacturing Small and Medium Enterprises." *Academy of Entrepreneurship Journal* 26, no. 4 (2020): 1–18.

Niederprüm, D. "Factors Affecting the Adoption of Sustainable Process Technologies: The Manufacturing Industries in the German Context *Journal of Cleaner Production*, 197 (2018): 1328–1337.

Notter, D. A., Gauch, M., & Widmer, R. "Life Cycle Assessment of a Lithium-Ion Battery Vehicle Pack." Environmental Science & Technology 44, no. 17 (2010): 6550–6556.

Panghal, A., P, A., Vern, P., & Mor, R. S. "Adoption Barriers to Green Logistics in the Indian Food Industry: A Circular Economy Perspective." *International Social Science Journal* 73 no. 247–248 (2023): 133–149.

Pouresmaieli, M., Ataei, M., Qarahasanlou, A. N., & Barabadi, A. "Integration of Renewable Energy and Sustainable Development with Strategic Planning in the Mining Industry." *Results in Engineering* 20 (2023): 101412.

Ribeiro, R. A., Ferreira, J. M., Barbosa-Póvoa, A. P., & Novais, A. Q. "Sustainable Manufacturing: Challenges and Opportunities." Procedia CIRP 17 (2014): 1–6.

Ritala, P., & Huotari, P. "Enhancing Sustainability through Collaborative Innovation: A Systematic Literature Review." *Sustainability* 13, no. 5 (2021): 2474.

Sarkis, J., Zhu, Q., & Lai, K. H. "Circular Economy: Closing the Loop on Sustainable Supply Chain Management." *International Journal of Production Economics* 130, no. 1 (2011): 1–15.

Sarkis, J., Zhu, Q., & Lai, K. H. "Sustainability in Manufacturing and Design: Principles, Practices, and Future Directions." *Journal of Manufacturing Systems* 32, no. 1 (2013): 1–19.

Shahzad, M., Qu, Y., Rehman, S. U., & Zafar, A. U. "Adoption of Green Innovation Technology to Accelerate Sustainable Development among Manufacturing Industry." *Journal of Innovation & Knowledge* 7, no. 4 (2022): 100231.

Song, J. Y., Lim, H. J., & Kang, D. H. "Life Cycle Assessment of Greenhouse Gas Emissions from Natural Gas Hydrate Production and Utilization for Power Generation." Energies 12, no. 3 (2019): 446.

Suresh, M., & Dharunanand, R. "Factors Influencing Sustainable Maintenance in Manufacturing Industries." *Journal of Quality in Maintenance Engineering* 29, no. 1 (2023): 94–113.

Tao, F., Zhang, H., Liu, A., & Nee, A. Y. C. "Smart Manufacturing: Characteristics, Technologies, and Enablers." Procedia CIRP 78 (2018): 4–9.

Tao, F., Zhang, H., Liu, A., & Nee, A. Y. C. "The Internet of Things in Manufacturing: Key Issues and Potential Applications." *Industrial Informatics, IEEE Transactions on* 10, no. 2 (2014): 799–809.

Ullah, S., Khan, F. U., & Ahmad, N. "Promoting Sustainability Through Green Innovation Adoption: A Case of Manufacturing Industry." *Environmental Science and Pollution Research* 29 (2022): 1–21.

Wang, H., Zhang, Y., & Zhang, Y. "Enhancing Supply Chain Transparency: A Blockchain and IoT Integration Approach." Sustainability 11, no. 17 (2019): 4639.

World Economic Forum. "Industry 4.0: The Future of Productivity and Growth in Manufacturing Industries." *World Economic Forum White Paper* (2017).

Wuest, T., Weimer, D., & Irgens, C. "Transition towards Sustainable Manufacturing: Key Challenges and Opportunities." Procedia CIRP 37, (2015): 849–854.

Yazdanpanah, M., Hayati, D., Hochrainer-Stigler, S., & Zamani, G. H. "Understanding Farmers' Intention and Behavior Regarding Water Conservation in the Middle-East and North Africa: A Case Study in Iran." *Journal of Environmental Management* 135 (2014): 63–72.

Yin, S., Zhang, N., & Li, B. "Enhancing the Competitiveness of Multi-Agent Cooperation for Green Manufacturing in China: An Empirical Study of the Measure of Green Technology Innovation Capabilities and Their Influencing Factors." *Sustainable Production and Consumption* 23 (2020): 63–76.

Zeng, S., Xie, X., & Tam, C. M. "Open Innovation in Sustainable Manufacturing: A Systematic Review." *Sustainability* 13, no. 8 (2021): 4181.

Zhang, H., Qu, T., & Gregory, M. "Sustainable Manufacturing and Design: Concepts, Practices and Needs." Procedia CIRP 60 (2017): 165–170.

Zhang, J., Wei, J., & Liu, L. "Collaborative Innovation and Sustainable Development of Manufacturing Industry." *Journal of Cleaner Production* 287 (2021): 125238.

Zhang, L., Wu, Q., & Chen, Y. "Renewable Energy Integration in Manufacturing: Challenges and Opportunities." *Renewable and Sustainable Energy Reviews* 82 (2018): 3963–3974.

APPENDIX A

A series of questionnaires were developed to survey experts from three distinct sustainable manufacturing units, aiming to evaluate the factors influencing innovative technologies in sustainable manufacturing. Experts were selected from various departments, possessing significant technical knowledge and extensive experience in sustainable field. Tables A1 and A2 provide the questionnaire utilized to gather experts' opinions and the profiles of the identified experts. The collected survey data was exclusively utilized for research purposes and was not shared with any external entities. Experts were asked to rate the factors using the linguistic scale values outlined in Table A3.

Kindly respond to the following inquiries

a) Name: b) Role: c) Experience (in years):

We value your perspective on the selection of crucial factors in sustainable manufacturing. Kindly provide your opinion on the factors listed below.

Table A1 Survey employed to gather expert perspectives

S.no	*Factors*	*Yes/no*
1.	Energy efficiency (F1)	
2.	Renewable energy integration (F2)	
3.	Material efficiency and recycling (F3)	
4.	Smart manufacturing and IoT (F4)	
5.	Advanced robotics and automation (F5)	
6.	Water conservation and management (F6)	
7.	Supply chain transparency (F7)	
8.	Life cycle assessment (LCA) (F8)	
9.	Collaborative and open innovation (F9)	
10.	Regulatory compliance and standards (F10)	

Table A2 Experts' profiles

Experts	*Domain*	*Designation*	*Experience (years)*
1.	Industry executives and leaders	Chief Operations Officer (COO)	30
2.	Researchers and academics	Principal investigator	26
3.	Government regulatory bodies	Regulatory affairs director	33
4.	Environmental and sustainability advocates	Environmental policy director	28
5.	Technology providers	Head of research and development	25
6.	Supply chain partner	Logistics manager	17
7.	Employees and labor unions	Labor union president	14
8.	Investors and financial institutions	Financial officer	16
9.	Consumers	Consumer advocate	12
10.	Educational institutions	Dean of the school	18

Table A3 Fuzzy linguistic scale used in this study

Five-point score for decision makers' preference	*Linguistic constants and their description*	*Trapezoidal fuzzy numbers (TrapFN) for equivalent scores*
0	No influence	(0, 0, 0.1, 0.2)
1	Very low influence	(0.1, 0.2, 0.3, 0.4)
2	Low influence	(0.3, 0.4, 0.5, 0.6)
3	High influence	(0.5, 0.6, 0.7, 0.8)
4	Very high influence	(0.7, 0.8, 0.9, 1)

APPENDIX B

The experts have furnished fuzzy assessment data aimed at comparing crucial factors associated with innovative technologies in sustainable manufacturing. Utilizing this data, several matrices have been computed, encompassing the fuzzy initial direct relationship matrix, the normalized direct relationship matrix, and the total relationship matrix. These matrices can be referenced in Appendix Tables B1 to B3.

Table B1 Fuzzy initial direct relation matrix (FDRM)

	F1	F2	F3	F4	F5	F6	F7	F8	F9	F10
F1	0	3.1	2.6	2.6	1.8	1.8	2	1.7	1.5	2.4
F2	1	0	2.9	2.3	1.8	1.6	1.1	1.7	1	1.9
F3	1.5	1.2	0	2	2	2.1	2	1.9	2.1	2.9
F4	1.5	1.8	2.1	0	2.8	2.1	2	2.9	2.9	3
F5	2.2	2.2	2.1	1.2	0	1.8	2	2	2	2.9
F6	2.2	2.3	1.9	1.6	1.7	0	2	2	1	1.9
F7	2.2	1.9	2	1.7	2	2	0	1.8	1.9	2.4
F8	2.2	1.9	2.1	1.2	2.1	2.1	1.7	0	1.2	2.1
F9	2.3	1.8	2.2	1.4	2	1.6	2	1.9	0	2.8
F10	1.4	1.7	1.5	1.2	1.2	1.9	1.2	1.7	1	0

Table B2 Normalized direct relation matrix (NDRM)

	F1	F2	F3	F4	F5	F6	F7	F8	F9	F10
F1	0.0000	0.1390	0.1166	0.1166	0.0807	0.0807	0.0897	0.0762	0.0673	0.1076
F2	0.0448	0.0000	0.1300	0.1031	0.0807	0.0717	0.0493	0.0762	0.0448	0.0852
F3	0.0673	0.0538	0.0000	0.0897	0.0897	0.0942	0.0897	0.0852	0.0942	0.1300
F4	0.0673	0.0807	0.0942	0.0000	0.1256	0.0942	0.0897	0.1300	0.1300	0.1345
F5	0.0987	0.0987	0.0942	0.0538	0.0000	0.0807	0.0897	0.0897	0.0897	0.1300
F6	0.0987	0.1031	0.0852	0.0717	0.0762	0.0000	0.0897	0.0897	0.0448	0.0852
F7	0.0987	0.0852	0.0897	0.0762	0.0897	0.0897	0.0000	0.0807	0.0852	0.1076
F8	0.0987	0.0852	0.0942	0.0538	0.0942	0.0942	0.0762	0.0000	0.0538	0.0942
F9	0.1031	0.0807	0.0987	0.0628	0.0897	0.0717	0.0897	0.0852	0.0000	0.1256
F10	0.0628	0.0762	0.0673	0.0538	0.0538	0.0852	0.0538	0.0762	0.0448	0.0000

Table B3 Total relation matrix (TRM)

	F1	**F2**	**F3**	**F4**	**F5**	**F6**	**F7**	**F8**	**F9**	**F10**
F1	0.2832	0.4323	0.4392	0.3738	0.3751	0.3726	0.3593	0.3763	0.3183	0.4769
F2	0.2714	0.2453	0.3829	0.3074	0.3158	0.3064	0.2706	0.3164	0.2495	0.3839
F3	0.3245	0.3329	0.3015	0.3232	0.3547	0.3575	0.3356	0.3563	0.3177	0.4622
F4	0.3709	0.4048	0.4413	0.2824	0.4329	0.4047	0.3798	0.4425	0.3882	0.5264
F5	0.3576	0.3802	0.3984	0.3020	0.2792	0.3535	0.3419	0.3665	0.3185	0.4711
F6	0.3354	0.3611	0.3665	0.2983	0.3288	0.2566	0.3212	0.3444	0.2619	0.4044
F7	0.3539	0.3646	0.3898	0.3173	0.3583	0.3570	0.2565	0.3556	0.3125	0.4475
F8	0.3354	0.3452	0.3725	0.2815	0.3424	0.3418	0.3095	0.2608	0.2684	0.4112
F9	0.3576	0.3606	0.3972	0.3058	0.3578	0.3419	0.3386	0.3588	0.2337	0.4628
F10	0.2515	0.2796	0.2879	0.2316	0.2531	0.2793	0.2384	0.2759	0.2133	0.2546

Chapter 10

International standards and cross-cultural communication in manufacturing

Priyanka Ramesh and Prabhakaran Jaganathan

10.1 INTRODUCTION

In today's dynamic manufacturing landscape, the importance of standardized practices cannot be overstated, especially in light of the increasing interconnectedness of global markets (Das & Dey, 2021). As organizations traverse these global markets, adherence to international standards becomes imperative, serving as a cornerstone for ensuring product quality, sustainability, and seamless integration on a global scale (Ninduwezuor-Ehiobu et al., 2023). This introduction sets the stage for our exploration of the intricate interplay between international standards, cross-cultural communication, and organizational dynamics in the manufacturing sector.

Within this context, green manufacturing emerges as a critical focal point, presenting both challenges and opportunities for organizations aiming to adopt environmentally sustainable practices. By delving into the challenges and applications of green manufacturing, we underscore the pivotal role of international standards in guiding organizations toward environmentally responsible practices while maintaining competitiveness in the global marketplace (Appolloni et al., 2022). Amidst the globalized landscape, the complexities of cross-cultural communication within manufacturing settings cannot be overlooked. Cultural nuances significantly impact decision-making processes, collaborative endeavors, and operational efficiency (Nguyen et al., 2021). Through a comprehensive examination of these complexities, we aim to shed light on the challenges and opportunities inherent in cross-cultural communication, offering insights into effective strategies for fostering inclusivity and maximizing productivity.

The contemporary manufacturing landscape is marked by a multifaceted interplay of factors, ranging from rapid technological advancements and expansive globalization to the rich tapestry of cultural diversity (Ashima et al., 2021). Within this dynamic milieu, the convergence of international standards and cross-cultural communication emerges as a pivotal determinant of organizational triumph and competitive prowess. As enterprises traverse the intricate pathways of global markets and navigate the diverse cultural landscapes therein, a nuanced comprehension of the intricate dynamics

DOI: 10.1201/9781003470342-10

interlinking these factors becomes indispensable for fostering sustainable and high-caliber manufacturing processes (Maddikunta et al., 2022).

10.1.1 The significance of international standards

International standards serve as bedrock frameworks that not only guide organizational practices but also ensure uniformity, dependability, and seamless interoperability across industries and geographical frontiers (Martínez-García et al., 2021). Particularly for manufacturing entities, adherence to globally recognized standards such as ISO 9001 for quality management and ISO 14001 for environmental stewardship transcends mere regulatory compliance—it assumes the mantle of a strategic imperative (Kawalkar et al., 2022). Through steadfast alignment with these internationally acclaimed benchmarks, organizations stand poised to augment product quality, streamline operational processes, mitigate inherent risks, and carve a distinctive niche within the cutthroat arena of the global marketplace.

10.1.2 Navigating cross-cultural communication: challenges and opportunities

In the domain of manufacturing, effective cross-cultural communication emerges as a linchpin for successfully traversing the labyrinthine corridors of multicultural work environments. It serves as the lifeblood for fostering harmonious collaboration and propelling operational efficiency to unprecedented heights (Aririguzoh, 2022). However, the confluence of disparate cultural norms, communication styles, and value systems poses formidable hurdles, often precipitating misunderstandings, intercultural frictions, and suboptimal outcomes (Obradović et al., 2021). Consequently, organizations are compelled to cultivate a robust arsenal of cultural intelligence and communication strategies that espouse inclusivity, engender mutual understanding, and underscore reverence for diversity (Kim & Williams, 2022). By embracing the rich tapestry of cultural diversity and harnessing it as a wellspring of innovation and creativity, manufacturing enterprises can unlock the latent potential of their workforce and orchestrate a symphony of heightened performance across all echelons of operation.

10.1.2.1 Research questions and objectives

Against this backdrop, this study seeks to address the following research questions:

RQ1. How do international standards influence organizational practices and performance outcomes in the manufacturing sector?
RQ2. What are the key challenges and opportunities associated with cross-cultural communication in manufacturing settings?

RQ3. How do cultural dimensions, language proficiency, use of automated manufacturing technologies, product quality metrics, and adherence to international standards collectively shape organizational performance in manufacturing?

Aligned with these research questions, the objectives are as follows:

RO1. To examine the role of international standards in shaping manufacturing practices and performance metrics.
RO2. To identify the challenges and opportunities associated with cross-cultural communication in the manufacturing context.
RO3. To analyze the impact of cultural dimensions, language proficiency, use of automated manufacturing technologies, and product quality metrics on organizational performance.
RO4. To evaluate how adherence to international standards influences organizational effectiveness, market competitiveness, and sustainability impact in manufacturing.

To achieve these objectives, we adopt a mixed-methods approach, combining fuzzy-set qualitative Comparative analysis (fsQCA) and artificial neural network (ANN) analysis. Through fsQCA, we aim to explore the complex causal configurations underlying the relationship between international standards, cross-cultural communication, and organizational performance in manufacturing. Additionally, ANN analysis enables us to model and predict the predictive power of various input variables on organizational dynamics.

10.2 LITERATURE REVIEW

The contemporary manufacturing landscape operates within a complex framework shaped by various factors, including rapid technological advancements, globalization, and cultural diversity. This literature review delves into seminal works and recent research findings to elucidate the multifaceted dynamics inherent in this domain, incorporating all variables outlined in Table 10.1 to provide a comprehensive understanding.

10.2.1 International standards in manufacturing

The significance of international standards in the manufacturing sector cannot be overstated. These standards serve as foundational frameworks that guide organizational practices, ensuring consistency, reliability, and compatibility across diverse industries and geographical boundaries. Key standards, such as ISO 9001 for quality management and ISO 14001 for environmental stewardship, are pivotal in shaping manufacturing operations. They go beyond mere regulatory compliance and assume strategic importance in driving organizational excellence (Shao et al., 2023).

Table 10.1 Framework of variables for analyzing organizational dynamics in manufacturing

Variable 1—Cultural dimensions			
Item 1	CD1	Organizational values regarding quality and sustainability	Margherita and Braccini (2023)
Item 2	CD2	Cultural diversity in the workforce	Lee et al. (2021)
Item 3	CD3	Attitudes toward change and innovation	Gajdzik and Wolniak (2022)
Item 4	CD4	Leadership style and approach to standardization	Tortorella et al. (2021)
Variable 2—Language proficiency across teams and stakeholders			
Item 6	LP1	Use of interpreters or translation services	Javaid et al. (2021)
Item 7	LP2	Cross-cultural training programs for employees	Dalal and Akdere (2023)
Item 8	LP3	Communication channels (formal vs. informal)	Chaudhry et al. (2021)
Item 9	LP4	Frequency of cross-cultural communication issues	Al-Hakimi et al. (2022)
Variable 3—Use of automated manufacturing technologies			
Item 10	AT1	Degree of supply chain integration across cultures	Papulová et al. (2022)
Item 11	AT2	Implementation of lean manufacturing principles	Rizvi et al. (2021)
Item 12	AT3	Degree of flexibility in manufacturing processes	Upadhyay et al. (2023)
Item 13	AT4	Adoption of sustainable manufacturing practices	Jamwal et al. (2021)
Variable 4—Product quality metrics			
Item 15	PQ1	Manufacturing efficiency (throughput, cycle time)	Dogan and Birant (2021)
Item 16	PQ2	Customer satisfaction scores across cultures	Peng et al. (2022)
Item 17	PQ3	Employee satisfaction and engagement levels	Alam et al. (2024)
Item 18	PQ4	Financial performance indicators (profit margins, return on investment)	Dogan and Birant (2021)
Variable 5—International standards			
Item 18	IS1	ISO 9001 certification	Martínez-García et al. (2021)
Item 19	IS2	Compliance with ISO 14001 environmental management standards	Kawalkar et al. (2022)
Item 20	IS3	Adoption of industry-specific international standards	Shao et al. (2023)
Item 21	IS4	Frequency of international standards updates	Yusr et al. (2020)
Item 22	IS5	Degree of alignment with international quality standards	Awasthi et al. (2020)
Output variable 8—Organizational performance			
Item 30	OP1	Overall performance effectiveness	
Item 31	OP2	Market competitiveness	
Item 32	OP3	Sustainability impact	

Adherence to international standards offers a multitude of benefits to manufacturing entities. Research by Amutabi (2024) has demonstrated that organizations complying with ISO standards experience enhanced operational efficiency. By establishing standardized processes and procedures, these standards facilitate smoother workflows, reduce errors, and optimize resource utilization. Consequently, organizations can achieve higher levels of productivity and cost-effectiveness.

Moreover, adherence to international standards is closely linked to improved product quality. ISO 9001 certification, in particular, emphasizes quality management systems that ensure consistency and reliability in product delivery. By adhering to rigorous quality standards, manufacturers can mitigate defects, minimize rework, and enhance overall product performance. This, in turn, leads to increased customer satisfaction and loyalty, as customers perceive higher value in products that meet stringent quality criteria (Ahmed et al., 2022).

In addition to operational efficiency and product quality, organizations certified under ISO standards demonstrate greater resilience in volatile market conditions. Yusr et al. (2020) highlight the role of robust quality management systems in providing organizations with the agility to adapt to changing market dynamics. By adhering to internationally recognized standards, organizations can anticipate market shifts, identify potential risks, and implement proactive measures to mitigate disruptions. This proactive approach not only safeguards organizational reputation but also enhances market competitiveness and sustains long-term profitability. International standards play a transformative role in shaping the manufacturing landscape (Awasthi et al., 2020). They empower organizations to achieve operational excellence, deliver superior products, and navigate the complexities of global markets with confidence. As manufacturing continues to evolve in an increasingly interconnected world, adherence to international standards will remain essential for driving innovation, fostering sustainability, and maintaining a competitive edge in the global marketplace.

10.2.2 Cross-cultural communication challenges and opportunities

In the dynamic and globalized landscape of manufacturing, effective cross-cultural communication is not just a desirable skill but a fundamental necessity. The sector operates within multicultural work environments and intricate global supply chains, where cultural nuances, communication styles, and value systems significantly influence organizational dynamics. As noted by Veile et al. (2020), these factors often pose formidable challenges, including misinterpretations, conflicts, and barriers to collaboration. In multicultural teams, diverse cultural backgrounds can lead to misunderstandings, hindering productivity and stifling innovation.

Yet, within these challenges lie opportunities for growth and enhancement. Jaeger and Upadhyay (2020) emphasize that adept navigation of cross-cultural communication challenges can foster innovation by leveraging diverse perspectives. Different cultural viewpoints bring unique insights and approaches to problem-solving, leading to creative solutions and novel approaches to manufacturing processes. By embracing and capitalizing on cultural diversity, manufacturing enterprises can cultivate an environment of inclusivity and openness, where employees feel valued and empowered to contribute their perspectives. Moreover, effective cross-cultural communication enhances operational agility within manufacturing organizations. In today's fast-paced and interconnected world, the ability to adapt quickly to changing market conditions is essential for maintaining competitiveness. Bonsu and Twum-Danso (2018) highlight that organizations adept at navigating cultural diversity can respond more effectively to market shifts, customer demands, and emerging trends. By fostering an environment where communication flows freely across cultural boundaries, manufacturing enterprises can improve decision-making processes, streamline operations, and seize opportunities for growth.

Embracing cultural diversity as a catalyst for creativity and resilience can drive organizational success. Khoo et al. (2018) argue that diverse teams bring a wealth of experiences, perspectives, and talents to the table, enabling organizations to innovate and excel in a rapidly evolving marketplace. By valuing and celebrating cultural differences, manufacturing enterprises can harness the full potential of their workforce, fostering a culture of collaboration, respect, and mutual understanding.

The integration of automated manufacturing technologies (AMTs) in the manufacturing sector has transformed traditional production processes, offering new opportunities and challenges for organizations.

10.2.3 Use of automated manufacturing technologies (AMTs)

The adoption of AMTs has become increasingly prevalent in manufacturing industries worldwide. These technologies encompass a range of automated systems and processes, including robotics, artificial intelligence, and computer-controlled machinery. Research by Javaid et al. (2021) highlights the transformative impact of AMTs on production efficiency, quality, and flexibility. By automating repetitive tasks and streamlining production workflows, AMTs enable organizations to achieve higher levels of productivity and cost-effectiveness.

Supply chain integration involves the seamless coordination of activities across different stages of the supply chain, from raw material sourcing to final product delivery. In a multicultural manufacturing environment, achieving supply chain integration across cultures presents unique challenges related to language barriers, cultural differences, and varying business practices. However, research by Ni and Sun (2019) suggests that organizations leveraging AMTs can overcome these challenges by facilitating real-time communication, data sharing, and collaboration among supply chain partners. By integrating AMTs into their supply chain management systems, organizations can enhance transparency, efficiency, and responsiveness, regardless of cultural differences.

Lean manufacturing principles emphasize the elimination of waste, continuous improvement, and value creation in manufacturing processes. The integration of AMTs aligns closely with lean principles by enabling organizations to streamline production workflows, minimize downtime, and optimize resource utilization. Research by Kafuku (2019) demonstrates that organizations adopting lean manufacturing principles alongside AMTs experience significant improvements in operational efficiency, product quality, and customer satisfaction. By embracing lean manufacturing practices, organizations can leverage AMTs to drive continuous improvement and competitive advantage in multicultural manufacturing settings (Prasad et al., 2020). Flexibility in manufacturing processes is essential for organizations to respond quickly to changing market demands, customer preferences, and unforeseen disruptions. AMTs offer inherent flexibility by allowing for rapid reconfiguration of production lines, adaptive scheduling, and agile manufacturing practices. Research by Jimeno-Morenilla et al. (2021) suggests that organizations leveraging AMTs can achieve higher levels of flexibility in manufacturing processes, enabling them to adapt to diverse cultural contexts, market conditions, and production requirements. By integrating flexible AMTs into their operations, organizations can enhance their ability to meet customer needs, optimize resource allocation, and maintain competitiveness in multicultural environments.

Sustainable manufacturing practices aim to minimize environmental impact, conserve resources, and promote social responsibility throughout the production lifecycle. The integration of AMTs presents opportunities for organizations to adopt sustainable manufacturing practices by reducing energy consumption, waste generation, and emissions. Research by Jamwal et al. (2021) highlights the role of AMTs in enabling organizations to achieve sustainability goals through improved resource efficiency, product recyclability, and eco-friendly production processes (Ali & Johl, 2023).

10.3 RESEARCH METHODOLOGY

10.3.1 Data collection process

For this study, data collection and sampling were conducted systematically to ensure the reliability and validity of the findings. A mixed-methods approach was employed to gather comprehensive data on the variables under investigation. Firstly, quantitative data was collected through surveys distributed to 90 manufacturing organizations operating in diverse cultural contexts. The surveys were designed to capture information on the use of automated manufacturing technologies, degree of supply chain integration across cultures, implementation of lean manufacturing principles, degree of flexibility in manufacturing processes, and adoption of sustainable manufacturing practices. Sampling was carried out using stratified random sampling techniques to ensure representation from various industry sectors and geographical regions. Additionally, qualitative data was obtained through in-depth interviews with 20 key stakeholders, including manufacturing managers, supply chain professionals, and lean manufacturing experts. These interviews provided valuable insights into the nuances of cross-cultural communication and the challenges faced in implementing standardized practices across diverse cultural landscapes.

10.3.2 Data analysis

10.3.2.1 Reliability and validity test

Reliability and validity testing were conducted to ensure the robustness and accuracy of the measurement scales used in this study across various constructs, including cultural dimensions (CD), language proficiency across teams and stakeholders (LP), use of automated manufacturing technologies (AT), product quality (PQ) metrics, international standards (IS), and organizational performance (OP). Results indicate high internal consistency reliability (ICR) across all constructs, with values ranging from 0.721 to 0.956, exceeding the acceptable threshold. Additionally, average variance extracted (AVE) values were above 0.5 for all constructs, affirming satisfactory convergent validity, while composite reliability (CR) values surpassed 0.7, indicating robust construct reliability. These findings validate the measurement scales' accuracy in capturing the intended constructs, ensuring the reliability and validity of the data collected for analysis (Table 10.2).

10.3.2.2 Fuzzy-set qualitative comparative analysis (fsQCA)

fsQCA stands as a methodological approach utilized for analyzing intricate causal relationships among variables within qualitative research frameworks. Developed as an extension of conventional qualitative comparative analysis (QCA), fsQCA addresses the inherent ambiguity and vagueness often present in social science data by integrating fuzzy logic principles (Kumar et al., 2022). This method enables researchers to explore how combinations of variables, rather than individual variables in isolation, contribute to specific outcomes or conditions. fsQCA is well-suited for uncovering complex causal configurations and identifying necessary and sufficient conditions for desired outcomes within sets of cases or observations (Stekelorum et al., 2021) (Figure 10.1).

Table 10.2 Reliability and validity testing

Condition and outcome	*Abbreviation*	*Item combinations*	*Factor analysis*
Cultural dimensions	CD	CD1 to CD4	ICR = 0.780 AVE = 0.63456 CR = 0.92249
Language proficiency across teams and stakeholders	LP	LP1 to LP4	ICR = 0.776 AVE = 0.65434 CR = 0.92378
Use of automated manufacturing technologies	AT	AT1 to AT5	ICR = 0.821 AVE = 0.56734 CR = 0.97854
Product quality metrics	PQ	PQ1 to PQ3	ICR = 0.721 AVE = 0.62557 CR = 0.91287
International standards	IS	IS1 to IS5	ICR = 0.839 AVE = 0.56784 CR = 0.95475
Organizational performance	OP	OP1 to OP4	ICR = 0.956 AVE = 0.63851 CR = 0.92963

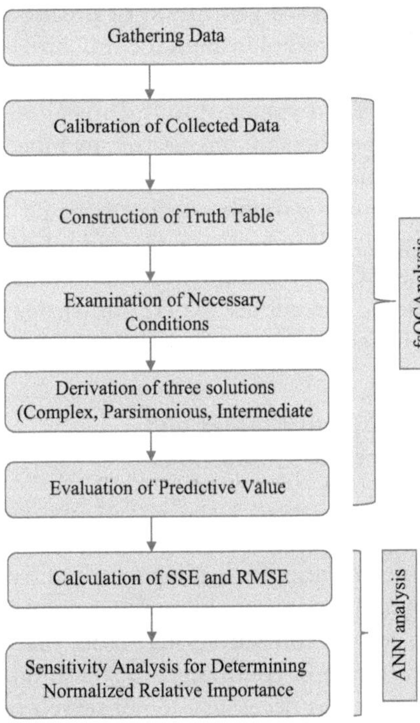

Figure 10.1 Framework of fsQCA-ANN analysis.

Prior to commencing the fsQCA process, the data gathered from 20 measurements is structured into 5 constructs. This is achieved by computing the mean of the measurements grouped under each construct. Subsequently, these constructs are slated for utilization within the fsQCA software. Once the mean values are determined, the data is exported from SPSS to the fsQCA software in the format of a.CSV file.

10.3.2.3 Data calibration

The calibration of data involves the systematic adjustment and standardization of variables to ensure comparability and accuracy in quantitative analysis. In the context of this study, data calibration encompasses refining and preprocessing collected data to enhance its quality and reliability before conducting subsequent analyses such as fsQCA and ANN analysis. Calibration methods may include data cleaning, normalization, transformation, and outlier detection. By calibrating the data, researchers can mitigate potential biases, enhance the robustness of their findings, and facilitate meaningful comparisons across variables and cases (Pappas & Woodside, 2021).

In the initial stage of the fsQCA method, the data undergoes calibration to convert it into fuzzy sets, which are depicted through binary values of 0s and 1s. Calibration stands as a pivotal step in fsQCA, facilitating the accurate interpretation of the data and the generation of precise results. To execute calibration, users establish threshold values for full, cross-over, and non-full membership, typically determined via percentiles based on the study's data. In this investigation, threshold values of four, three, and two were, respectively, assigned for full membership, cross-over membership, and non-full membership. The calibration process in fsQCA unfolds through specific steps: Analyze>Compute the variable>Assign a name to the target variable>Calibrate (x, n1, n2, n3), where n1, n2, and n3 denote the threshold values. It is notable that fsQCA provides diverse approaches for determining threshold values, including percentiles, expert knowledge, or predefined values, thereby allowing for enhanced representation of the data.

10.3.2.4 Construction of truth tables and analysis of necessary conditions

The construction of truth tables and analysis of necessary conditions are fundamental steps in the fsQCA methodology, aimed at systematically examining combinations of variables leading to specific outcomes or conditions of interest. Truth tables represent all possible combinations of selected variables and their respective configurations across cases. Through logical minimization techniques like Boolean algebra, truth tables are condensed to identify core configurations necessary for observed outcomes. Analysis of necessary conditions entails identifying minimal combinations of variables consistently associated with outcomes, elucidating essential pathways or conditions leading to desired results.

10.3.2.5 Artificial neural network (ANN) analysis

ANN analysis serves as a computational modeling technique inspired by biological neural networks, employed in this study to model and predict complex relationships between variables and outcomes within the manufacturing sector. ANNs consist of interconnected nodes (neurons) organized in layers, capable of learning patterns and relationships from data through iterative processes. By training ANN models on historical data, researchers can uncover hidden patterns, predict future trends, and simulate various scenarios to inform decision-making. ANN analysis complements fsQCA by providing a quantitative framework for understanding the intricate dynamics and predictive power of identified causal configurations. Through ANN analysis, researchers can gain deeper insights into drivers of OP, market competitiveness, and sustainability impact in manufacturing settings.

10.4 RESULTS FROM fsQCA-ANN ANALYSIS

10.4.1 fsQCA results

A truth table is constructed using calibrated data to represent the possible outcomes of a logical expression. Each variable is assigned binary values of 0 and 1, and the resulting value for the entire expression is determined. After analyzing the necessary conditions (Table 10.3), it is observed that no condition has a consistency value below 0.9, indicating a lack of sufficient influence on improved start-up performance. Therefore, sufficiency conditions are explored next. Three types of solutions—complex, parsimonious, and intermediate—are obtained by setting minimum case frequency and raw consistency benchmarks and conducting specific standard analysis. Complex solutions, involving a

Table 10.3 Examination of necessary conditions

	Organizational performance (OP)		*~ Organizational performance (~OP)*	
Conditions tested	*Consistency*	*Coverage*	*Consistency*	*Coverage*
CD	0.673287	0.584937	0.574839	0.659303
~CD	0.535459	0.583938	0.543399	0.540326
LP	0.643623	0.647383	0.658493	0.606980
~LP	0.564332	0.849300	0.746389	0.623593
AT	0.616757	0.657383	0.294755	0.501791
~AT	0.547367	0.763839	0.548403	0.634984
PQ	0.543679	0.336294	0.789322	0.655660
~PQ	0.642261	0.564893	0.748399	0.643017
IS	0.627552	0.743729	0.370102	0.456961
~IS	0.632246	0.843824	0.789608	0.657943

larger number of causal conditions, are often disregarded due to potential interpretational challenges. Constructs present in both intermediate and parsimonious solutions are identified as core, while those exclusive to the intermediate solution are considered peripheral. The analysis of these solutions is summarized in Table 10.4, showcasing various combinations of constructs to enhance start-up performance. Configurations are represented with black circles (●) indicating presence and crossed-out circles (⊗) indicating absence. Large circles represent key components, while small circles denote less significant elements. Blank spaces signify conditions that may be present or absent without impacting the outcome. The table provides set-theoretic consistency values exceeding 0.75 for each configuration and the overall solution, indicating robust approximations of subset relations. These results offer insights into potential strategies for improving start-up performance, achieving an overall consistency of 0.922182.

CD exhibit relatively high consistency values, ranging from 0.535 to 0.673, indicating a consistent influence on OP across various conditions tested. The coverage values for CD are also noteworthy, ranging from 0.584 to 0.659, suggesting that CD is prevalent across a significant proportion of cases. Conversely, Automated Manufacturing (AT) demonstrates lower consistency and coverage values compared to CD, indicating a comparatively weaker association with OP. These findings underscore the importance of CD in driving OP within the manufacturing sector, while also highlighting the nuanced relationships between different variables and organizational outcomes.

In the fsQCA results (Table 10.4), several configurations stand out for their ability to explain OP effectively. Configurations with consistency values exceeding 0.90, such as Configurations 2 and 3, are particularly notable for their robustness in explaining OP. Additionally, these configurations exhibit moderate to high raw and unique coverage values, indicating their ability to capture a diverse range of cases while maintaining internal consistency. The overall solution coverage (0.585) and consistency (0.922) further affirm the effectiveness of the identified configurations in explaining OP within the manufacturing context.

Regarding the SSE and RMSE values from the ANN analysis (Table 10.5), lower values indicate better model fit and predictive accuracy. Notably, ANN models 1, 3, and 7 demonstrate particularly low SSE and RMSE values, suggesting superior performance in predicting OP compared to other models. These findings highlight the effectiveness of certain ANN models in capturing the complex relationships between variables and predicting OP with a high degree of accuracy.

In the sensitivity analysis (Table 10.6), each variable's relative importance in predicting OP within the ANN models is assessed. Notably, language proficiency (LP) consistently emerges as crucial across all ANN models, with relative importance values approaching 1. This indicates

Table 10.4 fsQCA results

	Solution							
Configuration	1	2	3	4	5	6	7	8
Cultural dimension	⊗	⊗	⊗	●	⊗	●	●	⊗
Language proficiency	●	●	⊗	●	●	●	⊗	●
Automated manufacturing	●	●	⊗	⊗	●	●	●	●
Product quality metrics	⊗	●	⊗	●	●	⊗		●
International standards	⊗	⊗	●		⊗	⊗	⊗	●
Consistency	**0.93254**	**0.94373**	**0.95371**	**0.91274**	**0.92835**	**0.98613**	**0.91465**	**0.91626**
Raw coverage	**0.14566**	**0.12356**	**0.18693**	**0.17543**	**0.12694**	**0.16495**	**0.13864**	**0.09489**
Unique coverage	**0.04567**	**0.41368**	**0.54368**	**0.38178**	**0.11124**	**0.29652**	**0.24986**	**0.36891**

Overall solution coverage: 0.585565

Overall solution consistency: 0.922182

Table 10.5 SSE and RMSE values

Network	*SSE (training)*	*SSE (testing)*	*RMSE (training)*	*RMSE (testing)*
ANN1	8.546	1.341	0.464	0.457
ANN2	10.347	1.347	0.535	0.651
ANN3	5.218	1.472	0.671	0.431
ANN4	1.424	5.236	0.572	0.662
ANN5	5.859	2.845	0.356	0.542
ANN6	7.852	3.025	0.572	0.339
ANN7	6.436	4.34	0.389	0.561
ANN8	8.568	4.13	0.431	0.673
ANN9	6.856	3.577	0.596	0.445
ANN10	6.257	5.36	0.484	0.566
Mean	7.132	3.268	0.446	0.632
Std dev	2.4355	1.5632	0.1276	0.1240

that LP plays a pivotal role in predicting OP within the manufacturing sector. Conversely, CD display varying importance across different ANN models, underscoring its nuanced impact on OP prediction. Collectively, the sensitivity analysis highlights LP's dominant influence, constituting approximately 63% of the overall relative importance across all variables tested. This emphasizes LP's significance in shaping OP outcomes within the context of the study.

Table 10.6 Sensitivity analysis

	ANN1	*ANN2*	*ANN3*	*ANN4*	*ANN5*	*ANN6*	*ANN7*	*ANN8*	*ANN9*	*ANN10*	*Average relative importance*	*Normalized relative importance*
CD	0.32	0.39	0.08	0.45	0.19	0.29	0.57	0.11	0.30	0.55	0.32	33%
LP	0.95	1.00	1.00	0.92	1.00	1.00	0.93	1.00	1.00	1.00	0.98	100%
AT	1.00	0.13	0.58	0.72	0.55	0.32	1.00	0.53	0.41	0.92	0.62	63%
PQ	0.26	0.24	0.09	1.00	0.22	0.21	0.72	0.36	0.24	0.33	0.37	38%
IS	0.12	0.12	0.13	0.20	0.04	0.05	0.19	0.05	0.20	0.11	0.12	12%

10.5 DISCUSSION

The configurations presented in Table 4 offer valuable insights into the diverse pathways that lead to improved performance within the manufacturing sector. Each configuration represents a unique combination of constructs, shedding light on the complex interplay of factors influencing organizational effectiveness, market competitiveness, and sustainability impact.

Configuration 1 underscores the significance of CD and LP in driving manufacturing performance. The presence of both CD and LP suggests that fostering a culturally inclusive environment and facilitating effective communication practices are essential for optimizing operational efficiency and collaboration within manufacturing enterprises. This configuration highlights the importance of cultural alignment and communication strategies in navigating the diverse workforce and global market dynamics.

In Configuration 2, the inclusion of automated manufacturing technologies (AT) alongside CD and LP emphasizes the role of technological innovation in enhancing manufacturing performance. By leveraging automated manufacturing processes, organizations can streamline production, improve quality control, and meet the demands of a rapidly evolving market. This configuration underscores the importance of embracing technological advancements to drive productivity, innovation, and competitiveness in the manufacturing sector.

Configuration 3 introduces PQ metrics in conjunction with CD and LP, highlighting the critical role of quality management practices in manufacturing success. By prioritizing product quality and customer satisfaction, organizations can build a strong reputation, foster customer loyalty, and gain a competitive edge in the market. This configuration emphasizes the significance of quality assurance processes and continuous improvement initiatives in driving manufacturing excellence and market differentiation.

Configuration 4 emphasizes the importance of IS alongside CD and LP in shaping manufacturing performance. Adherence to IS ensures regulatory compliance, enhances product quality, and facilitates global market acceptance. This configuration underscores the value of standardization and quality assurance frameworks in building trust with customers, suppliers, and regulatory authorities, ultimately contributing to organizational success and sustainability.

10.6 PRACTICAL IMPLICATIONS

10.6.1 Cultural alignment and communication strategies

Manufacturing organizations should prioritize fostering a culturally inclusive environment and implementing effective communication strategies.

This includes promoting cross-cultural understanding among employees, providing language training programs, and leveraging technology to facilitate seamless communication across diverse teams. By promoting cultural alignment and effective communication, organizations can enhance collaboration, decision-making processes, and overall operational efficiency.

10.6.2 Investment in technological innovation

Embracing automated manufacturing technologies is crucial for enhancing productivity, quality control, and competitiveness within the manufacturing sector. Organizations should invest in advanced manufacturing technologies such as robotics, artificial intelligence, and Internet of Things (IoT) devices to streamline production processes, reduce operational costs, and meet evolving customer demands. By embracing technological innovation, manufacturing enterprises can stay ahead of the curve and maintain a competitive edge in the market.

10.6.3 Focus on quality management and continuous improvement

Prioritizing product quality metrics and implementing robust quality management systems are essential for manufacturing success. Organizations should establish rigorous quality assurance processes, conduct regular audits, and gather feedback from customers to ensure that products meet or exceed quality standards. Additionally, a culture of continuous improvement should be fostered, encouraging employees to identify areas for enhancement and implement corrective actions to drive operational excellence and customer satisfaction.

10.6.4 Adherence to international standards and regulations

Compliance with IS and regulatory requirements is paramount for manufacturing organizations operating in global markets. Organizations should ensure strict adherence to standards such as ISO 9001 for quality management and ISO 14001 for environmental management to demonstrate commitment to quality, sustainability, and regulatory compliance. By aligning with IS, manufacturing enterprises can enhance market acceptance, mitigate risks, and build trust with stakeholders.

10.6.5 Strategic decision-making and resource allocation

The findings of this study provide valuable insights for strategic decision-making and resource allocation within manufacturing organizations.

By identifying the key factors that drive OP, leaders can prioritize investments, allocate resources effectively, and develop targeted strategies to address areas of improvement. This includes investing in training and development programs, upgrading infrastructure, and fostering a culture of innovation and continuous learning.

10.7 CONCLUSION

This study has provided valuable insights into the intricate dynamics of IS and cross-cultural communication in the manufacturing sector. Through a comprehensive examination of key variables such as CD, language proficiency, use of automated manufacturing technologies, product quality metrics, and adherence to IS, we have elucidated the critical factors influencing OP, market competitiveness, and sustainability impact. The findings underscore the importance of cultural alignment, effective communication strategies, technological innovation, and quality management practices in driving manufacturing success. By embracing these factors, organizations can enhance operational efficiency, foster innovation, and maintain competitiveness in today's global marketplace. Moving forward, it is imperative to delve into the integration of emerging technologies, such as artificial intelligence and advanced robotics, to bolster productivity and efficiency in manufacturing. Additionally, further exploration of cross-cultural training programs, green manufacturing practices, and supply chain resilience strategies will be instrumental in fostering sustainable growth and resilience in the manufacturing sector. Moreover, continued research into the implications of digital transformation and evolving global market dynamics will provide invaluable insights for shaping strategic decisions and operational frameworks in manufacturing enterprises.

REFERENCES

Ahmed, S., Yaacob, M. R., & Ibrahim, M. A. H. (2022, November). Measuring the reliability of the sustainable performance instrument in ISO 14001-certified manufacturing firms in Malaysia. In *International Conference on Entrepreneurship, Business and Technology* (pp. 531–537). Springer Nature Singapore.

Alam, S., Zhang, J., Shehzad, M. U., Boamah, F. A., & Wang, B. (2024). The inclusive analysis of green technology implementation impacts on employees age, job experience, and size in manufacturing firms: empirical assessment. *Environment, Development and Sustainability*, 26(2), 4467–4486.

Al-Hakimi, M. A., Al-Swidi, A. K., Gelaidan, H. M., & Mohammed, A. (2022). The influence of green manufacturing practices on the corporate sustainable performance of SMEs under the effect of green organizational culture: A moderated mediation analysis. *Journal of Cleaner Production*, 376, 134346.

Ali, K., & Johl, S. K. (2023). Driving forces for industry 4.0 readiness, sustainable manufacturing practices and circular economy capabilities: does firm size matter? *Journal of Manufacturing Technology Management*, 34(5), 838–871.

Amutabi, C. (2024). Total factor productivity in Kenyan manufacturing firms: The role of ISO certification and direct vs indirect export intensity. *South African Journal of Economics*, 92(2), 260–285.

Appolloni, A., Jabbour, C. J. C., D'Adamo, I., Gastaldi, M., & Settembre-Blundo, D. (2022). Green recovery in the mature manufacturing industry: the role of the green-circular premium and sustainability certification in innovative efforts. *Ecological Economics*, 193, 107311.

Aririguzoh, S. (2022). Communication competencies, culture and SDGs: effective processes to cross-cultural communication. *Humanities and Social Sciences Communications*, 9(1), 1–11.

Ashima, R., Haleem, A., Bahl, S., Javaid, M., Mahla, S. K., & Singh, S. (2021). Automation and manufacturing of smart materials in Additive Manufacturing technologies using Internet of Things towards the adoption of Industry 4.0. *Materials Today: Proceedings*, 45, 5081–5088.

Awasthi, A., Saxena, K. K., & Arun, V. (2020). Sustainability and survivability in manufacturing sector. In *Modern Manufacturing Processes* (pp. 205–219). Woodhead Publishing.

Bonsu, S., & Twum-Danso, E. (2018). Leadership style in the global economy: A focus on cross-cultural and transformational leadership. *Journal of Marketing and Management*, 9(2), 37–52.

Chaudhry, I. S., Paquibut, R. Y., & Tunio, M. N. (2021). Do workforce diversity, inclusion practices, & organizational characteristics contribute to organizational innovation? Evidence from the UAE. *Cogent Business & Management*, 8(1), 1947549.

Dalal, R., & Akdere, M. (2023). Examining the relationship between talent management and employee job-related outcomes: The case of the Indian manufacturing industry. *Human Resource Development Quarterly*, 34(2), 201–226.

Das, A., & Dey, S. (2021). Global manufacturing value networks: Assessing the critical roles of platform ecosystems and Industry 4.0. *Journal of Manufacturing Technology Management*, 32(6), 1290–1311.

Dogan, A., & Birant, D. (2021). Machine learning and data mining in manufacturing. *Expert Systems with Applications*, 166, 114060.

Gajdzik, B., & Wolniak, R. (2022). Smart production workers in terms of creativity and innovation: The implication for open innovation. *Journal of Open Innovation*: Technology, Market, and Complexity, 8(2), 68.

Jaeger, B., & Upadhyay, A. (2020). Understanding barriers to circular economy: Cases from the manufacturing industry. *Journal of Enterprise Information Management*, 33(4), 729–745.

Jamwal, A., Agrawal, R., Sharma, M., & Kumar, V. (2021). Review on multi-criteria decision analysis in sustainable manufacturing decision making. *International Journal of Sustainable Engineering*, 14(3), 202–225.

Javaid, M., Haleem, A., Singh, R. P., & Suman, R. (2021). Significance of Quality 4.0 towards comprehensive enhancement in manufacturing sector. *Sensors International*, 2, 100109.

Javaid, M., Haleem, A., Singh, R. P., & Suman, R. (2021). Significance of Quality 4.0 towards comprehensive enhancement in manufacturing sector. *Sensors International*, 2, 100109.

Jimeno-Morenilla, A., Azariadis, P., Molina-Carmona, R., Kyratzi, S., & Moulianitis, V. (2021). Technology enablers for the implementation of Industry 4.0 to traditional manufacturing sectors: A review. *Computers in Industry*, 125, 103390.

Kafuku, J. M. (2019). Factors for effective implementation of lean manufacturing practice in selected industries in Tanzania. *Procedia Manufacturing*, 33, 351–358.

Kawalkar, R., Dubey, H. K., & Lokhande, S. P. (2022). A review for advancements in standardization for additive manufacturing. *Materials Today: Proceedings*, 50, 1983–1990.

Khoo, N. K., Hussin, H., & Abdullah, N. (2018). Managing occupational safety and health (OSH) culture practices at small and medium (S&M) Malaysia manufacturing sector. *Journal of Human Capital Development (JHCD)*, 11(1), 79–90.

Kim, M., & Williams, E. A. (2022). Emotional sustainability in human services organizations: Cultural and communicative paths to dealing with emotional work. *Sustainability*, 14(22), 15470.

Kumar, S., Sahoo, S., Lim, W. M., Kraus, S., & Bamel, U. (2022). Fuzzy-set qualitative comparative analysis (fsQCA) in business and management research: A contemporary overview. *Technological Forecasting and Social Change*, 178, 121599.

Lee, C. H., Liu, C. L., Trappey, A. J., Mo, J. P., & Desouza, K. C. (2021). Understanding digital transformation in advanced manufacturing and engineering: A bibliometric analysis, topic modeling and research trend discovery. *Advanced Engineering Informatics*, 50, 101428.

Maddikunta, P. K. R., Pham, Q. V., Prabadevi, B., Deepa, N., Dev, K., Gadekallu, T. R.,... & Liyanage, M. (2022). Industry 5.0: A survey on enabling technologies and potential applications. *Journal of Industrial Information Integration*, 26, 100257.

Margherita, E. G., & Braccini, A. M. (2023). Industry 4.0 technologies in flexible manufacturing for sustainable organizational value: reflections from a multiple case study of Italian manufacturers. *Information Systems Frontiers*, 25(3), 995–1016.

Martínez-García, A., Monzón, M., & Paz, R. (2021). Standards for additive manufacturing technologies: Structure and impact. In J. Pou, A. Riveiro, & J. P. Davim (Eds.), *Additive Manufacturing* (pp. 395–408). Elsevier. https://doi.org/10.1016/C2018-0-00910-X

Nguyen, H., Onofrei, G., & Truong, D. (2021). Supply chain communication and cultural compatibility: performance implications in the global manufacturing industry. *Business Process Management Journal*, 27(1), 253–274.

Ni, W., & Sun, H. (2019). The effect of sustainable supply chain management on business performance: Implications for integrating the entire supply chain in the Chinese manufacturing sector. *Journal of Cleaner Production*, 232, 1176–1186.

Ninduwezuor -Ehiobu, N., Tula, O. A., Daraojimba, C., Ofonagoro, K. A., Ogunjobi, O. A., Gidiagba, J. O., … & Banso, A. A. (2023). Exploring innovative material integration in modern manufacturing for advancing us competitiveness in sustainable global economy. *Engineering Science & Technology Journal*, 4(3), 140–168.

Obradović, T., Vlačić, B., & Dabić, M. (2021). Open innovation in the manufacturing industry: A review and research agenda. *Technovation*, 102, 102221.
Pappas, I. O., & Woodside, A. G. (2021). Fuzzy-set Qualitative Comparative Analysis (fsQCA): Guidelines for research practice in Information Systems and marketing. *International Journal of Information Management*, 58, 102310.
Papulová, Z., Gažová, A., & Šufliarský, Ľ. (2022). Implementation of automation technologies of industry 4.0 in automotive manufacturing companies. *Procedia Computer Science*, 200, 1488–1497.
Peng, Y., Wang, W., Li, S., & Veglianti, E. (2022). Competition and cooperation in the dual-channel green supply chain with customer satisfaction. *Economic Analysis and Policy*, 76, 95–113.
Prasad, M. M., Dhiyaneswari, J. M., Jamaan, J. R., Mythreyan, S., & Sutharsan, S. M. (2020). A framework for lean manufacturing implementation in Indian textile industry. *Materials Today: Proceedings*, 33, 2986–2995.
Rizvi, A. T., Haleem, A., Bahl, S., & Javaid, M. (2021). Artificial intelligence (AI) and its applications in Indian manufacturing: A review. Current Advances in Mechanical Engineering: Select Proceedings of ICRAMERD, 2020, 825–835.
Shao, G., Frechette, S., & Srinivasan, V. (2023, June). An analysis of the new ISO 23247 series of standards on digital twin framework for manufacturing. In *International Manufacturing Science and Engineering Conference* (Vol. 87240, p. V002T07A001). American Society of Mechanical Engineers.
Stekelorum, R., Laguir, I., Gupta, S., & Kumar, S. (2021). Green supply chain management practices and third-party logistics providers' performances: A fuzzy-set approach. *International Journal of Production Economics*, 235, 108093.
Tortorella, G. L., Fettermann, D., Fogliatto, F. S., Kumar, M., & Jurburg, D. (2021). Analysing the influence of organisational culture and leadership styles on the implementation of lean manufacturing. Production Planning & Control, 32(15), 1282–1294.
Upadhyay, A., Balodi, K. C., Naz, F., Di Nardo, M., & Jraisat, L. (2023). Implementing industry 4.0 in the manufacturing sector: Circular economy as a societal solution. *Computers & Industrial Engineering*, 177, 109072.
Veile, J. W., Kiel, D., Müller, J. M., & Voigt, K. I. (2020). Lessons learned from Industry 4.0 implementation in the German manufacturing industry. *Journal of Manufacturing Technology Management*, 31(5), 977–997.
Yusr, M. M., Salimon, M. G., Mokhtar, S. S. M., Abaid, W. M. A. W., Shaari, H., Perumal, S., & Saoula, O. (2020). Green innovation performance! How to be achieved? A study applied on Malaysian manufacturing sector. *Sustainable Futures*, 2, 100040.

Chapter 11

Industry 4.0

A step toward achieving the Sustainable Development Goals (SDGs)

Sambandhamoorthy Senthilrajan, Narayanan Venkateshwaran, Ragavendran Giri, and Thangaraju Tamilaban

11.1 INTRODUCTION

The objectives of the fourth industrial revolution include increased automation, output, and operational efficiency. Industry 4.0 is closely related to performance outcomes because of its significant role in manufacturing and services. Manufacturing depends on the industrialization process, and companies that have not embraced IT are finding it difficult to remain competitive. Organizations in every industry today have a similar perspective on the future and want to stay ahead of the curve. Consequently, because Industry 4.0 can produce high-quality customized products, many sectors are using it. Industry 4.0 has many advantages, but it still consumes a lot of energy, raw materials, resources, and data, which presents problems for environmental sustainability [1]. The implementation of the fourth industrial revolution in this industrial sector has primarily emphasized maximizing manufacturing and earnings, resulting in numerous challenges for other aspects. Natural resource depletion, adverse ecological impacts, wealth inequality, and unfavorable working conditions are some of these problems. Ultimately, these factors may contribute to consumption patterns that are unsustainable from environmental, economic, and social perspectives [2]. The main goal of this study was to look into how the concept of Industry 4.0 might affect how resources, energy, and raw materials are used in manufacturing processes, as well as how waste and byproducts are produced once Industry 4.0 technologies are put into place. Additionally, it sought to determine whether these impacts would ultimately prove beneficial or detrimental to environmental sustainability. Through this analysis, the research hopes to provide a more thorough understanding of the related effects, which will help stakeholders and policymakers develop strategies that support the best possible outcomes that are expected from the use of these cutting-edge technologies [3]. In the present circumstances, the industry prioritizes production and profitability while neglecting other crucial factors. This approach leads to the depletion of both finite and renewable resources, resulting in climate instability, severe environmental consequences, and a decline in biodiversity [4]. Prior studies have emphasized the importance of these environmental issues and shown how integrating the principles

DOI: 10.1201/9781003470342-11

of Industry 4.0 with cutting-edge technology might improve manufacturing quality and sustainability [5]. The main focus of Industry 4.0 is on enhancing productivity and product quality to boost profits and market competitiveness. Currently, this industrial revolution emphasizes manufacturing processes rather than environmental sustainability. This is despite the numerous challenges the industry faces, including harmonizing regulations, standardizing organizational procedures, recruiting skilled personnel, and implementing a suitable legal framework [6,7]. In the introduction, we went over a lot of territory about the fourth industrial revolution and the technologies that enhance smart manufacturing. An outline of how to achieve growth in sustainable development was also given in this chapter.

11.2 INDUSTRY 4.0 AND ITS TECHNOLOGIES

We will explore the fundamentals of Industry 4.0 in this section, beginning with an outline and delving into the definition of essential words. Finally, we will discuss the ramifications of these approaches, as well as the operations and capabilities they produce. But Industry 4.0 is a complicated subject. Three viewpoints need to be examined: functional, structural, and qualitative. Specifically presented in a 2011 German government initiative as a plan for industrial output in the nation, the phrase "Industry 4.0" has received broad usage in recent times [8]. According to this perspective, the concept of Industry 4.0 is an integrated and linked component of a world transformed by the information and communication technology (ICT) revolution. Industry 4.0 refers to a technological transition caused by the Internet of Things (IoT) and the Internet of Services (IOS), which electronically connect industry through a supply chain network both internally and outside. As a result, the industry becomes "smart," and cyber-physical systems (CPSs) facilitate this. This essay focuses on CPS, which are systems that mix hardware and software components to achieve large-scale effects [9,10] (Figure 11.1).

The benefit provided to the ultimate client and the abilities of the manufacturing process are the two key ways in which Industry 4.0 introduces novel characteristics and opportunities to industry. This paper highlights the primary benefits (Table 11.1).

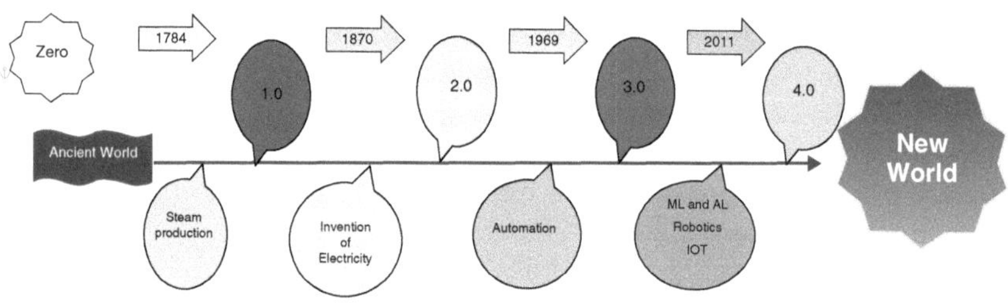

Figure 11.1 Evaluation of Industry 4.0.

Table 11.1 Significance of the fourth industrial revolution

S.no	*Benefits to clients*	*Extra benefits industries*
1	Increased personalization	Enhanced effectiveness
2	Enhanced client satisfaction	Increased endurance
3	Lower expenses	Improved communication with flow of information
4	Simplified adherence	Change every element in the list
5	Superior goods and a luxurious lifestyle	Analysis of intelligent development was encouraged
6.	Possibilities for employment established	Ability to react appropriately to various circumstances

11.3 KEY INNOVATIONS OF THE INDUSTRY 4.0

The fourth industrial revolution focuses on the modern industry. The smart factory is a component of, or dependent upon, every facet of the modern industrial revolution, including sales, machinery, and goods. Significant self-control and a flexible system that enables performance self-optimization over a wide variety of networks are features of smart factories. Self-adaptation also makes it possible to learn from events as they happen. They may also have the ability to oversee whole manufacturing processes alone [11].

As shown in Figure 11.2, the technologies adopted by these changes are known as the nine pillars of Industry 4.0.

1. IoT
2. Big data analysis (BDA)
3. Artificial intellect (AI).
4. Virtual reality (VR) and augmented reality (AR)

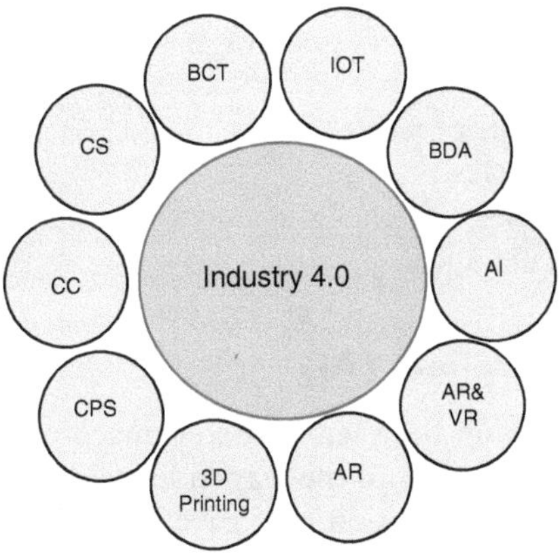

Figure 11.2 Core technologies of Industry 4.0.

5. Robotics (AR)
6. 3D printing
7. Cloud computing (CC)
8. Cybersecurity (CS)
9. Block chain technology (BCT)

11.3.1 IoT (Internet of Things)

"Internet of Things" refers to the method of connecting devices and sensors to the network in order to enable real-time data collecting and sharing. Industry 4.0 improves control, monitoring, and making choices by using the IoT to gather information regarding goods, equipment, and production processes [12].

Among the projected future IoT trends and advancements are:

1. A rise in the usage of low-energy communication tools.
2. Even stronger cloud integration.
3. Enhanced protection.
4. A deeper level of artificial intelligence integration.

11.3.2 Big data analysis (BDA)

Modern computing technology is applied to large data sets to find significant trends, correlations, patterns, and preferences that may aid organizations in making better informed decisions. This process is known as big data analytics. Industry 4.0 uses big data processing in certain areas, such smart factories, where sensor data from manufacturing equipment is analyzed to forecast when maintenance and replacements are required. Businesses that implement it gain from a self-service system and higher production efficiency. Familiarity with production management automation, predictive maintenance optimization, and real-time data are important [13].

It assists in enhancing business operations in the following ways:

1. Enhancing logistics processes.
2. Removal of obstacles.
3. Forecast demand.
4. Preventive maintenance.

11.3.3 Artificial intellect (AI)

A new type of connection between man and machine, fueled by artificial intelligence, has caused a shift in the way industry operates. A distinguishing feature of the industrialization that gave rise to Industry 4.0 is intelligent factories, where people and cyber-physical devices communicate in the cloud [14] (Figure 11.3).

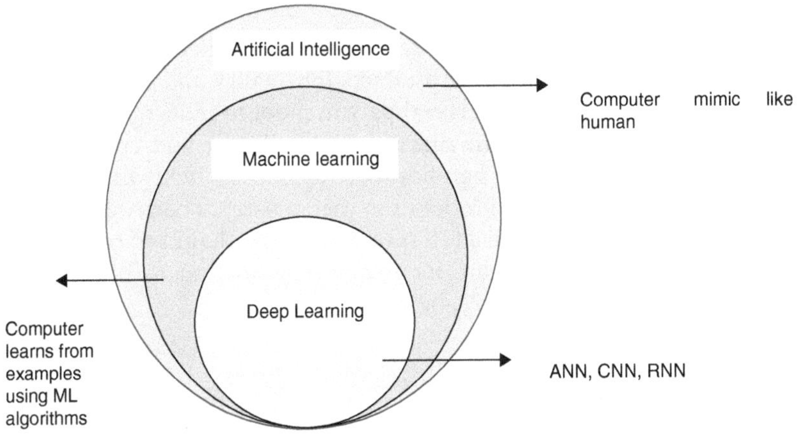

Figure 11.3 Artificial intelligence and its types.

In the 4.0 industry, its primary uses include the following:

1. Predictive maintenance and repair for overall equipment evaluation optimization.
2. Production quality is continuously improved through operational excellence, leading to quality 4.0.
3. Generative design using AI and automation algorithms that provide several workable design options for the same goal at the same time.
4. Robotics by means of cooperative, robotic devices that assist operators in being relieved of laborious, meticulous duties.

11.3.4 Virtual reality (VR) and augmented reality (AR)

The ever-evolving expectations of vendors and customers have presented manufacturing organizations with a number of hurdles in recent times. The fourth industrial revolution, which we are presently experiencing, was initiated by the digital revolution. The general availability of the more user-friendly Web and the application of robotics, intelligent parts, and technology are some of its distinctive features. The "smart factory," which combines digital and physical systems, is made possible by this transformation in order to accomplish "mass personalization" and "faster product development" [15]. A/V (augmented and virtual) worlds can be used for workforce training, which will improve human–machine interaction. Thus, they can: (i) expedite production line transformation; (ii) support shop floor operators; (iii) execute online instruction for component assembly; (iv) efficiently oversee the warehouse; (v) facilitate sophisticated tests integrated into the modules; and (vi) participate with the production environment for lowering exposure.

11.3.5 Robotics

Robots have been employed in industry for decades to do certain jobs, while humans work together to inspect products for quality and eliminate those with problems. Robots play a crucial part in modern industry, completing tasks intelligently while prioritizing safety, flexibility, and collaboration. Our civilization will undergo significant changes in the next years. The primary technologies are AI and robotics in the industry. The production and manufacturing phase has seen a full transformation, thanks to robotics and industrial automation. Both are focused on automation to boost output and the economy [16] (Figure 11.4).

11.3.5.1 Utilization of robotics in the fourth industrial revolution

Future generations will have a new standard of life due to the Modern Age, which will speed up global economic progress. Smart sensor human–machine interfaces will be used in the fourth industrial revolution to allow human and robots to work together collaboration on jobs [17].

The following are some of the many ways that robotics is being applied to Industry 4.0.

1. In Manufacturing.
2. Autonomous Vehicles.
3. In Health Care.
4. In Packaging.
5. In Mining.
6. Military and Public Safety Industry.
7. Supermarket and Malls.
8. Agriculture, etc.

Figure 11.4 Industry 4.0: utilization of robotics.

11.3.6 The 3D printing process

Mass manufacture of highly personalized objects is possible with 3D printing. Digitalizing inventories can result in considerable financial savings. It usually encompasses methods which are utilized to create specialized pieces, such as 3D printing. This innovation would benefit from Industry 4.0, which would also boost the production of personalized and customized items [18].

11.3.7 Cyber-physical framework

Systems used in production that communicate with the real world are referred to as cyber-physical systems (CPS). One popular method of integrating the digital and physical worlds is through the integration of CPS with the IoT. The goal of this partnership is to create smarter factories, which will improve production intelligence and enable faster, more effective digitization. The development of CPS technology will help numerous industries, such as metallurgy, heavy industry, logistics, transportation, and energy generation [19].

11.3.8 Cloud computing

A novel concept in computing, cloud computing offers consumers a multitude of Internet services at extremely little cost. They provide users with access to cloud resources, including hardware, software, and infrastructure resources, in accordance with their needs. They also function well with flexible memory allocation and utilization of resources. Real-time data transfer would be made possible by this, opening the door for the development of a cooperative, networked digital environment. Real-time data visibility and improved stakeholder interaction would enable reactive control of supply chains [20].

11.3.9 Cybersecurity

"Cybersecurity" refers to the area of computer technology that guards computer systems from hardware, software, data loss, and other components, as well as against service interruptions. The industry would be more productive and make it easier to see human–machine cooperation if interfaces between human and machine were implemented. This information may be utilized cruelly if it finds up in the wrong hands. Cybersecurity protocols are put in place to ensure that Industry 4.0 is safe for everyone to use [21].

11.3.10 Block chain technology

Block chain technology is utilized in banking and cryptocurrency applications to improve speed, security, usability, and secrecy. Several firms have built research centers to examine block chain technology's potential applications in various sectors [22].

11.3.10.1 Revolution of block chain technology

The figure depicts the detailed growth of block chain technology from its origin to the present.

a. **BLOCKCHAIN 1.0:** Block chain 1.0, the initial version of the technology, was launched in 2009 with the launch of the bitcoin network. The invention of the first cryptocurrency was introduced in this generation.
b. **BLOCKCHAIN 2.0:** In 2010, financial services for a range of applications and smart contracts were created, marking the next level of block chain technology.
c. **BLOCKCHAIN 3.0:** The shift toward decentralized applications was first seen in this generation of block chains. Decentralized applications were evaluated across several study domains, including health, governance, IoT, supply chain, business, and smart cities.
d. **BLOCKCHAIN 4.0:** The two main areas of focus for this generation were public records and distributed real-time databases. This level effortlessly integrates Industry 4.0 applications. With smart contract technology, it does not need [23] (Figure 11.5).

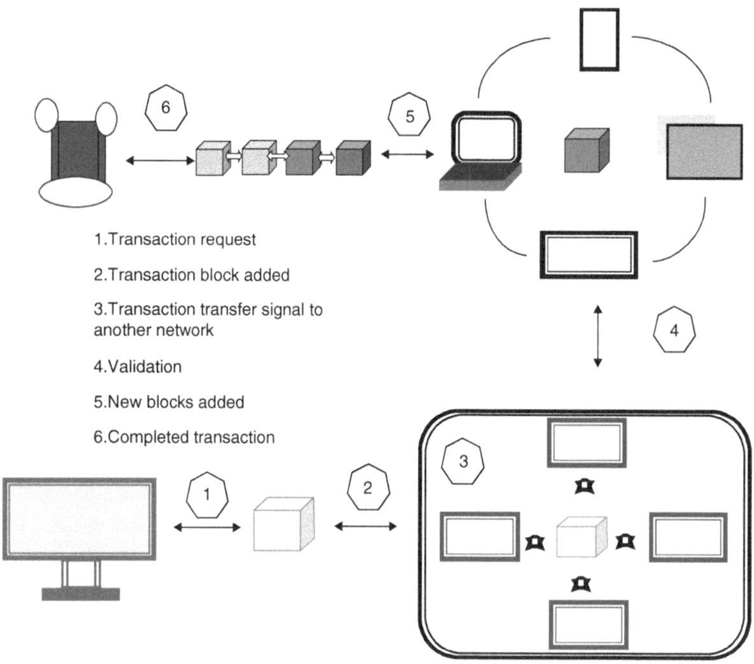

Figure 11.5 Block chain technology in Industry 4.0.

11.4 IMPACT OF INDUSTRY 4.0 ON SPECIFIC SUSTAINABLE DEVELOPMENT GOALS (SDGs)

The demand for a digital and intelligent manufacturing technique contributes to the manufacturing industry's current rapid migration to personalized manufacturing. "Industry 4.0," as the name suggests, relates to the latest change in the time, which has the power to provide inventive solutions to global problems faced in sustainable growth while also managing the entire lifecycle of a product. Throughout the creation of any new business model, there are broad challenges—17 in total—that must be overcome in order to have sustainable development. The goal of these challenges is to preserve a positive and harmonious interaction between people and the environment in order to achieve worldwide equitable and sustainable health [24].

A synopsis of the 17 primary challenges that sustainable development considers is provided below.

11.4.1 End poverty

Poverty is generally defined as the inability to provide for one's basic requirements, including food, housing, and clothing. One of the biggest challenges facing humanity is eradicating poverty in all of manifestations. The principal aim of the 2030 Sustainable Development Agenda is to guarantee the complete elimination of poverty in all its manifestations [25].

11.4.2 End hunger

The Global Status of Food and Nutrition Security Assessment was published. The number of people who are hungry has risen to one in nine during the past 3 years. Children who are hungry may suffer from severe malnourishment, which can restrict their growth, or they may be extremely underweight [26].

11.4.3 Ensure healthy lives

People are being impacted by the rise in pollution in all areas, including the water and air, leading to extreme health risks. One of the main causes of concern for worldwide is pollution, which creates a toxic environment that has an adverse effect on life. As a result, the need to guarantee healthy lives and advance age-appropriate well-being is crucial to the smooth operation of everything [27].

11.4.4 Comprehensive and perpetual education

Although more people have access to information than in the past, not everyone has profited from it globally. Education is a policy intervention that promotes self-sufficiency, economic growth, skill development, and improved employment possibilities. Everyone needs equal opportunity for training and education in order to benefit from specialized learning opportunities, regardless of source, financial or social status, physical or mental health, race, or age [28].

11.4.5 Attain gender equality

One important strategy for identifying the problem of gender equality appears to be emancipating women and girls via education, employment, and skill development. In addition to being a fundamental human right, gender equality is also necessary for a world that is viable and harmonious [29].

11.4.6 Clean water and hygiene for everyone

A healthy living environment requires access to pure, clean water. Even though fresh water is abundant on Earth, people's access to it is limited due to inadequate infrastructure, economic conditions, and hygiene and sanitation practices. In order to meet SGD targets, efforts must be made to decrease waste and water pollution [30].

11.4.7 Contemporary energy

An important issue facing the government is expanding the rural population's access to power. There is evidence that the great majority of people who prepare with bioenergy or lack access to electricity failed to meet the goals and policies of the past international order. A country may take into account these three energy tests: the majority of its population being energy-deprived and needs access to modern energy bearers [31].

11.4.8 Consistent economic development and profitable employment

A platform for productive work with equitable pay, safety for workers and their families in the workplace, good instruments for social integration and personal growth—all of these are indicators of good employment. Ensuring equitable participation of men and women in the workforce is also vital [32].

11.4.9 Facilities, creativity, and renewable industry

This is a complex issue. Facilities, industry, and creativity are the three interrelated words that form the foundation of Sustainable Development. One issue is that developing nations, especially locked development countries (LDCs), must compel more micro and small- and medium-sized enterprises (SMEs) to have Internet connectivity. Approximately 4 billion people worldwide, or 90% of the population, do not have access to the Internet [33].

11.4.10 Decrease discrimination

There is a global issue with income disparity that needs a global solution. Along with targeting national disparities, the goal also adds that linked characteristic and asks for the promotion of coordinated, safe, and orderly migration and people mobility [34].

11.4.11 Create sustainable cities and communities

People who work in remote areas to create and design towns and cities have been lauded for their contributions to centralized development. The industrial economy has resulted in air, water, and soil pollution, which has left cities with hazardous surroundings. In both rural and urban settings, to ensure that people who have impairments have the same rights to the outside world, information, travel, and communications, as well as other public facilities and services, local government must take reasonable measures [34].

11.4.12 Renewable manufacturing and consumption

The term "renewable manufacturing and consumption" refers to the process of creating and using goods and services in a way that is considerate of the economy, society, and the surroundings over their whole life period [35].

11.4.13 Tackle global warming

Right now, the biggest problem is about global warming, and things cannot continue the way they are. All countries on all continents are currently impacted by the changing climate. The average global temperature rose between 1880 and 2012 by 0.85°C. Grain yields have dropped by 5% for each degree Celsius that the temperature rises [35].

11.4.14 Maintain and protect oceans, seas, and marine resources

Seas, waterways, and coastal assets are necessary for sustainable growth as well as being vital to the ecosystem of the planet. They make up more than two-thirds of the planet's area and hold 97% of the total of its water. The oceans make up approximately 65% of the planet's area, contribute half of the oxygen in the atmosphere, and provide 16% of animal protein. The oceans are essential to human health and the global food chain. The world's coastal and marine resources support almost 3 billion people, and they generate a significant amount of jobs across a range of sectors, including tourism, fishing, shipping, and biotechnology [35].

11.4.15 Enhance biological and environment sustainability

The World Bank's biologically varied ecology offers significant economic benefits to the nation. An ecosystem is the whole of living things and abiotic elements that make up the area, such as air, water, and minerals. However, the notion of ecosystem management—which is based on four principles—was developed as a result of how different species interact with the environment and with one another [35], that is,

1. safeguards both the particular species and the whole environment,
2. discusses preserving each region's natural ecology,
3. addresses the management of resilience to disruptions, and
4. creates perimeters of safety around central reserves.

11.4.16 Promote friendly and happy societies

There has been inconsistent progress in the regions in advancing justice and peace and creating inclusive, accountable, and productive institutions. The Agenda for Sustainability has thoroughly encompassed harmony and safety concerns into the global development agenda, both at the operational and regulatory stages [35].

11.4.17 Promote international collaborations for sustainable growth

Partnerships between the government, business sector, and society are necessary for sustainable development to succeed. Immediately, mobilizing, redirecting, and transforming the power of affluent individuals is necessary to accomplish the ultimate goal [35].

11.5 SOLUTIONS TO CHALLENGES AND RISKS

- The goals are vital and comprehensive, but the hurdles may be overly optimistic or unattainable within the timeframe provided, necessitating active effort. The latest industrial revolution offers a plethora of ways to tackle and settle each of the previously described problems. Technology will need to be combined tenfold for the next industrial revolution, each with the capacity to address problems and promote fair growth. Possibly Useful Sector Industry the 4.0 Solution—to uphold the agricultural sector. More and more tasks are being handled by robotics, thanks to Industry 4.0 technologies such as detectors, unmanned aircraft, micro-robotics, and others. These technologies may also be utilized to construct export-oriented manufacturing projects in developing nations. Aside from helping with forest management and maintenance, deforestation prevention, and increasing tree planting and replanting, instruments like drone-based surveillance also provide current information. Additionally, IoT-capable hardware may make a significant impact on space and savings. Technologies like 3D printing, artificially intelligent systems, and automated manufacturing with additive software may cut waste and pollution. Smartphones might help developing nations spur greater innovation, which would boost the output and efficiency of their economies.

- They can also be used to train individuals to enhance their income levels. The supply chain can be strengthened by data analytics, smart cars, and simulation tools. These technologies may also increase demand estimate in supermarkets and restaurants and expand the food packaging business, which would reduce the disposal of food and transportation-related emissions.
- Through using fog computing, nanotechnologies, and other advances in food production, distribution, and preparation, there is a significant deal of potential to minimize food waste. Improvements are being made to solar and wind energy generation as viable sources of electrical energy.

11.6 MEASUREMENT OF SUSTAINABLE DEVELOPMENT

John Elkington invented the triple bottom line (TBL) concept in the year of 1990s in order to promote sustainability. An effective strategy that aims to provide overall financial results based on performance in individuals, revenues, and the environment is known as the triple bottom line, or 3BL. [36,37]. "Profit" or "loss" is commonly linked in the accounting system with the bottom line (TBL) of the revenue and expenses statement. The three parts of this accounting paradigm are Earth, profit, and society. Essentially, the goal of social work is to provide the public and workforce with corporate benefits. The TBL counts endangered species, evaluates environmental harm, assesses social welfare, compares company performance, and so on. Such a universally accepted method of evaluation does not exist. One of the difficult tasks in measuring the TBL is determining a standard unit of measurement. Using prices and indexes are the two methods that can be used as a standard gauge for the unit [38,39].

1. We can determine whether a product's price affects the bottom line by examining its price association with several items.
2. Another option is to employ an index, which would resolve the problem of incompatible units by permitting the application of generally recognized accounting principles.

3. TBL applications acknowledge the concepts of social conservation, ecological growth, and financial growth, which include businesses and governments. However, the methods by which these principles are measured may differ greatly. Let us examine the associated metrics for these three [40,41].
   a. **Measuring Economic Prosperity:** Economists generally focus on income when examining the growth of the economy.
   b. **Measuring Environmental Quality:** The measurement of natural resource availability is its area of focus.
   c. **Evaluating Equality and Social Assets:** Include sociological features of the community or region in your measurements.

These are the obstacles to putting the TBL into practice, which, if properly evaluated, may ultimately contribute to sustainable.

11.7 DIFFERENT FACTORS OF SUSTAINABLE GROWTH

11.7.1 The economic factor of sustainable development

The Industry 4.0 initiative increases productivity by enabling self-governing structures for growth in intelligent industrial facilities. The economic factor does not assess the company's financial position [42]. It focuses on four basic elements as follows:

1. The state of economy.
2. Visibility in the market.
3. The effects on the economy indirectly.
4. Procedures for purchasing.

11.7.2 The social factor of sustainable development

According to the social component of the environment, a company that wants to be viable needs to be supported and welcomed by consumers, staff, and the environment in which it works. The main objectives here are to treat the personnel well and to be a respectable local and international community member. Empathy for our coworkers and the society we live in is encouraged by the Industry 4.0 social sustainability pillar, which is illustrated in Figure 11.6. "Social sustainability" is having the capacity to ensure that people have access to security, wellness, schooling, and other aspects of life, and that these benefits are too evenly distributed across genders and levels of society [43].

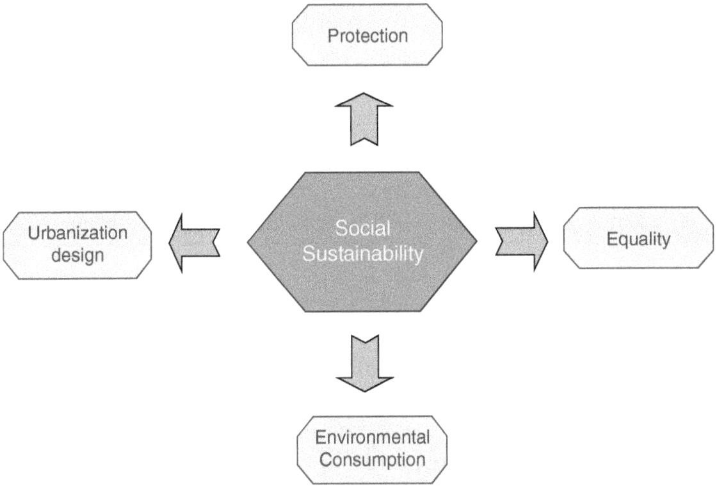

Figure 11.6 Social sustainability in Industry 4.0.

11.7.3 The aspect of sustainable growth related to the environment

The capacity to maintain the three main environmental functions over time is known as environmental sustainability. It consists of (i) direct utility; (ii) waste receiver; and (iii) availability of resources. The following are essential for environmental sustainability within a given region is the capacity to enhance the importance of nature while ensuring that the preservation and safety of natural resources and natural heritage are taken into consideration [44,45].

1. Decisions made by successive generations.
2. Prosperous environment.
3. Eliminate pollution.
4. Prioritize well-being ahead of gross domestic product.
5. Use more resources that are sustainable.

11.8 FUTURE PROSPECTS OF SUSTAINABLE DEVELOPMENT GOALS IN INDUSTRY 4.0

- Industry 4.0's opportunities are beginning to provide doors for long-term, steady growth. As per the hypothesis, the current economic and social dynamics are compatible with the perpetuation of natural resources and the advancement of living standards.

Table 11.2 Industry 5.0

S.no	*Benefits of Industry 5.0*
1	A greater focus on innovation and the creation of specifically tailored products
2	Increased communication among machines and people
3	Collaborative robotics (Cobots) are increasingly being used to enhance workforces
4	Robots-as-a-service (RaaS) enables smaller manufacturers to buy robots
5	The introduction of 5G technology will improve manufacturers' communication

- Proper approaches can be used in order to provide a more accurate estimate of Industry 4.0's potential ecological effect. For instance, energy accounting, life cycle assessment, or material intensity.
- The advancement of the idea of sustainable development is contingent upon the government's capacity to ensure an inclusive interaction of the (i) society, (ii) economy, and (iii) environment—three foundations of sustainability.
- All things considered, Industry 4.0 is the way of the future for a great deal, if not all, of life's goals, including sustainable development. Numerous portals linking the ever-expanding capabilities of technology and human creativity could lead to countless prospects for advancement and success in realizing humanity's goals. The future is already changing quickly, impacting many aspects of life and providing more hints about what is ahead.
- Industry 5.0 is an improvement on Industry 4.0, rather than a completely new concept. Table 11.2 presents the future Industry 5.0.

  Since Industry 5.0 robots will work together more, they are usually referred to as "cobots." Compared to previous generations, these newer robots are more affordable, transportable, and versatile. These cobots do not need safety cages to operate alongside humans in a secure manner.
- The idea of robotics-as-a service (RaaS) will permit the tiny enterprises pay-per-use without an upfront expenditure on robotics. The introduction of superfast 5G networks promises strong support for all of these, and when sensors become standard on the factory floor, artificial intelligence in manufacturing will reach new heights.

11.9 CONCLUSION

State and national governments can collaborate in a unified direction by using the Sustainable Development Goals, which offer a set of qualitative objectives. Sustainable development's five Ps are as follows:

- (1) People; (2) Planet; (3) Prosperity. Every aspect of maintaining the long-term links between the social, ecological, and economic systems is covered by (4) Peace and (5) Partnership.

- The 17 major challenges of sustainable development concentrate on eliminating hunger and enabling everyone to earn at least three meals a day. These challenges aim to end poverty by allowing everyone to have incomes above the poverty line.
- Ensuring a healthy life by mitigating hazardous diseases and creating, with the aid of the quick evolution of technology, spaces for inclusive and lifetime learning.
- In order to achieve gender equality, women and girls must be empowered via education, career opportunities, and skill development. Additionally, waste and pollution in the water must be reduced in order to provide fresh, pristine, and easily obtained water.
- Increasing energy through the generation of wind and solar power, promoting sustainable development in the economy, and creating good jobs by offering an excellent work environment.
- The ten Industry 4.0 technologies—sensors, unmanned aircraft, micro robots, cloud computing, augmented and virtual reality, additive manufacturing, data analytics, cybersecurity, and IoT—effectively address the problems facing sustainable growth.
- We are on the edge of entering the era of Industry 5.0 due to the tremendous improvements in technology, even though Industry 4.0 has not yet reached its full potential and companies remain in a state of transition of accepting this. The fifth version of Industry will emphasize human–machine cooperation and a greater emphasis on personalized production. It is anticipated that the new Industry 5.0 will undoubtedly overtake the fourth industrial revolution.

REFERENCES

1. McWilliams, A., Parhankangas, A., Coupet, J., Welch, E., &Barnum, D.T. (2016) Strategic decision making for the triple bottom line. *Business Strategy and the Environment*, *25*, 193–204.
2. Bonilla, S., Silva, H., Terra, M., Franco, G. R., &Sacomano, J. (2018). Industry 4.0 and sustainability implications: A scenario-based analysis of the impacts and challenges. *Sustainability*, *10*, 3740.
3. Zhang, Z, &Sutton, S. D. B. (2015). Linking Daly's proposition to policymaking for sustainable development: Indicators and pathways. *Journal of Cleaner Production*, *102*, 333–341.
4. Seddon, N., Mace, G. M., Naeem, S., Tobias, J. A., Pigot, A. L., Cavanagh, R., Mouillot, D., Vause, J., &Walpole, M. (2016). Biodiversity in the Anthropocene: Prospects and policy. *Proceedings of the Royal Society B: Biological Sciences*, *283*, 2094.
5. Szeremlei Keszi, A., &Magda, R. (2015). Sustainable production and consumption. *Visegrad Journal on Bioeconomy and Sustainable Development*, *4*, 57–61.
6. Xu, L. D., Xu, E. L., &Li, L. (2018). Industry 4.0: State of the art and future trends. *International Journal of Production Research*, *56*, 2941–2962.

7. Lentner, C., Vasa, L., Kolozsi, P. P., &Zéman, Z. (2019). New dimensions of internal controls in banking after the GFC. *Economic Annals-XXI, 176*, 38–48.
8. Ślusarczyk, B. (2018). Industry 4.0: Are we ready? *Polish Journal of Management Studies, 17*, 232–248.
9. Liao, Y., Deschamps, F., Loures, E. D. F. R., &Ramos, L. F. P. (2017). Past, present and future of Industry 4.0-a systematic literature review and research agenda proposal. *International Journal of Production Research, 55*, 3609–3629.
10. Da Silva, F. S. T., da Costa, C. A., Crovato, C. D. P., &da Rosa Righi, R. (2020). Looking at energy through the lens of Industry 4.0: A systematic literature review of concerns and challenges. *Computers & Industrial Engineering, 143*, 1–21.
11. Rojko, A. (2017). Industry 4.0 concept: Background and overview. *International Journal of Interactive Mobile Technologies (iJIM), 11* (5), 77–90.
12. Gilchrist, A. (2016). *Industry 4.0: The Industrial Internet of Things*. Apress.
13. Lu, Y. (2017) Industry 4.0: A survey on technologies, applications and open research issues. *Journal of Industrial Information Integration*, 6, 1–10. https://doi.org/10.1016/j.jii.2017.04.005
14. Timms, M. J. (2016). Letting artificial intelligence in education out of the box: Educational cobots and smart classrooms. *International Journal of Artificial Intelligence in Education, 26*(2), 701–712.
15. i-SCOOP. (2011). Industry 4.0: The fourth industrial revolution, https://www.i-scoop.eu/industry-4-0/
16. Zhou, K., Liu, T., & Zhou, L. (2015). Industry 4.0: Towards future industrial opportunities and challenges. *12th International Conference on Fuzzy Systems and Knowledge Discovery (FSKD)*, Zhangjiajie, 2015, pp. 2147–2152.
17. Rojko, A. (2017). Industry 4.0 concept: Background and overview. *International Journal of Interactive Mobile Technologies (iJIM), 11*(5), 77–90.
18 Lee, H. K., &Yang, S. (2014). Service Innovation and smart analytics for Industry 4.0 and big data environment. *Procedia CIRP*, 16, 3–8. https://doi.org/10.1016/j.procir.2014.02.001.
19. Wan, J., Tang, S., Shu, Z., Li, D., Wang, S., Imran, M., &Vasilakos, A. V. (2016). Software defined industrial Internet of Things in the context of Industry 4.0. *IEEE Sensors Journal, 16*(20), 7373–7380.
20. Luenendonk, M. (2019). Industry 4.0: Definition, design principles, challenges, and the future of employment. https://www.cleverism.com/industry-4-0/
21. Nayyar, A., Jain, R., Mahapatra, B., & Singh, A. (2019). Cyber security challenges for smart cities. In Management Association, Information Resources (Ed.), *Driving the Development, Management, and Sustainability of Cognitive Cities* (pp. 27–54). IGI Global. https://doi.org/10.4018/978-1-5225-8491-9.ch002.
22. Kara filoski, E., &Mishev, A. (2017, July). Blockchain solutions for big data challenges: A literature review. In *IEEE EUROCON 2017-17th International Conference on Smart Technologies* (pp. 763–768). IEEE.
23. Bodkhe, U., Tanwar, S., Parekh, K., Khanpara, P., Tyagi, S., Kumar, N., Alazab, M. (2020). Blockchain for Industry 4.0: A comprehensive review. *IEEE Access*. https://doi.org/10.1109/access.2020.2988579.

24. Yang, F., &Gu, S. (2021). Industry 4.0, a revolution that requires technology and national strategies. *Complex & Intelligent Systems*, 7, 1311–1325. https://doi.org/10.1007/s40747-020-00267-9
25. Beckerman, W. (2002). *A Poverty of Reason: Sustainable Development and Economic Growth*. The Independent Institute.
26. Belahsen, R. (2014). Nutrition transition and food sustainability. *Proceedings of the Nutrition Society*, *73*(3), 385–388.
27. Pandey, B. D. (2015). An appraisal of the global warming: Problems issues corrective measures. *Bonfring International Journal of Industrial Engineering and Management Science*, *5*(4), 133.
28. Pal, J. (2003). The developmental promise of information and communications technology in India. *Contemporary South Asia*, *12*(1), 103–119.
29. Johnsson-Latham, G. (2007). A study on gender equality as a prerequisite for sustainable development. Report to the Environment Advisory Council.
30. Rijsberman, F. R. (2006). Water scarcity: fact or fiction? *Agricultural Water Management*, *80*(1–3), 5–22.
31. Carney, D. (1998). *Sustainable Livelihoods. Sustainable Livelihoods: What Contribution Can We Make*. DFID.
32. Anker, R., Chernyshev, I., Egger, P., Mehran, F., &Ritter, J. A. (2003). Measuring decent work with statistical indicators. *International Labour Review*, *142*(2), 147–178.
33. Liedtke, C., Baedeker, C., Hasselkuß, M., Rohn, H., &Grinewitschus, V. (2015). User-integrated innovation in Sustainable LivingLabs: An experimental infrastructure for researching and developing sustainable product service systems. *Journal of Cleaner Production*, *97*, 106–116.
34. Doyle, M. W., & Stiglitz, J. E. (2014). Eliminating extreme inequality: A sustainable development goal, 2015–2030. *Ethics & International Affairs*, *28*(1), 5–13.
35 Tukker, A., Charter, M., Vezzoli, C., Stø, E., &Andersen, M. M. (Eds.). (2017). *System Innovation for Sustainability 1: Perspectives on Radical Changes to Sustainable Consumption and Production*. Routledge.
36. Hammer, J., &Pivo, G. (2017). The triple bottom line and sustainable economic development theory and practice. *Economic Development Quarterly*, *31*(1), 25–36.
37. Toth, F. L. (2003). *Ecosystems and Human Well-Being: A Framework for Assessment*. Island Press.
38. Bennett, M. I., & Giloth, R. P. (Eds.). (2012). *Economic Development in American Cities: The Pursuit of an Equity Agenda*. SUNY Press.
39. Sharma, A., &Ranjan, R. (2017) Software effort estimation using neuro fuzzy inference system: past and present. *International Journal on Recent Innovation Trends in Computing and Communication (IJRITCC)*, *5*(8), 78–83.
40. Kumar, A., Sharma, A., Sharma, S., &Kashyap, S. (2017, July). Performance analysis of keyword extraction algorithms assessing extractive text summarization. In *2017 International Conference on Computer, Communications and Electronics (Comptelix)* (pp. 408–414). IEEE.
41. Costanza, R., de Groot, R., Sutton, P., Van der Ploeg, S., Anderson, S. J., Kubiszewski, I., Farber, S., &Turner, R. K. (2014). Changes in the global value of ecosystem services. *Global Environmental Change*, *26*, 152–158.

42. Brook, J. W., & Pagnanelli, F. (2014). Integrating sustainability into innovation project portfolio management–a strategic perspective. *Journal of Engineering and Technology Management*, *34*, 46–62.
43. Kumar, A., & Joshi, A. (2017, March). Ontology driven sentiment analysis on social web for government intelligence. In *Proceedings of the Special Collection on eGovernment Innovations in India* (pp. 134–139). ACM.
44. Kumar, A., & Khorwal, R. (2017). Firefly algorithm for feature selection in sentiment analysis. In Bhabani Sankar Pradhan & Subhransu Sekhar Dash (Eds.), *Computational Intelligence in Data Mining* (pp. 693–703). Springer. https://doi.org/10.1007/978-81-322-3799-5_59
45. Singh, S. P., Nayyar, A., Kumar, R., & Sharma, A. (2019). Fog computing: From architecture to edge computing and big data processing. *The Journal of Supercomputing*, *75*(4), 2070–2105.

Chapter 12

Green manufacturing approaches

Case study

Sasi Kumar Mani, Kumaresan Mayakannan, Sindhu Selvaraj, Nanssy Vijayakumar, Dharani Subramani, and Vaishali Rajasekaran

12.1 INTRODUCTION

Green manufacturing, sometimes referred to as sustainable manufacturing or eco-friendly manufacturing, is a revolutionary method of industrial processes that focuses on reducing the harmful effects on the environment and preserving energy and natural resources. This cutting-edge approach incorporates sustainability concepts into every aspect of the production process, including design, sourcing, manufacturing, distribution, and disposal at the end of the product's life cycle. Green manufacturing, through its focus on efficiency, waste reduction, and utilisation of renewable resources, not only tackles environmental issues but also improves operational efficiency and supports long-term economic sustainability. With the increasing global awareness of environmental challenges, green manufacturing has become crucial in promoting a more sustainable future for enterprises and society worldwide [1]. This case study examines the evolution of green manufacturing practices across a variety of industries, as well as green management strategies. The advent of mass manufacturing techniques in the early 19th century caused a revolutionary change in industrial output. During this time, assembly lines and other forms of mechanisation greatly improved production efficiency. There was less regard for environmental implications, however, as the focus was squarely on boosting productivity and decreasing costs. A break from inflexible mass production paradigms was signalled by the 1970s paradigm shift towards flexible manufacturing systems. These systems were designed to be flexible and quick to respond to changes in the market, allowing for a wider range of products and more personalisation. Sustainable practices were also progressively acknowledged during this age as enterprises began to incorporate more eco-friendly methods into their operations. Thus, the idea of "Green Manufacturing" developed in the 20th century [2]. The green manufacturing (GM) technique allows for substantial economic improvements while also maintaining environmental integrity [3]. One definition of green productivity describes it as encompassing all actions aimed at reducing waste. They emphasised the impact that green productivity might have on the entire production performance.

DOI: 10.1201/9781003470342-12

The ISO, or International Organisation for Standardisation, released the initial global standard for Environmental Management Systems (EMS) called ISO14001. EMS can serve as a mechanism for implementing GM. One study showed that the implementation of GM (green management) can lead to improvements in economic, environmental, and social performance by reducing waste and costs [4]. A study demonstrated that companies that engage in green management and green innovation can enhance their environmental performance while meeting consumer demand, so enhancing their corporate image with regulators and the public [1].

12.2 BACKGROUND

During the period of the "13th Five-Year Plan", the Ministry of Industry and Information Technology placed a significant amount of focus on environmentally friendly manufacturing as an essential component of the development of environmentally friendly industries. A document titled "Green Manufacturing Engineering Implementation Guide (2016–2020)" was handed out by them. By the time the year 2021 came to a close, more than 300 significant GM projects had been organised and put into action. Furthermore, 184 GM system solution suppliers were issued, over 500 GM standards were developed, and 2,783 green factories, 223 green industrial parks, and 296 green supply chain companies were cultivated and created [5]. All of these accomplishments were made through the implementation of GM. These efforts have been extremely important in guiding the sector towards a transformation that is environmentally friendly and low in carbon emissions. The current scenario regarding environmental protection has resulted in certain regions adopting measures that are overly simplistic and even excessive. These regions have resorted to a "one-size-fits-all" approach, which has had some negative impacts on the social development and stability of the local community. This study broke the policy down into four distinct phases: the time of exploration, which lasted from 1949 to 1977; the period of formal establishment, which lasted from 1978 to 2001; the period of improvement and strengthening, which lasted from 2002 to 2011; and the period of comprehensive improvement, which spans from 2012 to the present day [6]. One of the main obstacles is the perceived greater upfront expenses linked to the adoption of green technologies and practices. Nevertheless, as time has passed, numerous environmentally friendly technologies have grown more financially viable as a result of progress and increased production efficiency [7]. Implementing GM practices generally necessitates modifications across the entire supply chain, encompassing activities such as procuring sustainable resources and assuring the use of environmentally friendly packaging and transportation methods. Ensuring the coordination of these endeavours among many vendors and partners can present difficulties. Staying abreast of changing environmental legislation and standards in many locations and jurisdictions can be intricate. Companies must ensure that they are well informed and adjust their manufacturing processes as necessary in order to maintain compliance. Consumers' demands for environmental responsibility are growing, compelling corporations to embrace more sustainable practices in order to stay competitive in the market. Nevertheless, properly conveying the environmental advantages of products to consumers can provide a challenge. Anticipating the precise requirements of future generations is a challenging task that demands flexibility and strategic foresight. Achieving a harmonious equilibrium among economic expansion, safeguarding the environment, and fostering social progress can be intricate and give rise to conflicts. Attaining sustainable development necessitates global collaboration to tackle worldwide challenges such as climate change and the depletion of resources. The notion of sustainable development was popularised by describing it as "development that satisfies the requirements of the current generation without jeopardising the capacity of future generations to satisfy their own requirements." Through international conferences, initiatives such as Agenda 21 and the Sustainable Development Goals (SDGs), I have played a crucial role in advancing sustainable development [8]. Green manufacturing refers to the implementation of innovative production techniques with the goal of minimising negative environmental effects through waste reduction, waste recycling, improved utilisation of natural resources, and other related initiatives [9]. The concept goes beyond the traditional "reduce, reuse, and recycle" strategy of 3R and adopts a 5R approach: "repair, reuse, refurbish, rebuild, and recycle". This technique allows for more efficient utilisation of resources and extends the lifespan of products [10]. Attaining a state of "zero waste" is indeed possible. For the sake of sustainable development, casting and forging are increasingly being phased out. It is necessary to find solutions to the economic issues that plague the manufacturing industry, which include low profit margins, high costs of capital, and unstable market conditions [11]. A German government group comprised of scientists and industry leaders introduced the "Industry 4.0" concept in 2011. It proposed ways in which the ideas and technology of the Internet of Things and Services may be used in the manufacturing sector, combining political ambitions with technological advances [1,2]. A key component of Industry 4.0 is the ongoing use of intelligent technology to automate conventional production processes. Cyber-Physical Systems that function in manufacturing [12]. Sustainability and ecological concerns are rapidly emerging as perhaps the primary focus for crucial business, management, manufacturing, and product development decisions. Green manufacturing emerged in the early 1990s as an innovative production method aimed at reducing environmental pollution and mitigating the negative impacts of resource utilisation. Several exploratory studies have been conducted to identify sustainable practices in small and medium-sized enterprises (SMEs) [6]. A model was created to establish a correlation between the L&G (Labour and General) approach and waste. The industrial industry has consistently faced the difficulty of a dearth of L&G competence. In order to optimise the application of the L&G approach in the sector, it is necessary to establish a methodical approach that enables employees to comprehend the advantages of L&G. The analysis reveals a deficiency in policy support and dedication from both the government and industry participants when it comes to adopting the L&G method. The L&G model emphasises personnel, machine, material, money, and environmental factors as the primary components of the concept [1]. Due to the rising consumer environmental awareness (CEA), there is a growing global demand for green products [13].

12.3 METHODOLOGY

The research methodology involves the formulation of effective methods and the creation of a benchmarking questionnaire focused on GM. The participants were asked to rate the extent to which their organisations implemented each dimension of green concept practices on a five-point scale, where 1 represents good, 2 represents average, 3 represents better, 4 represents outstanding, and 5 represents best. The research also includes describing the influence of environmental factors on design choices for manufacturing processes and products, evaluating a technological product to assess its appropriateness for environmentally friendly production, and implementing design modifications as necessary to achieve the desired technological result, guided by well-informed design judgement criteria. The study examines the precise requirements, such as acceptable variations, for the final product that will be produced considering the limitations and advantages of the manufacturing site and emphasises choosing an environmentally friendly manufacturing process while implementing stringent quality control measures to ensure that the units satisfy the specified standards and tolerances [14]. It also includes efficient managing resources and ensures precise execution of procedures in accordance with applicable rules of practice by supervising the production process and making necessary adjustments. The research further assesses the efficacy of the manufacturing process in fulfilling environmental requirements and evaluates the extent to which the manufacturing process has achieved success in meeting environmentally friendly criteria [2]. The concept of LCSA was initially discussed in literature around 15 years ago. The purpose of this analysis is to examine the pertinent literature on life cycle sustainability assessment (LCSA) and provide a comprehensive review of the current advancements in LCSA within the manufacturing sector. The overview will serve as a foundation for future scientific advancements in this field. In 2008, Klöpffer devised a conceptual formula for LCSA, which stands for life cycle sustainability assessment. The formula is as follows: LCSA=LCA+LCC+SLCA. The analysis reveals that the case studies provide distinct system boundaries when implementing the LCSA approach, particularly in terms of the life cycle phases that are taken into account. This literature evaluation yields two distinct outcomes. This study has examined and outlined the methodology of various methods to LCSA as applied in case studies within the industrial domain, highlighting significant disparities in their implementation. The number three is green. The process of methodically engineering a product has three stages: (i) conceptual, preliminary, and detailed design; (ii) production and construction; and (iii) operational use and system support. Green supply chain management (GSCM) encompasses practices such as reuse, remanufacturing, and recycling [15]. Agricultural production is on the rise in order to address the demands of food security, but it also presents significant environmental risks. Agricultural Green Productivity Technologies (AGPTs) are eco-friendly innovations that support the practice of sustainable agriculture. Adoption rates of advanced genetic testing procedures (AGPTs) are low in developing nations, notwithstanding their advantages. Behavioural variables are of utmost importance in determining farmers' decisions to embrace certain practices. This article introduces a new multi-attribute decision analysis (MADA) technique termed the probabilistic linguistic preference selection index (PL-PSI) method. The PL-PSI framework combines the PSI approach, which gives priority to competing qualities, with PLTSs to represent farmers' language judgements while accounting for ambiguity. This enables the recording of the subtle aspects of farmers' perspectives of the diverse elements that influence their decisions to adopt [16]. The GM methods are categorised in Table 12.1.

Table 12.1 Green manufacturing methods are categorised

Types of green manufacturing	*Functions*
Design for environment	• The methodology prioritises environmental factors from product creation to disposal. • Improves product functionality and reduces unnecessary parts can reduce material usage. • Utilise recyclable and biodegradable materials. Easily disassembled and remade. Product-use energy efficiency assessment.
Life cycle assessment (LCA)	• LCA evaluates the environmental effect of a product or process throughout its lifecycle, including resource extraction, material processing, manufacture, transportation, usage, and disposal/recycling. • LCA identifies improvement areas and prioritises eco-friendly approaches to reduce environmental impact.
Lean manufacturing	• While not exclusively environmental, Lean principles typically align with green activities. • Focusing on waste elimination and efficiency, Lean can have environmental benefits. • Process step reduction, inventory reduction, and product flow optimisation can reduce energy and resource utilisation.

(*Continued*)

Table 12.1 (Continued) Green manufacturing methods are categorised

Types of green manufacturing	*Functions*
Circular economy	• Promotes product lifecycle extension through disassembly, reuse, remanufacturing, and recycling. • To reduce waste and resource depletion, use materials longer. • Design goods for several lifecycles to follow circular economy ideas. Designing excellent reverse logistics for used goods collection. Instead of discarding, reuse or fabricate components.
Use of renewable energy	• Replacing fossil fuels with renewable energy sources like solar, wind, or geothermal electricity can drastically minimise the environmental effect of manufacturing operations. • Installing renewable energy infrastructure can save money over time.
Green supply chain management	• It prioritises sustainable procedures and works with green suppliers. • Consider supplier's environmental record when buying.
Continuous improvement	• Green production is ongoing. A commitment to continuous improvement is needed to implement these techniques. Monitor environmental performance and find solutions to lower the environmental footprint for long-term success.
Pollution prevention (P2)	• P2 aims to prevent pollution rather than address it following its occurrence. • Includes methods such as the following: • Utilising cleaner production technology to reduce waste. • Optimising procedures to save energy and materials. • Water conservation measures implemented. Replace dangerous materials with safer ones.

12.4 CASE STUDY ON GREEN MANUFACTURING IN VARIOUS INDUSTRIES

12.4.1 Green manufacturing in textile industry

This article employs a bibliometric analysis to investigate research trends and patterns in GM for the textile sector. This study examines the expansion of this discipline over the last three decades and highlights significant areas of research.

This article investigates the development of this topic over the course of the previous three decades and identifies important study areas [17].

12.4.1.1 Summary

GM is becoming increasingly popular. The textile sector has a notable environmental impact, but the adoption of GM processes is being seen as a solution. There has been a significant increase in the number of articles on this topic since 2013. Research areas: the research encompasses multiple disciplines, with engineering, environmental sciences, and business being the primary fields of focus. Key themes encompass life cycle evaluation, lean methodologies, sustainability, cleaner production, and waste management. The *Journal of Cleaner Production* is the foremost journal for conducting research on green production in the textile industry. China, the USA, and India are the foremost collaborators in this subject. Significant publications: the papers with the highest number of citations go into subjects such as the breakdown of dyes using photo catalysis and the management of a sustainable supply chain.

12.4.1.2 Environmental impact on textile industry

The textile sector has a substantial influence on environmental contamination, causing considerable effects on: Water pollution occurs as a result of dyeing and finishing operations that emit toxic chemicals. Air pollution is caused by the release of pollutants from industrial plants and energy production sources. Land degradation occurs as a result of the extensive cultivation of cotton and the creation of synthetic fibres. Waste production arises from the accumulation of textile remnants, discarded garments, and minuscule particles of plastic.

12.4.1.3 Trends and findings from the research

The quantity of articles on eco-friendly textile production has increased dramatically from 2013 onwards. The increasing concern about the industry's impact on the environment is evident from this trend. Engineers, environmental scientists, and business experts all work together in this field's multidisciplinary research approach. This emphasises how important it is to address sustainable textile production from every angle. Focus of investigation: the term "life cycle assessment" (LCA) refers to the process of analysing a product's effect on the environment from the time it is created until it is thrown away. Eliminating inefficiencies and maximising productivity is the goal of lean manufacturing. Integrating social, economic, and environmental factors into textile production techniques is what we mean when we talk about sustainability. Less pollution and hazardous waste: making production less polluting by using new technologies and methods. Management of waste: research and development of textile waste recycling, composting, and reuse technologies. Collaborating with top sources: one of the best places to get studies about environmentally friendly textile production is in the *Journal of Cleaner Production*. Research collaboration in this subject is spearheaded by China, the USA, and India. Most influential works: the development of strategies to break down toxic textile dyes has been the subject of highly referenced research. One such topic is photo catalytic degradation of dyes. Making the textile supply chain more ethical and less harmful to the environment is the goal of sustainable supply chain management.

12.4.1.4 Green manufacturing as solution

GM as a solution: resource reduction refers to the practice of minimising the use of resources, such as water and energy, throughout the production process. Material selection: opting for sustainable and recyclable materials whenever feasible. Pollution prevention is the implementation of cleaner industrial processes aimed at minimising both waste and emissions. Waste management include the practices of recycling, composting, and discovering inventive applications for textile waste. LCA is a method used to examine the environmental impact of a product from the moment its raw materials are extracted until its disposal. Lean manufacturing involves the optimisation of production processes with the aim of minimising waste and maximising efficiency.

12.4.1.5 Notes

The information was culled from the Scopus database in September 2023. Reviews and publications that have been peer-reviewed were the primary objects of the analysis. The authors call for additional studies to investigate digital and circular technologies in clothing using a variety of data sources.

12.4.1.6 Future research

In the future, researchers may look into new data sources and investigate developing technologies such as digital tools for maximising resource use and decreasing waste, or the concepts of the circular economy as they pertain to closed-loop textile production. For example: Figure 12.1 represents the case study on Arvind Ltd Textile Industry.

12.4.2 Green manufacturing in cement industry

GM in the cement sector is examined in this article. GM reduces environmental effect during production. Cement makers may reduce waste, energy use, and pollution by adopting greener techniques [18].

12.4.2.1 Objective

This study seeks to examine the correlation between the implementation of environmentally friendly manufacturing techniques and the efficiency of operations at a cement firm located in Kenya. This involves evaluating whether implementing environmentally friendly practices can result in reduced expenses or enhanced operational effectiveness. Notable features or significant aspects: the article outlines essential environmentally friendly methods that are applicable to cement production, such as effective waste management to reduce dust, maximising energy efficiency by utilising alternative fuels, and optimising processes by incorporating recycled materials.

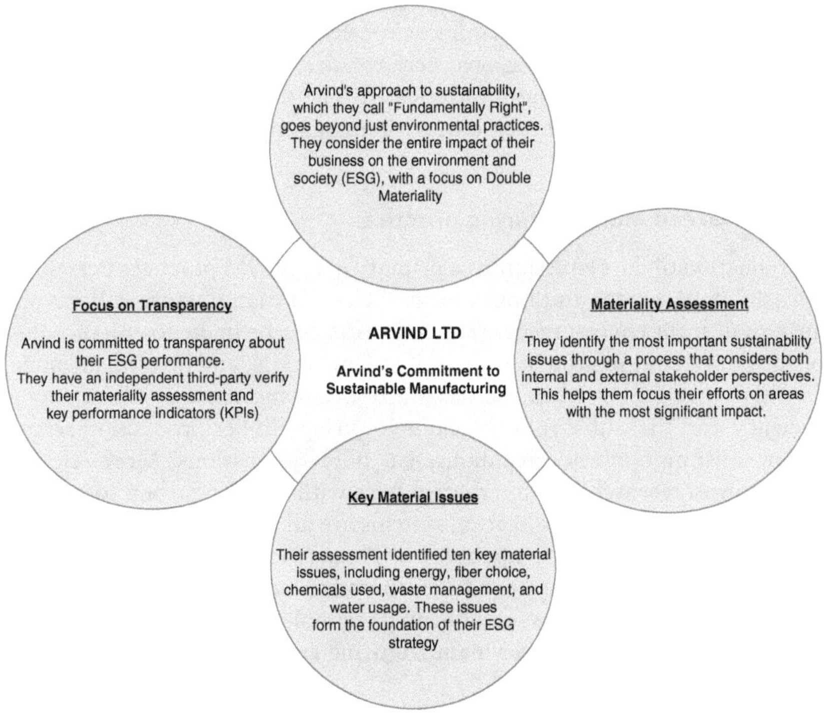

Figure 12.1 Case study on Arvind Ltd Textile Industry.

Analysis theoretical frameworks: the study examines stakeholder theory and institutional theory to gain insight into the factors that motivate corporations to embrace environmentally sustainable practices. Stakeholder concerns and external pressures arising from rules and social norms are taken into account.

12.4.2.2 Efficiency and effectiveness of operations

The study investigates the financial implications of burning clinker, which are influenced by the use of environmentally friendly production techniques. Minimising waste and reducing energy use can result in financial savings. Compliance with the environmental rules established by the National Environment Management Authority (NEMA) of the Kenyan government is of utmost importance for cement producing enterprises.

12.4.2.3 Research and knowledge gap

The research aims to investigate the correlation between environmentally friendly practices and operational efficiency through a singular case study, while also identifying any existing gaps in research and knowledge. Conducting additional study with a more extensive sample size could enhance the robustness of the results. The potential influence of environmentally friendly methods on other dimensions of operational performance, such as product quality or production speed, has not been investigated. The study does not explore particular green technology or their efficacy in mitigating environmental damage.

12.4.2.4 Green manufacturing practices

The reduction of dust emissions is a primary goal of GM practices pertaining to waste management and minimisation. Cement manufacture is known to cause significant environmental harm. Efforts can be made to enhance dust collecting systems, maximise the use of raw materials, and investigate potential alternative fuel sources such as waste oil or biomass. Goals of energy efficiency and fuel substitution include lowering clinker production-related energy consumption and greenhouse gas (GHG) emissions. More effective use of power, research into alternative fuels with smaller carbon footprints, and the installation of new heating systems are all ways to achieve this goal. Design, modifications, and upgrades of Panasonic processes: this includes enhancing current processes and incorporating new technology to reduce environmental effect. Innovative burning technologies and the use of recovered waste materials in clinker manufacturing are two examples.

12.4.2.5 Extra points

Stakeholder theory: sustainable manufacturing must include stakeholder issues like environmental regulations and public pressure, according to the study. Institutional theory: government rules and social norms might urge corporations to go green to compete.

12.4.2.6 Overall assessment

The study provides a valuable viewpoint on how the cement sector might adopt environmentally friendly methods. Companies can enhance their environmental sustainability and potentially enhance their financial performance by applying measures such as waste reduction, energy efficiency, and process optimisation. The article advocates for additional research to investigate the wider ramifications of GM on the cement sector. For example: Figure 12.2 represents the case study on JSW Cement Industry.

12.4.3 Green manufacturing in automobile industry

The study discusses automotive GSCM. GSCM promotes green techniques from raw material procurement to product disposal. The study discusses how carmakers might lower their environmental effect with GSCM [19].

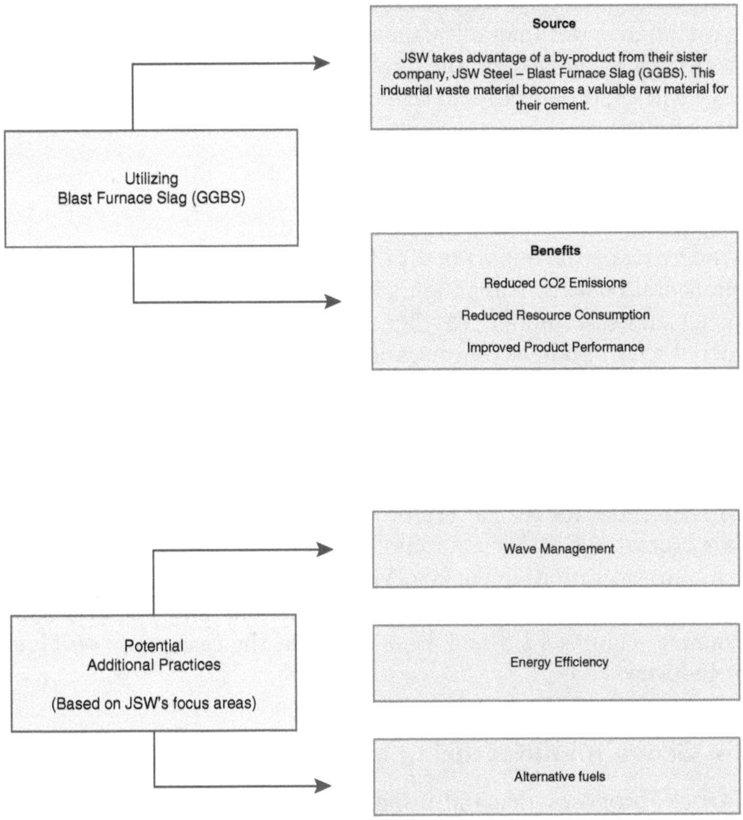

Figure 12.2 Case study on JSW Cement Industry.

12.4.3.1 GSCM in automobile

Uses recycled materials, LCA for environmental effect evaluation, and fuel-efficient technologies such as MPF-i, CRD-i, and hybrids. Green operations: eco-friendly production, logistics, and waste management. This includes GM and remanufacturing: using appropriate technology and materials and reusing components to reduce waste. Optimising reverse logistics and network design to recycle or reuse used autos and parts. Reduce waste by reducing sources and preventing pollution.

12.4.3.2 Barriers to the implementation of green soppy chain management (GSCM) in India

This study identifies the specific problems encountered by the Indian automobile sector when attempting to embrace GSCM methods. Some of these challenges include inadequate technological infrastructure, obsolete manufacturing methods, insufficient number of highly skilled workers, lack of enough governmental assistance, financial considerations, insufficient customer knowledge and low market interest in environmentally friendly vehicles, organisational reluctance to embrace change, and intense market competition.

12.4.3.3 Conclusion

This study stresses the importance of GSCM in the automotive industry for environmental reasons. It highlights how GSCM may reduce environmental impact and save money. Further research is needed on product selection tactics that maximise output and enhance brand value. Additionally, GSCM benefits must be communicated to the Indian auto sector.

12.4.3.4 Key takeaways

GSCM is essential for the car sector to go green. Green design and operations are crucial to GSCM in automobile production. Social, technical, and economic considerations make GSCM adoption in India difficult. A cleaner atmosphere requires overcoming these limitations and boosting GSCM. For example: Figures 12.3 and 12.4 represent the case study on Hyundai Motor Industry.

12.4.4 Green manufacturing in electronics industry

This article proposes demand-responsive data-driven sustainable intelligent manufacturing for energy-intensive businesses. Using data-driven approaches and demand responsiveness, GM improves energy efficiency and lowers costs. This requires optimising production processes and energy consumption with real-time data and modifying production schedules to match lower energy demand and cost [20].

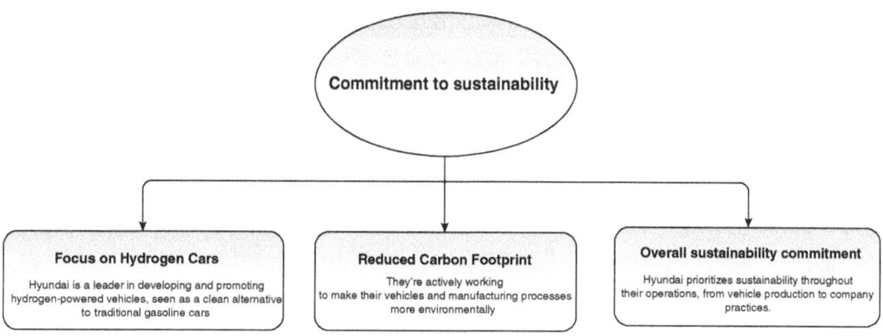

Figure 12.3 Case study on Hyundai Motor Automobile Industry.

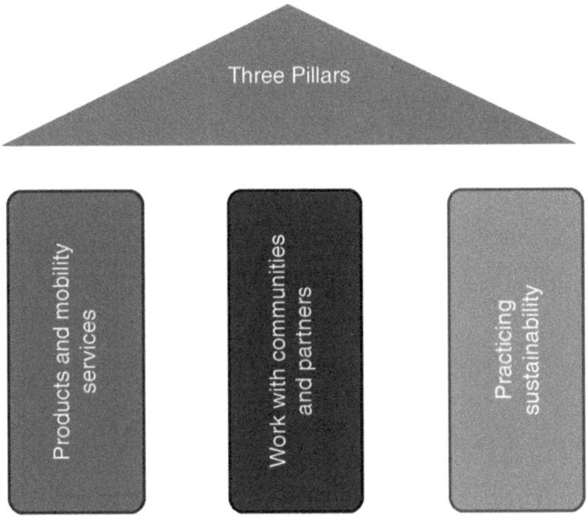

Figure 12.4 Three Pillars of Hyundai Motor Automobile Industry.

12.4.4.1 Proposed framework feature

Data collection: sensors and metres measure energy use in real time from machine, shop floor, and factory levels. Energy monitoring: data are checked for energy waste, equipment idle, and improvement opportunities. Energy efficiency assessment: products, teams, and equipment are assessed for energy efficiency improvements. Production schedules and equipment functioning are optimised using energy efficiency and demand response (DR) methods. DR: production schedules use more energy outside peak hours when electricity prices are lower.

Authors' study results: sustainable intelligent manufacturing in energy-intensive industries is achieved using data. They propose an enterprise-level energy efficiency optimisation framework and technology architecture. Using peak and off-peak electricity pricing, multi-level DR models optimise energy use for machines, shop floors, and factories.

12.4.4.2 Problem

Steel, cement, and chemical industries, which use a lot of energy, have a big influence on the environment and high costs as a result of their energy consumption. This study addresses both concerns by outlining a methodology that makes use of data analysis and astute changes to accomplish: the framework's goal is to reduce energy consumption by making process and equipment operation more efficient. The goal of the framework is to reduce energy costs by coordinating production with times when electricity prices are lower, a process known as demand response. Reduced energy consumption is one way the framework helps make manufacturing more sustainable and less of an environmental impact.

12.4.4.3 Impact

This research provides energy-intensive industry with useful tools: the data-driven framework helps to develop sustainable and cost-effective manufacturing methods. Technological architecture: serves as a blueprint for factory-wide energy optimisation framework design and implementation. Models with multiple DR levels allow for flexible implementation of DR techniques at multiple levels (machines, shop floors, and factories) based on energy usage trends. Less energy means reduced expenses and a smaller environmental impact. Enhanced production efficiency: energy efficiency optimises production. Cost savings: DR reduces energy costs.

12.4.4.4 Data and discussion

Optimisation analysis: production plans and equipment operation are carefully organised to use less energy and fulfil targets. This may involve sequencing activities: reduce machine switching and energy use by rearranging production activities. Avoid equipment breakdowns and energy waste with predictive maintenance based on real-time sensor data. Variable-speed drives: based on production demands, adjust motor and pump speeds to save energy during low demand. The DR technique uses dynamic electricity pricing. Production plans shift energy-intensive jobs to off-peak hours when electricity prices are lower. Imagine a factory scheduling more metal smelting at night when energy prices are lower.

12.4.4.5 Future research

Implementation: a greener industrial future is within reach, according to this research. What is needed to put the proposed framework into action is: Investing in data infrastructure and sensors: for the purpose of gathering and analysing data on energy consumption in real time. Acquiring proficiency in data analytics: to understand data and draw conclusions that can be used for improvement. Working together with utilities: gaining insight into and making the most of existing DR initiatives. For example: Figure 12.5 represents the case study on Philips Electronics Industry.

12.4.5 Green manufacturing in agriculture industry

This article examines emerging economy smallholder rice farmers' AGPT adoption decisions. It suggests a novel MADA method for analysing these elements [16].

12.4.5.1 Framework for study

The framework is utilised in this study to classify behavioural characteristics into the following categories: characteristics such as personality, risk tolerance, and environmental concerns are examples of dispositional traits.

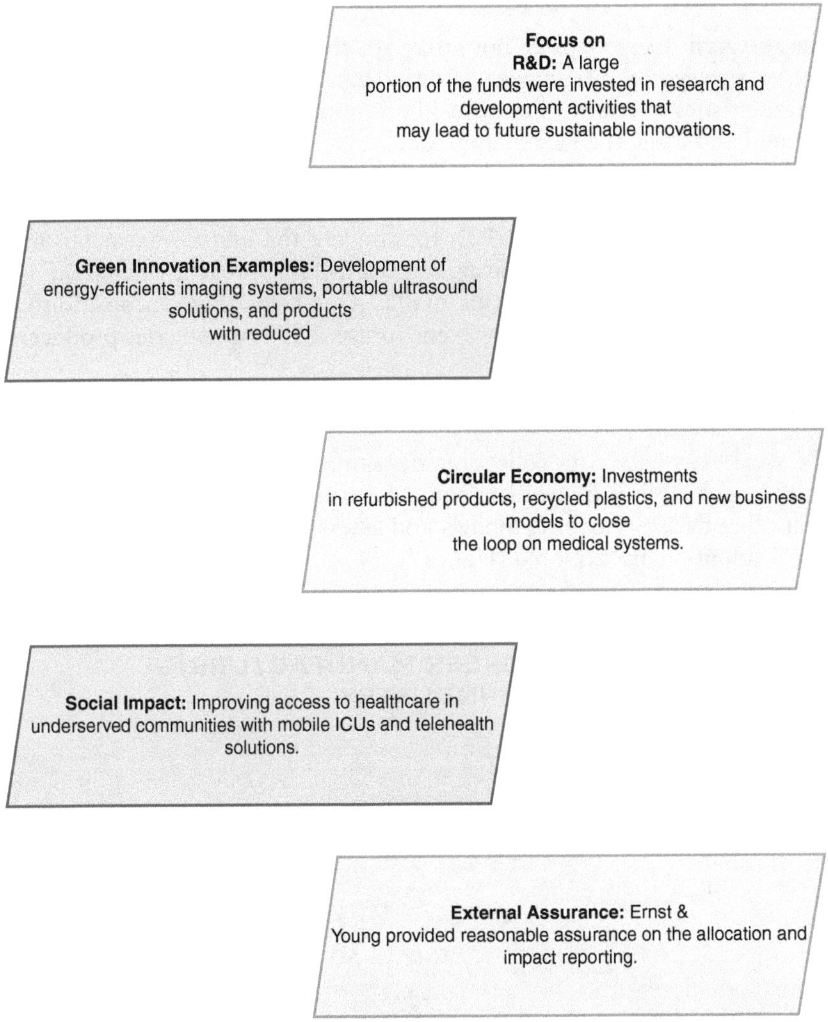

Figure 12.5 Case study on Philips Electronics Industry.

Relationships with other farmers and the influence of social norms are examples of social factors. The cognitive aspects of AGPTs include things like knowledge, perceived costs and benefits, and perceived risks.

12.4.5.2 Research gap

This study seeks to address the lack of information by investigating a wide range of behavioural characteristics that encourage the use of AGPT in rice cultivation. Probabilistic linguistic term sets (PLTSs) are utilised to tackle the uncertainties and ambiguities in decision-making.

12.4.5.3 Analyse and data collection

The research does not state how they got their data, but it was probably farmer surveys or interviews. In order to determine which behavioural characteristics are most important in promoting AGPT adoption, the data are analysed using the PL-PSI approach.

The study is expected to add to the existing body of knowledge by conducting an analysis of a thorough framework of behavioural factors that influence the adoption of AGPT. To simulate the unknowns in farmers' choices, we use PLTSs. By creating a new PL-PSI approach, we find out the factors affecting the adoption of AGPT. This helps those in positions of power to better understand how to encourage AGPTs among rice producers.

12.4.5.4 Future research

The work may have data collection and sample size constraints. Empirical PL-PSI applicability in real life could be studied. Comparison of PL-PSI with other decision-making models and assessment of policy initiatives for AGPT adoption are explored (Figure 12.6).

12.5 ADOPTATION OF GREEN MANUFACTURING TECHNOLOGIES BY INDUSTRY

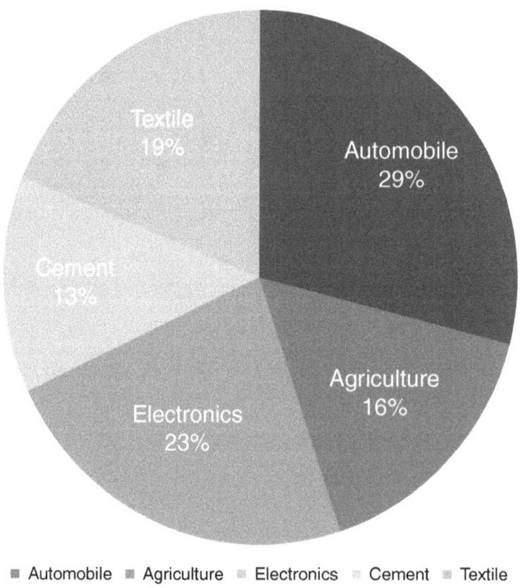

Figure 12.6 Represents the adaptation of green manufacturing technologies by various industries [18].

12.6 CONCLUSION

GM could help the manufacturing industry become more sustainable (Figure 12.6). GM helps businesses and the environment by reducing environmental impact, preserving resources, and improving output. As demand for eco-friendly products rises and environmental rules tighten, GM could help the manufacturing industry become more sustainable. GM helps businesses and the environment by reducing environmental impact, preserving resources, and improving output. As customer demand for eco-friendly products rises and environmental rules tighten, GM may become increasingly more important for industry success. Continuous innovation and collaboration will be needed to expand green practices across all industries. New technology, workforce training, and green supply chains are needed for this change. GM benefits enterprises, society, and the environment. GM could help the manufacturing industry become more sustainable. GM helps businesses and the environment by reducing environmental impact, preserving resources, and improving output. As customer demand for eco-friendly products rises and environmental rules tighten, GM may become increasingly more important for industry success. Continuous innovation and collaboration will be needed to expand green practices across all industries. New technology, workforce training, and green supply chains are needed for this change. GM benefits enterprises, society, and the environment. GM may become even more important for industry. Continuous innovation and collaboration will be needed to expand green practices across all industries. New technology, workforce training, and green supply chains are needed for this change.

REFERENCES

1. W. D. Leong, H. L. Lam, W. P. Q. Ng, C. H. Lim, C. P. Tan, and S. G. Ponnambalam, Lean and green manufacturing—A review on its applications and impacts. *Process Integration and Optimisation for Sustainability*, vol. 3, pp. 5–23.Mar. 15, 2019, Springer. doi:10.1007/s41660-019-00082-x
2. S. V. Ghinmine and D. I. Sangotra, Implementation of green manufacturing in industry-A case study, *International Journal of Research in Engineering and Technology*, vol. 04, pp. 42–45, 2015. [Online]. Available: https://www.ijret.org
3. C. W. Y. Wong, K. H. Lai, K. C. Shang, C. S. Lu, and T. K. P. Leung, Green operations and the moderating role of environmental management capability of suppliers on manufacturing firm performance, *International Journal of Production Economics*, vol. 140, no. 1, pp. 283–294, Nov. 2012, doi:10.1016/j.ijpe.2011.08.031
4. A. Ahmed and S. Vij, Industry 4.0 and green sustainable manufacturing: A smarter and effective process management, *Journal of Mountain Research*, vol. 17, no. 1, 2022, doi:10.51220/jmr.v17i1.9

5. M. Rinaldi, M. Caterino, M. Fera, P. Manco, and R. Macchiaroli, Technology selection in green supply chains - The effects of additive and traditional manufacturing, *Journal of Cleaner Production*, vol. 282, p. 124554, Feb. 2021, doi:10.1016/J.JCLEPRO.2020.124554
6. K. Sharma, R. Arora, R. Nangia, R. Singel, Dharmveer, and S. Dixit, Effects of green manufacturing and technological innovations on sustainable development, *Materials Today: Proceedings*, vol. 69, pp. 266–270, Jan. 2022, doi:10.1016/J.MATPR.2022.08.475
7. S. Ruidas, M. R. Seikh, and P. K. Nayak, A production inventory model for green products with emission reduction technology investment and green subsidy, *Process Integration and Optimisation for Sustainability*, vol. 6, no. 4, pp. 863–882, Dec. 2022, doi:10.1007/S41660-022-00258-Y/FIGURES/5
8. J. Mensah, Sustainable development: Meaning, history, principles, pillars, and implications for human action: Literature review, *Cogent Social Sciences*, vol. 5, no. 1, Jan. 2019, doi:10.1080/23311886.2019.1653531
9. M. Singh, K. Singh, and A. Singh, Green manufacturing practices in manufacturing industry: A state of art, *Asian Journal of Engineering and Applied Technology*, vol. 7, no. 2, pp. 52–56, Oct. 2018, doi:10.51983/ajeat-2018.7.2.956
10. J. Sarkis and A. Rasheed, Greening the manufacturing function, *Business Horizons*, vol. 38, pp. 17–22, 1995.
11. A. Haleem, M. Javaid, R. P. Singh, R. Suman, and M. A. Qadri, A pervasive study on green manufacturing towards attaining sustainability, *Green Technologies and Sustainability*, vol. 1, no. 2, p. 100018, May 2023, doi:10.1016/J.GRETS.2023.100018
12. J. Zamorano, M. Alfaro, V. M. de Oliveira, G. Fuertes, C. Durán, R. Ternero, J. Sabattin, and M. Vargas, New manufacturing challenges facing sustainability, *Manufacturing Letters*, vol. 30, pp. 19–22, Oct. 2021, doi:10.1016/J.MFGLET.2021.09.003
13. Q. Zhang, Q. Zhao, X. Zhao, and L. Tang, On the introduction of green product to a market with environmentally conscious consumers, *Computers & Industrial Engineering*, vol. 139, p. 106190, Jan. 2020, doi:10.1016/J.CIE.2019.106190
14. V. A. Kirik, S. Cheng, N. I. Vyunova, O. V. Galustyan, S. S. Gamisonija, and S. D. Galustyan, Education of future green engineers for achieving sustainable development in green manufacturing industry, *International Journal of Engineering Pedagogy*, vol. 11, no. 5, pp. 138–145, 2021, doi:10.3991/IJEP.V11I5.22165
15. I. D. Paul, G. P. Bhole, and J. R. Chaudhari, A review on green manufacturing: It's important, methodology and its application, *Procedia Materials Science*, vol. 6, pp. 1644–1649, Jan. 2014, doi:10.1016/J.MSPRO.2014.07.149
16. L. O. Asiedu-Ayeh, X. Zheng, K. Agbodah, B. S. Dogbe, and A. P. Darko, Promoting the adoption of agricultural green production technologies for sustainable farming: A multi-attribute decision analysis, *Sustainability (Switzerland)*, vol. 14, no. 16, Aug. 2022, doi:10.3390/su14169977
17. M. S. I. Sarker and I. Bartok, Global trends of green manufacturing research in the textile industry using bibliometric analysis, *Case Studies in Chemical and Environmental Engineering*, vol. 9, p. 100578, Jun. 2024, doi:10.1016/J.CSCEE.2023.100578

18. S. M. Eshikumo and S. O. Odock, Green manufacturing and operational performance of a firm: Case of cement manufacturing in Kenya, *International Journal of Business and Social Science*, vol. 8, pp. 106–120, 2017. [Online]. Available: www.ijbssnet.com
19. P. Sanghavi, Y. Rana, S. Shenoy, and R. Yadav, A review on green supply chain management in automobile industry, *International Journal of Current Engineering and Technology*, vol. 5, pp. 3697–3702, 2015. [Online]. Available: https://inpressco.com/category/ijcet
20. S. Ma, Y. Zhang, Y. Liu, H. Yang, J. Lv, and S. Ren, Data-driven sustainable intelligent manufacturing based on demand response for energy-intensive industries, *Journal of Cleaner Production*, vol. 274, Nov. 2020, doi:10.1016/j.jclepro.2020.123155

Chapter 13

Lean and green manufacturing

Gurusamy Seetharam, Sankaranarayanan Bathrinath, and Priyanka Ramesh

13.1 INTRODUCTION

Lean manufacturing, derived from the Toyota Production System, is a methodology designed to optimize production processes by minimizing waste and maximizing value [1]. It has gained widespread recognition for its effectiveness in enhancing operational efficiency, reducing costs, and improving customer satisfaction [2]. While the benefits of lean manufacturing are well-documented, its implementation presents distinct challenges, particularly for small and medium-sized enterprises (SMEs) [3]. Unlike larger organizations, SMEs often operate with limited resources, face significant financial constraints, and may lack in-house expertise, making the adoption of lean practices more complex [4].

The unique characteristics of SMEs, such as their smaller scale, tighter budgets, and constrained personnel, present specific hurdles to successful lean implementation [5]. Financial limitations can restrict their ability to invest in lean tools, technologies, and training. Additionally, resistance to change from employees and management, coupled with a lack of Lean expertise, can hinder the effective adoption of lean principles [6,7]. To address these challenges and fully harness the potential benefits of lean manufacturing, it is crucial to understand the factors that influence its implementation in the SME context [8,9]. This study aims to investigate the challenges SMEs face in adopting lean manufacturing practices and provide actionable insights for overcoming these barriers. The research objectives are as follows:

RO1. Identify and Categorize Key Challenges: To identify SMEs' primary challenges in implementing lean manufacturing and categorize them into cause-and-effect factors. Understanding these challenges will help in pinpointing the root causes that impede lean adoption and their subsequent effects on the implementation process.

RO2. Analyze Cause-and-Effect Relationships: To analyze the cause-and-effect relationships between various factors influencing lean adoption using the Pythagorean fuzzy decision-making trial and evaluation laboratory (DEMATEL) method. This approach will provide a detailed understanding of how different factors interact and impact the effectiveness of lean practices in SMEs.

DOI: 10.1201/9781003470342-13

RO3. Develop Practical Recommendations: Based on the analysis, develop practical recommendations for overcoming the identified challenges. These recommendations will be tailored to the unique needs of SMEs and aim to provide strategies for enhancing lean implementation, improving operational efficiency, and achieving sustainable growth.

RO4. Contribute to the Body of Knowledge: To contribute to the existing body of knowledge on lean manufacturing by focusing specifically on the SME context. The study will offer valuable insights into how SMEs can navigate the complexities of lean adoption and provide a foundation for future research and practice in this area.

13.2 LITERATURE REVIEW

Lean manufacturing is a widely adopted methodology designed to improve operational efficiency by eliminating waste and optimizing processes [10,11]. Originating from the Toyota Production System, lean principles focus on streamlining operations, enhancing value, and promoting continuous improvement. These principles, including value stream mapping, just-in-time production, and kaizen (continuous improvement), aim to reduce production costs, enhance quality, and increase efficiency. Successful implementation of lean practices has been shown to lead to significant improvements in both financial performance and customer satisfaction [12].

However, implementing lean manufacturing presents several challenges. Common barriers include resistance to change, insufficient training, and a lack of management support [13]. Resistance to change often stems from fear of disruption and uncertainty about the benefits of lean practices. Additionally, the lack of expertise and experience in lean methodologies can hinder effective implementation [14].

For SMEs, these challenges are often amplified by their specific characteristics. Financial constraints are a major barrier, as SMEs frequently struggle to allocate resources for lean tools, technologies, and training [15]. The smaller scale of operations and limited workforce can also restrict SMEs' ability to adopt comprehensive lean strategies. Moreover, the lack of in-house lean expertise and the need for customized solutions can complicate the implementation process [16].

In SMEs, workforce engagement and consistent leadership support are crucial for successful lean adoption [17]. Many SMEs face difficulties in fully engaging their workforce and securing unwavering leadership commitment. Additionally, the limited data collection systems and performance measurement tools typical of SMEs can affect their ability to track lean progress and measure success effectively [18].

The Pythagorean fuzzy DEMATEL method provides a valuable approach for analyzing complex cause-and-effect relationships, particularly in the context of lean manufacturing. This method offers a nuanced way to handle uncertainty and ambiguity in decision-making processes, enabling a deeper understanding of how various factors interact and influence lean implementation. Table 13.1 represents key challenges of adopting lean manufacturing in small and medium-sized enterprises (SMEs).

This literature review highlights that while lean manufacturing offers substantial benefits, its implementation in SMEs involves distinct challenges related to financial constraints, resistance to change, and expertise gaps. Understanding these challenges is essential for developing effective strategies to overcome barriers and achieve successful lean adoption. The insights gained from the Pythagorean fuzzy DEMATEL method can further illuminate these complexities, providing valuable guidance for SMEs seeking to enhance their lean practices.

Table 13.1 Key challenges of adopting lean manufacturing in small and medium-sized enterprises

S.no	*Factors*	*Description*	*Literature review(s)*
C1	Limited financial resources	SMEs often face budget constraints, making it difficult to invest in lean tools, technologies, and staff training.	[11]
C2	Resistance to change	Employees and management may resist adopting lean practices due to fear of disruption, additional workload, or uncertainty about the benefits.	[12]
C3	Lack of lean expertise	SMEs typically have limited in-house expertise in lean principles, making it harder to initiate and sustain lean transformations.	[13]
C4	Inadequate workforce engagement	Difficulty in gaining full workforce participation can hinder the success of lean projects, especially in environments where employees are not empowered.	[14]
C5	Time constraints	SMEs often operate with a lean workforce and cannot afford to allocate time for extensive process improvement initiatives.	[15]
C6	Inconsistent leadership support	Without strong and consistent support from leadership, lean initiatives may lose momentum and fail to deliver expected results.	[16]
C7	Limited data and metrics	Many SMEs lack robust data collection systems, making it challenging to track performance, identify waste, and measure the success of lean initiatives.	[17]
C8	Supply chain challenges	SMEs often rely on external suppliers, who may not adopt lean principles, leading to inefficiencies that SMEs cannot directly control.	[18]
C9	Customization of lean tools	Standard lean tools may not be fully applicable to the unique processes of SMEs, requiring customization that adds complexity to implementation.	[19]
C10	Short-term focus	SMEs often focus on short-term profitability and survival, making it difficult to invest in long-term lean initiatives that may not show immediate results.	[20]

13.3 METHODOLOGY

13.3.1 Sampling and data selection

The sampling for this study involved selecting a group of 33 experts from SMEs who have experience in lean manufacturing practices. The experts were carefully chosen based on their professional background, which included roles such as operations managers, production heads, process improvement specialists, and consultants who have been involved in lean implementations across various industries. The purposive sampling technique was used to ensure that only individuals with relevant expertise and knowledge in lean methodologies and SME environments were included in the study. A structured questionnaire was developed and utilized as the primary data collection tool. The questionnaire was designed to gather expert insights on the key challenges and opportunities related to lean adoption in SMEs.

13.3.2 Pythagorean fuzzy DEMATEL

In this section, we begin by providing a concise overview of the classical DEMATEL method, which serves as a basis for understanding the application of Pythagorean fuzzy DEMATEL. Pythagorean fuzzy DEMATEL is utilized to assess the causal relationships among the challenges that interrupt the lean and green manufacturing in SMEs. Figure 13.1 represents a framework for the Pythagorean fuzzy DEMATEL method.

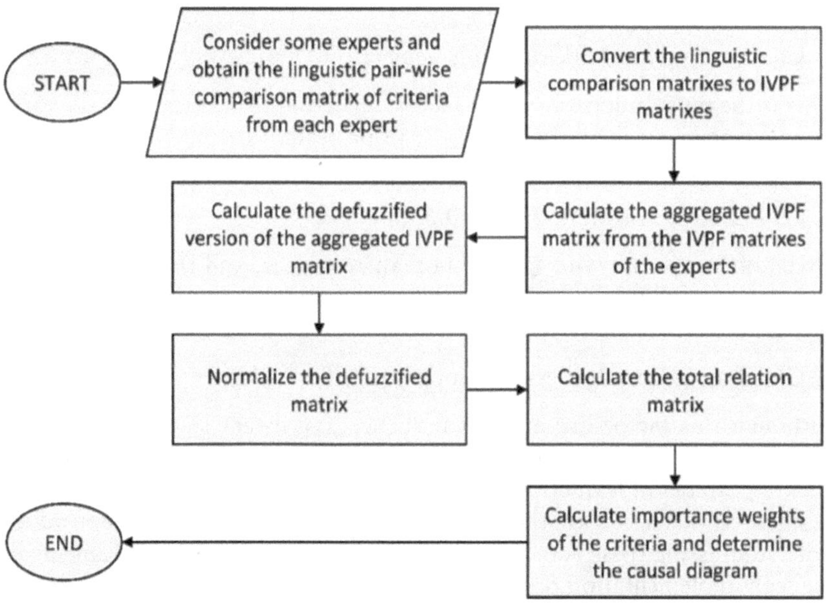

Figure 13.1 Framework for Pythagorean fuzzy DEMATEL [21].

13.4 RESULT AND ANALYSIS

The results of the Pythagorean fuzzy DEMATEL method are presented in [6], which provides a clear distinction between the cause-and-effect factors influencing lean manufacturing adoption in SMEs. These factors are determined based on the values of $(D - R)$, where positive values indicate cause factors and negative values indicate effect factors. Additionally, the values of $(D + R)$ represent the overall prominence of each factor in the system, highlighting their relative importance.

13.4.1 Cause factors

The following factors were identified as cause factors, meaning they have a strong driving influence on the system:

C1 – Limited financial resources $(D-R=0.157)$

SMEs often face budget constraints, making it difficult to invest in lean tools, technologies, and staff training. This factor has a moderate positive influence and drives many of the challenges in lean implementation.

C2 – Resistance to change $(D-R=0.548)$

Resistance from employees and management, stemming from fear of disruption and uncertainty, significantly hinders lean adoption. This factor has a high net influence and must be addressed to ensure smooth implementation.

C4 – Inadequate workforce engagement $(D-R=0.750)$

This is the most influential cause factor, reflecting the critical role of workforce engagement in the success of lean projects. Low participation from employees can lead to the failure of lean initiatives.

C5 – Time constraints $(D-R=0.530)$

SMEs often operate with limited human resources, and this factor shows that the inability to allocate time for lean activities is a significant barrier to successful implementation.

C6 – Inconsistent leadership support $(D - R=0.014)$

Although this factor has a lower influence, consistent leadership support remains a crucial driver for sustaining lean practices. A lack of leadership backing can derail lean efforts.

These cause factors are the primary drivers of inefficiencies in lean adoption. Addressing these root causes can lead to substantial improvements in the lean implementation process.

13.4.2 Effect factors

The following factors were identified as effect factors, meaning they are outcomes influenced by the cause factors:

C3 – Lack of Lean expertise ($D-R=-0.173$)

SMEs typically have limited in-house lean expertise, making it difficult to sustain lean transformations. This factor is impacted by issues such as inconsistent leadership support and resistance to change.

C7 – Limited data and metrics ($D-R=-0.262$)

The absence of robust data systems in SMEs hampers the ability to track lean performance. This factor is a direct outcome of limited resources and time constraints.

C8 – Supply chain challenges ($D-R=-0.473$)

SMEs often depend on external suppliers who may not adopt lean principles. Supply chain inefficiencies are affected by limited engagement and financial constraints within the SME itself.

C9 – Customization of Lean tools ($D-R=-0.296$)

Many SMEs struggle with the applicability of standard lean tools to their specific processes. This effect is influenced by a lack of expertise and financial resources to adapt tools effectively.

C10 – Short-term focus ($D-R=-0.972$)

SMEs tend to prioritize short-term profitability over long-term lean initiatives. This is the most heavily impacted effect factor, influenced by financial constraints, time pressures, and inconsistent leadership.

These effect factors represent the results of the interactions between the driving forces (cause factors). To mitigate the negative impacts on these areas, SMEs must focus on addressing the root causes, particularly financial limitations, workforce engagement, and leadership support. Table 13.2 represents cause/effect analysis.

13.4.3 Causal diagram interpretation

The causal diagram generated from the Pythagorean fuzzy DEMATEL analysis helps visualize the relationships between the cause-and-effect factors. It positions cause factors (C1, C2, C4, C5, C6) as the driving elements of the system, while the effect factors (C3, C7, C8, C9, C10) reflect the outcomes of the system's dynamics.

Cause factors like limited financial resources (C1) and resistance to change (C2) are central to initiating lean implementation challenges.

Effect factors like lack of Lean expertise (C3) and short-term focus (C10) are heavily influenced by the driving causes, resulting in barriers to lean transformation. Figure 13.2 represents the causal diagram.

Table 13.2 Cause/effect analysis

Factors	*D*	*R*	*D + R*	*D − R*	*Cause/effect*
F1	2.501	2.344588	4.846	0.157	Cause
F2	2.611	2.062778	4.673	0.548	Cause
F3	2.264	2.437161	4.701	−0.173	Effect
F4	2.840	2.090069	4.930	0.750	Cause
F5	2.222	1.692438	3.915	0.530	Cause
F6	1.996	1.981679	3.977	0.014	Cause
F7	1.889	2.15122	4.041	−0.262	Effect
F8	1.090	1.563496	2.654	−0.473	Effect
F9	2.157	2.453522	4.611	−0.296	Effect
F10	1.595	2.567175	4.162	−0.972	Effect

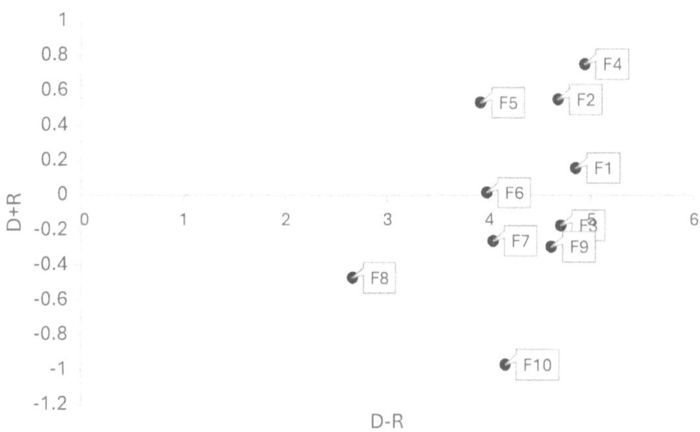

Figure 13.2 Causal diagram.

13.5 DISCUSSION

The results of this study offer critical insights into the challenges and factors influencing lean manufacturing implementation in SMEs. Using the Pythagorean fuzzy DEMATEL method, we identified key cause-and-effect factors that both drive and hinder lean adoption. These findings provide a deeper understanding of how SMEs can navigate the complexities of lean implementation, where limited resources and organizational resistance often pose significant barriers.

13.5.1 Key challenges in lean implementation

The analysis highlights several major challenges faced by SMEs when adopting lean practices. Among these, limited financial resources (C1), resistance to change (C2), and inadequate workforce engagement (C4) emerged as critical cause factors driving lean implementation difficulties. The results suggest that SMEs struggle to allocate adequate funds to invest in lean tools, training, and technologies. Furthermore, resistance from employees and management due to fear of disruption and uncertainty is a significant barrier, impeding the smooth adoption of lean principles.

Another essential challenge is time constraints (C5), as SMEs often operate with leaner workforces and face intense pressure to prioritize daily operations over continuous improvement initiatives. This inability to dedicate sufficient time to lean projects can delay or even derail the implementation process. Additionally, Inconsistent Leadership Support (C6), while a moderate cause factor, indicates that without strong and committed leadership, lean initiatives can lose momentum and fail to deliver their intended outcomes.

13.5.2 Effect factors and their implications

The effect factors, such as lack of Lean expertise (C3), limited data and metrics (C7), and short-term focus (C10), highlight the consequences of the challenges mentioned above. For instance, the lack of Lean expertise within SMEs is a direct result of insufficient investment in training and knowledge-building initiatives. Many SMEs lack the internal capabilities to design and sustain lean transformations, making them reliant on external consultants or incapable of fully leveraging lean tools.

Similarly, Limited Data and Metrics (C7) reflect the inadequate infrastructure within SMEs to collect and analyze performance data, which is vital for tracking lean progress. Without proper data systems, SMEs struggle to measure waste reduction, process efficiency, and overall lean success. The short-term focus (C10) further compounds this problem, as SMEs tend to prioritize immediate profitability over long-term lean investments. This short-sightedness, influenced by financial constraints, inhibits SMEs from reaping the full benefits of lean manufacturing, which often requires a sustained, long-term commitment.

13.5.3 Practical implications for SMEs

The findings from this study suggest several practical implications for SMEs seeking to adopt lean manufacturing practices. First, addressing the cause factors should be the primary focus for SMEs. This includes securing necessary financial resources, fostering a culture that embraces change, and ensuring strong leadership support for lean initiatives. By tackling these root causes, SMEs can significantly reduce the negative impacts seen in the effect factors, such as lack of expertise and limited data systems.

Second, workforce engagement is a pivotal element in lean success. SMEs must invest in employee training and empowerment to overcome resistance and enable a more collaborative lean environment. Employees at all levels should be encouraged to participate in lean initiatives and contribute to continuous improvement efforts.

Additionally, SMEs should consider customizing lean tools and strategies to fit their specific operational needs. The one-size-fits-all approach of standard lean tools may not be fully applicable to SMEs, requiring tailored adaptations to make them more effective in smaller business environments.

13.6 CONCLUSION

This study provides a detailed analysis of the challenges and opportunities for lean manufacturing implementation in SMEs. Key cause factors such as limited financial resources, resistance to change, and inadequate workforce engagement drive many of the difficulties SMEs face in adopting lean practices. Addressing these root issues is crucial for overcoming barriers like lack of Lean expertise and limited data and metrics, which are identified as key factors. To improve lean adoption, SMEs must focus on long-term goals, invest in employee training, and develop robust data systems, despite their financial and operational constraints. By tackling these challenges strategically, SMEs can enhance efficiency and operational performance. The study emphasizes the importance of balancing immediate business needs with sustained lean efforts. Future research should consider larger sample sizes and technological advancements to further explore ways to support SMEs in their lean journey.

REFERENCES

1. Kumar, N., Hasan, S. S., Srivastava, K., Akhtar, R., Yadav, R. K., & Choubey, V. K. (2022). Lean manufacturing techniques and its implementation: A review. *Materials Today: Proceedings*, *64*, 1188–1192.
2. Deshmukh, M., Gangele, A., Gope, D. K., & Dewangan, S. (2022). Study and implementation of lean manufacturing strategies: A literature review. *Materials Today: Proceedings*, *62*, 1489–1495.
3. Palange, A., & Dhatrak, P. (2021). Lean manufacturing a vital tool to enhance productivity in manufacturing. *Materials Today: Proceedings*, *46*, 729–736.
4. Maware, C., Okwu, M. O., & Adetunji, O. (2022). A systematic literature review of lean manufacturing implementation in manufacturing-based sectors of the developing and developed countries. *International Journal of Lean Six Sigma*, *13*(3), 521–556.
5. Antony, J., Psomas, E., Garza-Reyes, J. A., & Hines, P. (2021). Practical implications and future research agenda of lean manufacturing: A systematic literature review. *Production Planning & Control*, *32*(11), 889–925.
6. Buer, S. V., Semini, M., Strandhagen, J. O., & Sgarbossa, F. (2021). The complementary effect of lean manufacturing and digitalisation on operational performance. *International Journal of Production Research*, *59*(7), 1976–1992.
7. Santos, D. M. C. D., Santos, B. K. D., & Santos, C. G. D. (2021). Implementation of a standard work routine using lean manufacturing tools: A case study. *Gestão & Produção*, *28*, e4823.

8. Rahamneh, A., Alrawashdeh, S., Bawaneh, A., Alatyat, Z., Mohammad, A., & Al-Hawary, S. (2023). The effect of digital supply chain on lean manufacturing: A structural equation modelling approach. *Uncertain Supply Chain Management*, *11*(1), 391–402.
9. Ciliberto, C., Szopik-Depczyńska, K., Tarczyńska-Łuniewska, M., Ruggieri, A., & Ioppolo, G. (2021). Enabling the circular economy transition: A sustainable lean manufacturing recipe for Industry 4.0. *Business Strategy and the Environment*, *30*(7), 3255–3272.
10. Johan, A., & Soediantono, D. (2022). Literature review of the benefits of lean manufacturing on industrial performance and proposed applications in the defense industries. *Journal of Industrial Engineering & Management Research*, *3*(2), 13–23.
11. Schmitt, T., Wolf, C., Lennerfors, T. T., & Okwir, S. (2021). Beyond "Leanear" production: A multi-level approach for achieving circularity in a lean manufacturing context. *Journal of Cleaner Production*, *318*, 128531.
12. Gebeyehu, S. G., Abebe, M., & Gochel, A. (2022). Production lead time improvement through lean manufacturing. *Cogent Engineering*, *9*(1), 2034255.
13. Touriki, F. E., Benkhati, I., Kamble, S. S., & Belhadi, A. (2021). An integrated smart, green, resilient, and lean manufacturing framework: A literature review and future research directions. *Journal of Cleaner Production*, *319*, 128691.
14. Vlachos, I. P., Pascazzi, R. M., Zobolas, G., Repoussis, P., & Giannakis, M. (2023). Lean manufacturing systems in the area of Industry 4.0: A lean automation plan of AGVs/IoT integration. *Production Planning & Control*, *34*(4), 345–358.
15. Shahin, M., Maghanaki, M., Hosseinzadeh, A., & Chen, F. F. (2024). Improving operations through a lean AI paradigm: A view to an AI-aided lean manufacturing via versatile convolutional neural network. *The International Journal of Advanced Manufacturing Technology*, *133*(11), 5343–5419.
16. Mulugeta, L. (2021). Productivity improvement through lean manufacturing tools in Ethiopian garment manufacturing company. *Materials Today: Proceedings*, *37*, 1432–1436.
17. Sadiq, S., Amjad, M. S., Rafique, M. Z., Hussain, S., Yasmeen, U., & Khan, M. A. (2021). An integrated framework for lean manufacturing in relation with blue ocean manufacturing-A case study. *Journal of Cleaner Production*, *279*, 123790.
18. Naeemah, A. J., & Wong, K. Y. (2022). Positive impacts of lean manufacturing tools on sustainability aspects: A systematic review. *Journal of Industrial and Production Engineering*, *39*(7), 552–571.
19. Debnath, B., Shakur, M. S., Bari, A. M., & Karmaker, C. L. (2023). A Bayesian best–worst approach for assessing the critical success factors in sustainable lean manufacturing. *Decision Analytics Journal*, *6*, 100157.
20. Marinelli, M., Deshmukh, A. A., Janardhanan, M., & Nielsen, I. (2021). Lean manufacturing and Industry 4.0 combinative application: Practices and perceived benefits. *IFAC-PapersOnLine*, *54*(1), 288–293.
21. Torkzadeh, J., Pamucar, D., & Niroomand, S. (2023). Interval-Vvalued Pythagorean fuzzy extension of DEMATEL for prioritizing and casualty analysis of environmental criteria of organizational behaviour in higher education sector. Informatica, 1–26. https://doi.org/10.15388/23-INFOR541

Chapter 14

Additive manufacturing for sustainable design and production

Eganathan Venkatesaperumal, Jayakrishna Kandasamy, Paulchamy Jeyapandiarajan, and S. Arulvel

14.1 INTRODUCTION

The manufacturing industry is a major contributor to environmental concerns, responsible for 20% of global carbon emissions, 15% of energy consumption, and 30%–40% of global material use. Rapid growth further strengthens these issues, generating waste and pollutants that drain resources and destroy the environment [1]. Subtractive manufacturing produces huge waste during production. In contrast, additive manufacturing (AM) has the possibility to reduce this waste by a notable amount of 90%. Conversely, AM is not entirely waste-free. AM produces less waste, including support structures needed for complex shapes, excess material particles, and waste from printing errors [2]. AM is a technique to produce products efficiently and reduce material waste by depositing material layer by layer using 3D model data. AM methods are classified into seven groups: material jetting, directed energy deposition, powder bed fusion, sheet lamination, vat photopolymerization, material extrusion, and binder jetting [3]. Benefits of additive manufacturing over subtractive manufacturing include the utilization of recyclable materials, the absence of tools, jigs, and fixtures, improved functionality of components, greater material efficiency, and reduced supply chain costs. However, the use of AM in manufacturing industries has been hindered due to its poor surface quality, dimensional exactness, mechanical characteristics, and microstructure compared to subtractive manufacturing (SM) [4].

In the present world, product design entails real-time data sharing and collaboration on essential design concepts. The early design phases (EDS) have crucial steps of processes where assessments about a product drastically impact a substantial portion of future expenses across design, production, assembly, maintenance, and disassembly. Interactive product design plays a vital role in the improvement of different industrial products and practices, involving cost-effective and strategic value. Interactive design is to facilitate the process of knowledge modeling in the initial stages of design. Interactive design is influenced by three key criteria that limit product production: the designer's

 DOI: 10.1201/9781003470342-14

expertise, the end user's satisfaction, and the implementation of desired functionalities [5]. In the product design and production processes, professionals from various disciplines, including engineering, ergonomics, and design, collaboratively work together to achieve optimal product environmental performance. These professionals must possess the requisite skills and equipment to account for all business constraints, including eco-design restrictions [6].

Sustainable manufacturing brings up the integration of processes and systems that can produce superior products and services with less material utilization and optimized energy resources. It also prioritizes the safety of employees, customers, and the surrounding communities, intending to minimize environmental and social impacts throughout the product life cycle [7]. A sustainable manufacturing system must satisfy all the components of the triple bottom line objectives. The three pillars of sustainability are the economy, society, and environment. Several factors are included in each pillar, which are used to evaluate and supervise the sustainable process. Environmental pillars consist of three critical components: waste management, pollution control, and resource utilization. The social pillar is composed of a variety of components, including copyright and patent protection, public acceptance, healthcare enhancement, ethical progress, improvement in society, labor advancement, and superior product quality. The economic aspect encompasses the administration of the supply chain, market development, production expenses, both direct and indirect expenditures, and productivity [8].

14.2 THE SUSTAINABILITY OF ADDITIVE MANUFACTURING TO SUBTRACTIVE MANUFACTURING

SM involves the removal of material from a solid block to achieve the desired shape, while AM involves the addition of material to form the desired shape. AM is frequently implemented in real-world applications to generate intricate and small-scale components with lesser material content in industries such as aerospace, healthcare, and automotive [9]. AM has been extensively employed in the biomedical sector, particularly in the development of customized prosthetics, implant dentistry, organ and tissue structures, anatomy models, and medical devices. Metals, polymers, and ceramics are the three primary categories of biomaterials that are utilized in the production of these implants and human organs [10]. AM has been employed in numerous applications as shown in Figure 14.1 (Table 14.1).

Watson and Taminger [11] implemented a computational model that assessed sustainability performance by examining resource consumption. This model functioned as a decision-support instrument and established a critical volume fraction threshold that is contingent upon resource utilization. Both additive and subtractive methods were equally effective in the product's fabrication, as indicated by the framework. AM methods are more suitable than SM methods, based on smaller volume fractions.

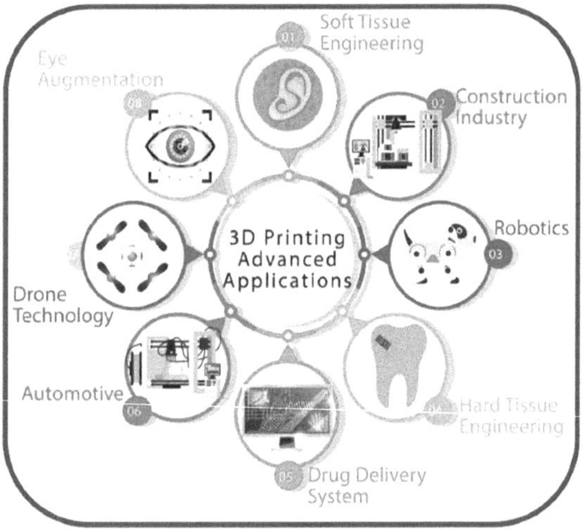

Figure 14.1 Additive manufacturing applications [10].

Table 14.1 Comparison of subtractive and additive manufacturing

Process	*Material compatibility*	*Design complexity*	*Waste*	*Energy consumption*	*Cost*
Subtractive manufacturing	Wider range (metals, wood, plastics, etc.)	Complex designs are challenging	High waste generation by removing excess material	High energy consumption, especially for hard materials	Cost-effective for high volume
Additive manufacturing (3D printing)	Limited range (specific plastics, metals, composites)	Excels at complex geometries	Reduced waste due to layer-by-layer building	Energy-intensive for LPBF processes	Expensive for mass production

Jayawardane et al. [12] introduced a concept called "techno-eco-efficiency" that combines technical, economic, and environmental factors into a unified framework for assessing the sustainability of AM and SM parts. The environmental effects of AM parts are significantly reduced in comparison to traditional SM parts, as evidenced by the implementation of the decision-support framework. Likewise, AM components have been accepted as environmentally friendly, as the increased operating costs associated with AM is offset by the reduced environmental consequences.

Özceylan et al. [13] studied the supply chain cost of orthopedic insoles, comparing traditional manufacturing (TM) with AM technologies, found that AM offered a significant cost benefit. AM offered a 25.75% decrease in supply chain expenses in comparison to TM. The benefit arises from various aspects, including a 26.33% decrease in production costs, a 10.17% decrease in delivery costs, and a 20.44% decrease in inventory holding costs for AM in comparison to TM.

Lyons et al. [14] evaluated the environmental impact of producing a knee implant's femoral component from Ti-6Al-4V alloy. The study compared electron beam melting (EBM), an AM process, to a conventional milling approach. For both techniques, a life cycle assessment from cradle to grave was conducted, with a focus on primary energy consumption (PEC) and carbon dioxide emissions. EBM provides significant environmental benefits in comparison to conventional manufacturing (CM) for producing intricate components such as femoral knee implants. EBM results in a significant 45.5% reduction in primary energy consumption (PEC) and an impressive 68.2% reduction in CO_2 emissions. This research emphasizes the environmental feasibility of AM to produce geometrically intricate components.

14.3 ECO-DESIGN AND ENERGY EFFICIENCY OF ADDITIVE MANUFACTURING

The conceptual framework of eco-design for AM is made up of three fundamental elements: design, environmental sustainability, and AM. The eco-design for AM is represented by the areas where these three dimensions' overlap. The AM dimension investigates respective AM techniques, such as fused deposition modeling (FDM) and laser powder bed fusion (LPBF), their technological characteristics, and the pros and cons associated with them. The design dimension focuses on addressing issues such as generative design, topology optimization, and material design to improve the functional performance of products. The dimension of ecological sustainability examines the environmental implications of AM, such as CO_2 emissions, resource and energy efficiency, and other ecological aspects. The eco-design for AM integrates the three dimensions of design, ecological sustainability, and AM pointed out in Figure 14.2. Its purpose is to optimize the design advantages of AM while minimizing its environmental effects [15].

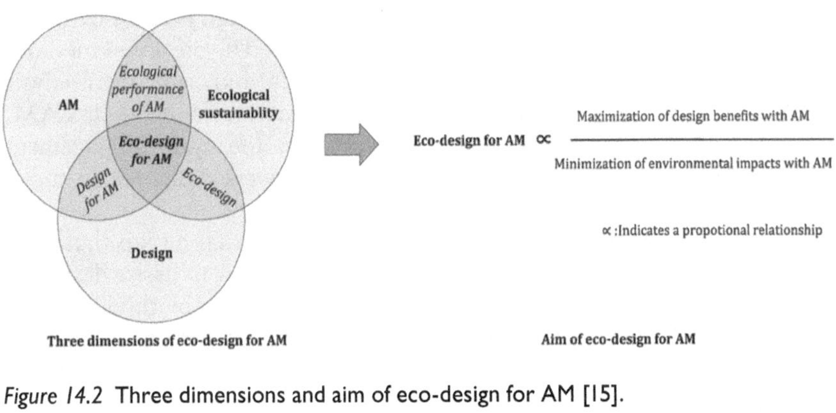

Figure 14.2 Three dimensions and aim of eco-design for AM [15].

Yi et al. [15] developed a methodology for eco-design in AM that integrates energy performance assessment (EPA). To ensure the ecological advantages of AM, it is effective to implement eco-design methodologies in the life cycle assessment (LCA) to assess and mitigate the environmental impacts of AM. However, the application of LCA is restricted to the later phase of the design process, after most of the design and decision activities have been completed, as it requires comprehensive process and inventory data for the full life cycle. The proposed method utilizes a comprehensive framework that includes three components: a simulation tool for predicting energy consumption, an evaluation model for evaluating energy performance, and an overall workflow for eco-design in AM.

Wang et al. [16] introduced a novel concept called eco-design for additive manufacturing (EcoDfAM). This approach integrates the sustainability principles in the preliminary stages of product development for AM. To streamline information reuse for AM design sustainability analysis, they developed the EcoDfAM ontology using the Web Ontology Language (OWL). This ontology focuses on the LPBF method and incorporates terms like performance, eco-parameters, and eco-properties. The effectiveness of the EcoDfAM ontology has been shown in retrieving critical data, including carbon emissions, cost, eco-parameters, and eco-design suggestions, for ecologically friendly and cost-effective AM solutions.

Santiago-Herrera et al. [17] conduct an ex-ante life cycle assessment (LCA) on AM technology. They use a systematic methodology with parametrized modeling to enable a fair comparison with the TM method. The study focuses on a gearbox component and compares the direct energy deposition (WAAM) and casting processes to evaluate their suitability for producing a gearbox. The study findings indicate that the competitiveness of AM technology can be enhanced through process design optimization and upscaling. This can result in a 94% reduction in climate change impacts, with emissions falling from 4,520 to 264 kg CO_2 eq./kg. Additionally, the produced component experiences a significant weight reduction, leading to further fuel savings during its use phase.

Chtioui et al. [18] developed a collaborative eco-design process by incorporating eco-design tools into various design stages. The proposed methodology utilizes the sustainable-failure modes, effects, and criticality analysis (S-FMECA) eco-designing tool to facilitate interaction with CAD, CAM, LCA, topology optimization (TO), and product life cycle management (PLM) software. This integrated approach empowers designers to make environmentally conscious decisions.

Markou et al. [19] developed a method for conceptual design that integrates expert guidance through brainstorming sessions to assist designers in making eco-friendly selections for AM projects. This method employs a customized lifecycle design strategies (LiDS) wheel that is specifically adapted for AM. The AM information is classified according to parameters such as AM category, technique, material state (powder or liquid), material

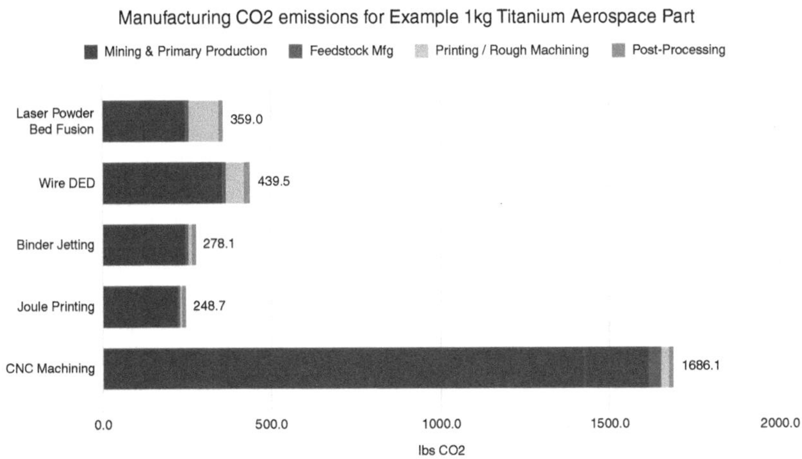

Figure 14.3 CO_2 emission of different AM methods for 1 kg of titanium material [20].

capability (metal or polymer), inert gas requirements, energy consumption rate (ECR) (kWh/kg), post-processing processes, and other requirements. This organized approach has proven successful in promoting environmentally conscious decision-making in AM design.

Figure 14.3 presents the carbon dioxide (CO_2) emissions of different AM methods used to manufacture a 1 kg titanium aerospace component in comparison to computer numerical control (CNC) machining. AM is a highly promising manufacturing process. It has the advantage of significantly reducing CO_2 emissions by using fewer materials during production [20].

Priarone et al. [21] conducted a study to determine the most suitable methods for identifying the production path that produces the least amount of CO_2 emissions and requires the minimum amount of energy. The study focuses on the important process parameters and the resulting impacts of redesign in AM to produce titanium alloy components.

14.4 GENERATIVE DESIGN AND TOPOLOGY OPTIMIZATION IN AM

Generative design is an essential tool in the manufacturing industry, as it allows designers to produce parts that provide substantial value additions. By utilizing computing power, it optimizes the effectiveness of parts and assemblies. It is a technology that operates in a collaborative cloud-based environment, is driven by data from traditional design processes, and relies on automated activities. Generative design employs evolutionary algorithms to identify novel designs, thereby allowing designers to develop high-value components [22].

Coulthard and Wang [23] investigate the impact of combining TO, generative design, and AM on car suspension upright optimization. The researchers found that both topological optimization and generative design techniques resulted in a reduction of volume by 65.17% and 88.13%, respectively, as compared to the original design. The prototype components were fabricated using the FDM AM technique, utilizing acrylonitrile butadine styrene (ABS) material, to serve as a demonstration of the idea. TO resulted in a more robust component, whereas generative design resulted in a refined design with less organic material usage. Research has revealed that the generative process yields substantial and beneficial sustainability outcomes (Figure 14.4).

AM processes have the potential to substantially address the limitations of TM methods in terms of topological design complexities. TO is a technique that optimizes an objective parameter against constraints to generate the optimal material or structural layout. As AM methods improved, TO was extensively integrated into part design workflows to produce intricate shapes and structures [25] (Figure 14.5).

Figure 14.6 demonstrates that the most used density methods in TO software tools are gradient-based, specifically solid isotropic material with penalization (SIMP) and rational approximation of material properties

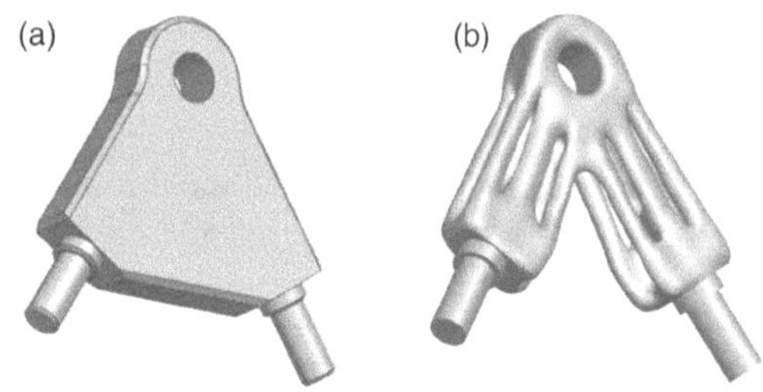

Figure 14.4 Variants of the A-arm: (a) conventional; (b) generative design [24]

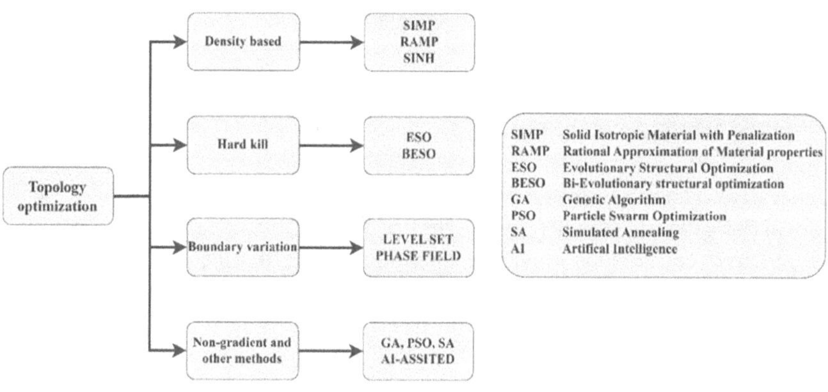

Figure 14.5 Topology optimization methods.

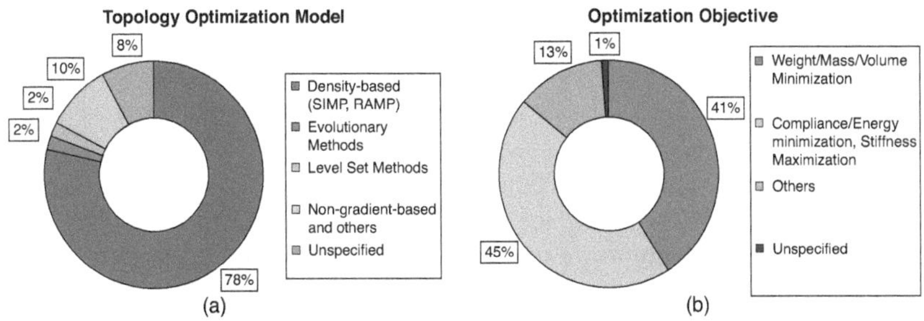

Figure 14.6 Usage of topology-optimization model and optimization objective [25].

(RAMP), with non-gradient approaches accounting for 10% of the total usage. Both volume and compliance minimization are equally valued optimization objectives, and over 43% of case studies use Ti6Al4V material as an increased preference for these approaches.

Mikulikova et al. [26] examined the topology-optimized clutch lever produced through AM. They employed finite element method (FEM) analysis both before and after the TO procedure to attain an equivalent safety factor and stiffness. The optimized component witnessed substantial modifications, such as decreased component weight, enhanced rigidity, and a more attractive design. The utilization of the selective laser melting (SLM) manufacturing technique led to a decrease in the number of components, combining multiple-part assemblies into a single component featuring a novel design and shiny surfaces (Figure 14.7).

Okorie et al. [27] utilized a density-based approach and level-set methodology to enhance the efficiency of an industrial bracket designed for aerospace components. A FDM process was employed to produce the parts, and a compression and tension test was done to compare them. The improved designs successfully achieved a 20% decrease in weight while preserving the original component compression displacement. The findings demonstrate that the utilization of topologically optimized parts may greatly improve the design of actual components, especially those used for weight-sensitive applications in industries.

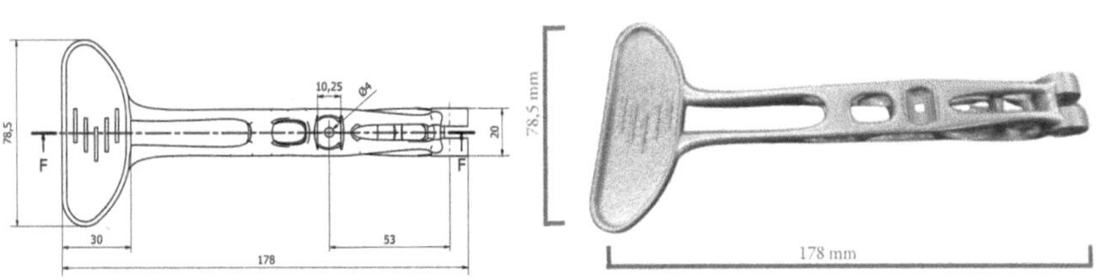

Figure 14.7 Topology-optimized clutch design [26].

Gebisa and Lemu [28] employ a topologically optimal design technique for AM, emphasizing the lightweight design of a jet engine bracket. The decrease in weight results in significant savings in material, processing energy, and cost. The final design achieved a weight reduction of 65%, lowering the original product from 2.067 to 0.72 kg.

Kayacan and Alisihai [29] conducted research to reduce the weight of a steering pump housing for an aircraft's nose landing gear. They achieved this by employing two lattice structures: gyroid and Voronoi. The Gyroid lattice reduced the component's weight by 32.15% and had a maximum Von Mises stress value of 343.85 MPa, while the Voronoi lattice reduced the weight by 38.30% and had a maximum Von Mises stress value of 344.86 MPa. The use of gyroid and Voronoi lattices effectively reduced the component's weight while preserving its structural integrity.

14.5 AM CIRCULAR ECONOMY AND SUSTAINABILITY STRATEGIES

A circular economy (CE) emphasizes optimizing resource utilization and minimizing waste generation for disposal. This approach strategically employs waste reduction and recycling through principles like reusing, redesigning, reclaiming, and remanufacturing [30]. The CE approach emphasizes using materials compatible with AM systems, including recycled, biodegradable, and renewable materials. Meticulous material selection and procurement are crucial throughout the entire product lifecycle, from primary production to repair, maintenance, refurbishment, and remanufacturing. Safety, cost, and origin of both vendors and materials are vital considerations [31]. Additionally, recycling specific polymers like ABS, PP, and polylactic acid (PLA) has gained traction in AM to enhance the CE. While fossil fuel-based virgin polymers are still used, particularly for 3D-printed nylon fabrics, there is a growing shift toward bio-based and recycled plastics. However, the mechanical properties of recycled plastics can be limited in AM due to degradation during recycling [32] (Figure 14.8).

Plastics and their related composites are widely used in various industries because of their adaptability in creating customized parts. The significant increase in plastic usage has made it extremely difficult to manage plastic garbage at the end of its life cycle. Each year, a significant amount of plastic garbage is disposed of in landfills. This waste has the potential to enter the environment, causing severe harm to ecosystems. There is a need for innovative methods to reuse, recycle, and reutilize waste plastic to tackle these global issues [33]. Volatile organic compounds (VOCs) such as styrene, cyclohexanone, butanol, and ethylbenzene are released by ABS, PLA, and nylon, which are regularly employed in FDM [34].

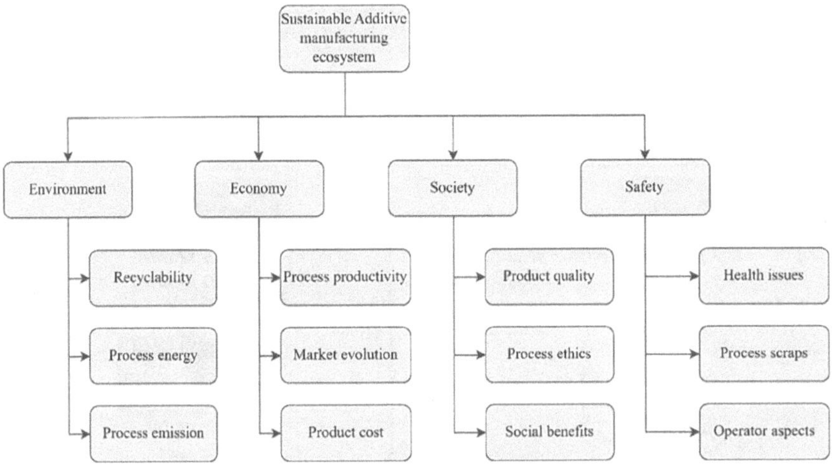

Figure 14.8 Sustainable additive manufacturing ecosystem.

14.5.1 Recycle

Rashid and Koç [33] analyzed the utilization of recycled PET (rPET) derived from waste plastic bottles as a feedstock for fused filament fabrication (FFF). In addition, they examined filaments that blend rPET with virgin carbon fiber-reinforced polyamide-6 (PA6-CF). The chemical and thermal properties of the rPET-based filaments were analyzed, afterwards they underwent mechanical testing on 3D-printed specimens. The study revealed that the rPET/PA6-CF composite holds a greater tensile strength than pure rPET. Finally, these findings have the potential to support a CE by transforming the plastic waste into feedstock for FFF through a straightforward process, thereby achieving sustainable production and consumption (Figure 14.9).

Olawumi et al. [35] conducted an analysis to compare the efficacy of virgin plastic and recycled plastic waste in 3D printing. The study investigates the environmental consequences of utilizing recycled materials in 3D printing through a sustainability analysis and a LCA. The findings highlight benefits such as enhanced resource efficiency, reduced waste, and opportunities for customization. The process of recycling plastic waste involves collecting, sorting, cleaning, and shredding the material into smaller fragments. Subsequently, the waste is melted and transformed into a filament for the purpose of mechanical recycling. Chemical recycling methods such as depolymerization or pyrolysis are used to transform waste into usable raw materials. After that, the test specimens are fabricated using both traditional and recycled materials, and their performance is assessed for mechanical testing.

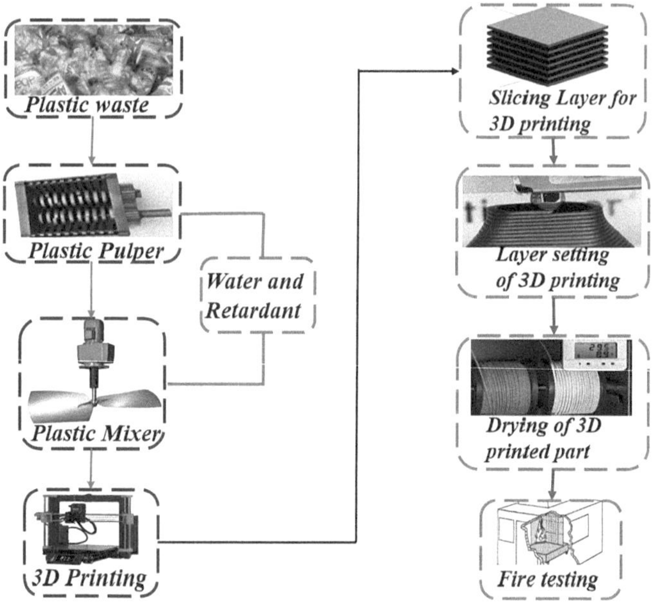

Figure 14.9 Plastic waste to 3D printed product [35].

14.5.2 Reuse

Metal AM is an emerging discipline that allows for greater design freedom and produces novel materials that would be difficult to manufacture using traditional production methods. Effective monitoring is critical for successful powder bed fusion procedures because it directly affects the characteristics and performance of the final parts. This brings up the idea of reusing powder, which is beneficial for business and the environment. Powder reuse methods consist of single batch and collective aging, topping up, and refreshing. Single batch and collective aging methods provide excellent traceability, but they lead to fast deterioration of both powder and component characteristics. Top-up refers to straightforward techniques for handling powder, but it also has the potential for applications that do not need mechanical processes. The refreshing technique reduces waste and maintains the quality of the parts. It is not extensively used due to the difficulty of handling and the limited traceability of the powder [36].

Contaldi et al. [37] investigated the effects of precipitation hardening stainless steel using direct metal laser sintering with both virgin and recycled powder. The researchers conducted a comparison of powder characteristics and static and fatigue behavior and discovered that the mechanical properties of martensitic stainless steel were not affected by the reuse of powder, except for the elongation at break. The characteristics of austenitic stainless steel were significantly impacted at the yield strength and elongation at break.

Mohd Yusuf et al. [38] studied the characteristics of recycled 316L stainless steel and AlSi10Mg powders for laser powder bed fusion additive manufacturing (LPBF AM). The findings revealed that both recycled powders increased particle dimensions and altered the particle size distribution (PSD) toward greater magnitudes in comparison to virgin powders. The recycled 316L SS powder particles maintained their spherical and almost spherical shapes, but the recycled AlSi10Mg powder exhibited more irregular shapes. The recycled powder mean circularity decreased by around 2%, whereas the AlSi10Mg powder lost 17%.

14.5.3 Circular design strategies prefer product integrity across their lifecycle

1. **Design for Attachment and Trust:** The objective of this strategy is to develop products that consumers value and maintain for an extended period. AM has the potential to significantly contribute to this attempt by facilitating personalization and customization.
2. **Design for Reliability and Endurance:** Products must perform consistently over a definite lifespan with appropriate maintenance.
3. **Design for ease of Maintenance and Repair:** Products designed for easy repair prolong their working life. AM has the capability for auxiliary part production.
4. **Design for Upgradability and Adaptability:** Products should be able to accommodate modifications to enhance functionality or meet current needs.
5. **Design for Standardization and Compatibility:** The use of standardized components across products enables future repairs and adaptability.
6. **Disassembly and Reassembly Design:** Products that are designed for effortless disassembly facilitate the efficient recovery of materials and the reuse of components.
7. **Design for Recyclability:** Product designs that facilitate material recovery, reduce waste, and conserve natural resources. But make sure the recycled materials retain their original characteristics.

AM facilitates the purpose of circular design techniques, which involve making improvements and repairs, even on products that were not originally designed for such reasons. However, to fully utilize AM capacity for a CE, various obstacles must be addressed. The main challenge is to create materials that are both long-lasting and capable of being recycled at the end of a product's lifespan. Furthermore, monolithic and geometrically complex parts designed for recyclability might hinder efficient material recovery [39]. To optimize the potential of AM in a CE, designers should evaluate how AM might facilitate various product lifecycles. Incorporating AM concerns into the design process can help to develop new types of products that perform in a CE.

14.6 CHALLENGES IN SUSTAINABLE ADDITIVE MANUFACTURING SYSTEMS

AM presents an opportunity for sustainable production through on-demand manufacturing and reduced waste. The environmental benefits of the product are hindered because of limited eco-friendly materials, energy-intensive manufacturing processes, the generation of powder waste, and the absence of standardized design for sustainability [40]. There is a lack of design guidelines and tools that are precisely designed to adapt the unique features such as support structures, material properties, and wall thickness of AM processes. The AM process requires a lot of energy, which makes it less environmentally friendly. To mitigate this, more energy-efficient printing processes must be developed to protect the environment. Powder waste results from the lack of ability to reuse residual materials from the AM process, recycling of powder waste must be done to reduce the cost of material. Robust methods are essential to evaluate the environmental impacts and lifespan costs of components and products that are manufactured using AM. In the present day, there is a lack of qualified trainers who possess a complete understanding of AM technologies and processes. The future of AM is supported by the development of a competent workforce.

14.7 CONCLUSION

The sustainable possibility of AM can be enhanced through eco-design and TO. Eco-design contributes to a sustainable future by integrating strategic part design, optimized printing for lower energy consumption, and reusable materials. These benefits are measured using LCA and the eco-efficiency method. Generative and TO have become the future of product design for AM parts, in which different design models are generated through AI technology, and the designer chooses a suitable design for the application without compromising the product's strength and stiffness. AM presents a transformative opportunity to adopt the principles of a CE, with the potential to reduce environmental impact and material usage by up to 90% compared to conventional methods. Materials that reveal greater circularity, reusability, or recyclability will be more practical for future use in AM as national legislation keeps pushing the industry toward eco-friendly materials and methods. Nevertheless, there are still challenges, including the development of efficient recycling methods for specific materials and reducing energy consumption in certain AM processes. By engaging in research, innovation, and collaboration, the AM industry can overcome these obstacles and establish a sustainable future.

REFERENCES

1. C. Gao, S. Wolff, S. Wang, Eco-friendly additive manufacturing of metals: Energy efficiency and life cycle analysis, *J. Manuf. Syst.* 60 (2021) 459–472. https://doi.org/10.1016/j.jmsy.2021.06.011
2. T. Peng, K. Kellens, R. Tang, C. Chen, G. Chen, Sustainability of additive manufacturing: An overview on its energy demand and environmental impact, *Addit. Manuf.* 21 (2018) 694–704. https://doi.org/10.1016/j.addma.2018.04.022
3. Y. Huang, M.C. Leu, J. Mazumder, A. Donmez, Additive manufacturing: Current state, future potential, gaps and needs, and recommendations, *J. Manuf. Sci. Eng. Trans. ASME* 137 (2015). https://doi.org/10.1115/1.4028725
4. R. Huang, M. Riddle, D. Graziano, J. Warren, S. Das, S. Nimbalkar, J. Cresko, E. Masanet, Energy and emissions saving potential of additive manufacturing: The case of lightweight aircraft components, *J. Clean. Prod.* 135 (2016) 1559–1570. https://doi.org/10.1016/j.jclepro.2015.04.109
5. X. Fischer, J.-P. Nadeau, Research in interactive design, *Res. Interact. Des.* 3 (2011). https://doi.org/10.1007/978-2-8178-0169-8
6. F. Laverne, F. Segonds, N. Anwer, M. Le Coq, Assembly based methods to support product innovation in design for additive manufacturing: An exploratory case study, *J. Mech. Des.* 137 (2015). https://doi.org/10.1115/1.4031589
7. A. Moldavska, T. Welo, The concept of sustainable manufacturing and its definitions: A content-analysis based literature review, *J. Clean. Prod.* 166 (2017) 744–755. https://doi.org/10.1016/j.jclepro.2017.08.006
8. B. Purvis, Y. Mao, D. Robinson, Three pillars of sustainability: In search of conceptual origins, *Sustain. Sci.* 14 (2019) 681–695. https://doi.org/10.1007/s11625-018-0627-5
9. G. Ingarao, P.C. Priarone, Y. Deng, D. Paraskevas, Environmental modelling of aluminium based components manufacturing routes: Additive manufacturing versus machining versus forming, *J. Clean. Prod.* 176 (2018) 261–275. https://doi.org/10.1016/j.jclepro.2017.12.115
10. Z.U. Arif, M.Y. Khalid, R. Noroozi, M. Hossain, H.T.H. Shi, A. Tariq, S. Ramakrishna, R. Umer, Additive manufacturing of sustainable biomaterials for biomedical applications, *Asian J. Pharm. Sci.* 18 (2023). https://doi.org/10.1016/j.ajps.2023.100812
11. J.K. Watson, K.M.B. Taminger, A decision-support model for selecting additive manufacturing versus subtractive manufacturing based on energy consumption, *J. Clean. Prod.* 176 (2018) 1316–1322. https://doi.org/10.1016/j.jclepro.2015.12.009
12. H. Jayawardane, I.J. Davies, G. Leadbeater, M. John, W.K. Biswas, "Techno-eco-efficiency" performance of 3D printed impellers: An application of life cycle assessment, *Int. J. Sustain. Manuf.* 5 (2021) 44–80. https://doi.org/10.1504/IJSM.2021.116871
13. E. Özceylan, C. Çetinkaya, N. Demirel, O. Sabırlıoğlu, Impacts of additive manufacturing on supply chain flow: A simulation approach in healthcare industry, *Logistics* 2 (2017) 1. https://doi.org/10.3390/logistics2010001
14. R. Lyons, A. Newell, P. Ghadimi, N. Papakostas, Environmental impacts of conventional and additive manufacturing for the production of Ti-6Al-4V knee implant: A life cycle approach, *Int. J. Adv. Manuf. Technol.* 112 (2021) 787–801. https://doi.org/10.1007/s00170-020-06367-7

15. L. Yi, M. Glatt, P. Sridhar, K. de Payrebrune, B.S. Linke, B. Ravani, J.C. Aurich, An eco-design for additive manufacturing framework based on energy performance assessment, *Addit. Manuf.* 33 (2020). https://doi.org/10.1016/j.addma.2020.101120
16. Y. Wang, T. Peng, Y. Xiong, S. Kim, Y. Zhu, R. Tang, An ontology of eco-design for additive manufacturing with informative sustainability analysis, *Adv. Eng. Informatics* 60 (2024). https://doi.org/10.1016/j.aei.2024.102430
17. M. Santiago-Herrera, E. Igos, J.M. Alegre, S. Martel-Martín, R. Barros, Ex-ante life cycle assessment of directed energy deposition based additive manufacturing: A comparative gearbox production case study, *Sustain. Mater. Technol.* 39 (2024). https://doi.org/10.1016/j.susmat.2023.e00819
18. N. Chtioui, R. Gaha, S. Chatti, A. Benamara, S-FMECA based collaborative design proposal for additive manufacturing methodology, *Ann. "Dunarea Jos" Univ. Galati, Fascicle XII, Weld. Equip. Technol.* 34 (2023) 36–48. https://doi.org/10.35219/awet.2023.04
19. F. Markou, F. Segonds, M. Rio, N. Perry, A methodological proposal to link design with additive manufacturing to environmental considerations in the early design stages, *Int. J. Interact. Des. Manuf.* 11 (2017) 799–812. https://doi.org/10.1007/s12008-017-0412-1
20. A. Huckstep, Economics of metal additive manufacturing, *Digit. Alloy. Guid. Met. Addit. Manuf.* (2019). https://www.digitalalloys.com/blog/economics-metal-additive-manufacturing
21. P.C. Priarone, G. Ingarao, V. Lunetto, R. Di Lorenzo, L. Settineri, The role of re-design for additive manufacturing on the process environmental performance, *Procedia CIRP* 69 (2018) 124–129. https://doi.org/10.1016/j.procir.2017.11.047
22. C. Hyunjin, A study on application of generative design system in manufacturing process, *IOP Conf. Ser. Mater. Sci. Eng.* 727 (2020). https://doi.org/10.1088/1757-899X/727/1/012011
23. J. Coulthard, C.J. Wang, Generative design and topology optimisation of products for additive manufacturing, *Adv. Transdiscipl. Eng.* 25 (2022) 81–86. https://doi.org/10.3233/ATDE220570
24. S. Junk, N. Rothe, Lightweight design of automotive components using generative design with fiber-reinforced additive manufacturing, *Procedia CIRP* 109 (2022) 119–124. https://doi.org/10.1016/j.procir.2022.05.224
25. O. Ibhadode, Z. Zhang, J. Sixt, K.M. Nsiempba, J. Orakwe, A. Martinez-Marchese, O. Ero, S.I. Shahabad, A. Bonakdar, E. Toyserkani, Topology optimization for metal additive manufacturing: Current trends, challenges, and future outlook, *Virtual Phys. Prototyp.* 18 (2023). https://doi.org/10.1080/17452759.2023.2181192
26. A. Mikulikova, J. Mesicek, J. Karger, J. Hajnys, Q.P. Ma, A. Sliva, J. Smiraus, D. Srnicek, S. Cienciala, M. Pagac, Topology optimization of the clutch lever manufactured by additive manufacturing, *Materials (Basel)* 16 (2023). https://doi.org/10.3390/ma16093510
27. O. Okorie, A. Perveen, D. Talamona, K. Kostas, Topology optimization of an aerospace bracket: Numerical and experimental investigation, *Appl. Sci.* 13 (2023). https://doi.org/10.3390/app132413218
28. A.W. Gebisa, H.G. Lemu, A case study on topology optimized design for additive manufacturing, *IOP Conf. Ser. Mater. Sci. Eng.* 276 (2017). https://doi.org/10.1088/1757-899X/276/1/012026

29. Y. Kayacan, M. Alshihabi, M. Engineering, A. Sciences, K. Tarihi, Adaptive topology optimization for additive manufacturing in aerospace applications Havacılık uygulamalarında eklemeli imalat için uyarlanabilir topo, *Pamukkale Üniversitesi Mühendislik Bilimleri Dergisi, Pamukkale University J. Eng. Sci.* 30 (2024) 145–154. https://doi.org/10.5505/pajes.2023.52578
30. K.L. Ameta, V.S. Solanki, V. Singh, A.P. Devi, R.S. Chundawat, S. Haque, Critical appraisal and systematic review of 3D & 4D printing in sustainable and environment-friendly smart manufacturing technologies, *Sustain. Mater. Technol.* 34 (2022). https://doi.org/10.1016/j.susmat.2022.e00481
31. M. Kravchenko, D.C.A. Pigosso, T.C. McAloone, Circular economy enabled by additive manufacturing: Potential opportunities and key sustainability aspects, *DS 101: Proceedings of NordDesign 2020*, Lyngby, Denmark (2020). https://doi.org/10.35199/norddesign2020.4
32. H.A. Colorado, E.I.G. Velásquez, S.N. Monteiro, Sustainability of additive manufacturing: the circular economy of materials and environmental perspectives, *J. Mater. Res. Technol.* 9 (2020) 8221–8234. https://doi.org/10.1016/j.jmrt.2020.04.062
33. A. Al Rashid, M. Koç, Additive manufacturing for sustainability, circularity and zero-waste: 3DP products from waste plastic bottles, *Compos. Part C Open Access* 14 (2024). https://doi.org/10.1016/j.jcomc.2024.100463
34. S. Wojtyła, P. Klama, T. Baran, Is 3D printing safe? Analysis of the thermal treatment of thermoplastics: ABS, PLA, PET, and nylon, *J. Occup. Environ. Hyg.* 14 (2017) D80–D85. https://doi.org/10.1080/15459624.2017.1285489
35. M.A. Olawumi, B.I. Oladapo, O.M. Ikumapayi, J.O. Akinyoola, Waste to wonder to explore possibilities with recycled materials in 3D printing, *Sci. Total Environ.* 905 (2023). https://doi.org/10.1016/j.scitotenv.2023.167109
36. J.H. Warner, S.P. Ringer, G. Proust, Strategies for metallic powder reuse in powder bed fusion: A review, *J. Manuf. Process.* 110 (2024) 263–290. https://doi.org/10.1016/j.jmapro.2023.12.066
37. V. Contaldi, F. Del Re, B. Palumbo, A. Squillace, P. Corrado, P. Di Petta, Mechanical characterisation of stainless steel parts produced by direct metal laser sintering with virgin and reused powder, *Int. J. Adv. Manuf. Technol.* 105 (2019) 3337–3351. https://doi.org/10.1007/s00170-019-04416-4
38. S.M. Yusuf, E. Choo, N. Gao, Comparison between virgin and recycled 316l ss and alsi10mg powders used for laser powder bed fusion additive manufacturing, *Metals (Basel)* 10 (2020) 1–18. https://doi.org/10.3390/met10121625
39. M. Sauerwein, E. Doubrovski, R. Balkenende, C. Bakker, Exploring the potential of additive manufacturing for product design in a circular economy, *J. Clean. Prod.* 226 (2019) 1138–1149. https://doi.org/10.1016/j.jclepro.2019.04.108
40. W. Liu, X. Liu, Y. Liu, J. Wang, S. Evans, M. Yang, Unpacking additive manufacturing challenges and opportunities in moving towards sustainability: An exploratory study, *Sustainability* 15 (2023). https://doi.org/10.3390/su15043827

Chapter 15

Future of green manufacturing

Senthil Kumar Shanmugakani, Rathinavel Subbiah, Senthilkumar Thangarajan Sivasankaran, and Angshuman Chattopadhyay

15.1 INTRODUCTION

Sustainability has emerged as a major driving force behind the dramatic shift of the manufacturing sector in recent years. Global industry adoption of green manufacturing processes has been driven by the need to minimize environmental impacts, reduce resource consumption, and mitigate climate change.

Clean manufacturing, environmentally conscious manufacturing, ecologically benign manufacturing, environmentally responsible manufacturing, and sustainable manufacturing are some other names for genetically modified organisms. Whatever the nomenclature, the objective is still the same: creating and supplying goods that, in their manufacture, use, and disposal, have the fewest adverse effects possible on the environment [1].

One of the strongest forces behind this transition is the necessity to reduce climate change. Manufacturing processes that need a lot of energy, logistics involved in shipping, and the preparation of raw materials all contribute significantly to the world's greenhouse gas emissions. There is a general feeling of urgency to cut emissions and switch to low-carbon industrial techniques as the effects of climate change become more apparent. By utilizing sustainable materials and technology [2], maximizing resource efficiency, and utilizing renewable energy sources, green manufacturing processes provide a way to accomplish this goal.

The main principle of green manufacturing is resource consumption reduction in addition to mitigating climate change. Linear "take-make-dispose" models are a common characteristic of traditional manufacturing processes, which deplete limited resources and produce significant waste [3]. Manufacturers may reduce resource inputs, extend product life, and make it easier for materials to be recovered and reused at the end of their life cycle by adopting the concepts of the circular economy. This change to a more regenerative approach helps the manufacturing ecosystem become more resilient and innovative while also conserving precious resources.

In this regard, green manufacturing has a bright future ahead of it as it develops more in response to demands on the environment throughout the world. Manufacturers may lead the path toward a future that is more resilient, egalitarian, and sustainable for future generations by adopting

 DOI: 10.1201/9781003635789-15

sustainability as a guiding principle and valuing innovation, collaboration, and responsible stewardship [4].

This chapter examines the course that green manufacturing will take going forward, emphasizing significant developments, tactics, and trends that will influence the sector in the years to come.

15.2 EVOLUTION OF GREEN MANUFACTURING

The development of green manufacturing is a dynamic process propelled by the need to adopt sustainable methods and an increasing awareness of environmental concerns. Green manufacturing has changed dramatically from its early phases to its current importance, altering industrial paradigms and reshaping entire industries. This chapter delves into the evolutionary path of green manufacturing, charting its beginnings, emphasizing significant turning points, and analyzing the variables that have influenced its evolution throughout time.

15.2.1 The history of green manufacturing

In the late 1980s and early 1990s, the Germans came up with the idea for genetic modification. Waste reduction in production became the primary focus of sustainable manufacturing efforts starting in the 1980s. A worldwide manufacturing standard that mandates that any business hoping to compete on a global scale begin producing goods compliant with the European market's green regulations was developed [5]. Following this, the paradigm for sustainable manufacturing shifted from being process-oriented to being product-oriented, intending to minimize waste, emissions, and raw material consumption. Products are built for reusability and recycling.

The early environmental movement of the late 20th century, which brought attention to pollution, resource depletion, and the need for sustainable development, is where green manufacturing got its start [6]. As environmental concerns grew, forward-thinking businesses started looking at new technology and alternate production strategies that would lessen their negative effects on the environment. The basis for what would eventually be referred to as "green manufacturing" was established by these early initiatives.

15.3 REVIEW ON GREEN MANUFACTURING

An examination of the literature on green manufacturing indicates that conceptual comprehension and real-world application are given equal weight. While some studies add to the theoretical knowledge of the concepts behind green manufacturing, others offer useful techniques and tools for converting these concepts into strategies that can be implemented at different stages of the manufacturing process. When combined, these two types of labor help to develop green manufacturing techniques and make the industrial ecosystem more sustainable. Below is an overview of the main findings, which highlight the significance of green manufacturing.

Green productivity is defined as waste reduction initiatives in manufacturing operations [7]. They illustrate how green productivity can improve total production performance by decreasing resource consumption, lowering environmental impact, and increasing efficiency through case studies that illustrate waste elimination techniques.

Waste management and green manufacturing are closely related; thus, manufacturers can improve their environmental performance and achieve sustainable operations by determining and removing the causes of waste [8].

Green and sustainable manufacturing is a paradigm for the future, with a business plan focused on environmentally conscious design utilizing emerging nano, bio, and material technologies [9]. The authors stress that this new paradigm will boost competitiveness and innovation while meeting customer demand for more environmentally friendly products.

Environmental and social costs and values into economic activities are reviewed and found that they offer a thorough triple bottom line framework for tracking sustainability information at the company level [10]. This paradigm facilitates decision-making by offering an organized method for assessing how manufacturing procedures and goods affect sustainability.

Based on their research on small and medium-sized enterprises (SMEs) manufacturers who obtained ISO 14001 certification, offer a framework for sustainability [11]. This framework emphasizes the significance of environmental management systems and certification standards in advancing sustainability activities and offers an organized method for achieving green manufacturing goals.

The second group of research focuses on analytical tools and models for green manufacturing and includes the contribution of the following.

The green material requirements planning (MRP) tool incorporates environmental factors into production scheduling by modifying traditional MRP systems [12]. Green MRP helps reduce environmental impact and maximize resource usage in manufacturing operations by highlighting possible environmental concerns and waste management difficulties throughout the planning phase.

In order to support green manufacturing, a variety of analytical tools that emerged from research on product and process design need to be compiled. Risk analysis, screening techniques, design for the environment (DfE), and life cycle analysis (LCA) are a few examples [13]. These instruments offer methodical techniques for evaluating and enhancing the environmental performance of goods and procedures over the course of their lives.

Using fuzzy set theory and network analytic techniques, a model for evaluating environmental hazards in manufacturing was established [14]. This model examines the possible effects on the environment of various waste types generated during manufacturing operations, with an emphasis on risks to ecological health. The approach makes it possible to assess the total environmental impact on ecosystems by allocating numerical fuzzy weighting factors to impact categories.

The Environmental Value Systems Analysis technique was developed in order to assess the environmental performance of semiconductor processing at the machine level [15]. By creating equipment environmental models to characterize a system and evaluate its environmental impact, this tool uses a "bottom-up" analytical method to provide insights for enhancing environmental performance in industrial processes.

The aerospace industry mainly concentrates on burr reduction and cleanability in machine tool optimization [16]. They sought to maximize manufacturing efficiency in aerospace production processes and minimize environmental effects by analyzing these factors as optimization objectives and improving machine tool performance.

15.4 FUTURE OF GREEN MANUFACTURING

Reducing waste, carbon emissions, environmental contaminants, and resource consumption are the goals of green manufacturing. It is anticipated that the following fields will experience developments in the field of green manufacturing:

- Artificial intelligence and data analytics
- Sustainable product development
- Robotics and automation
- Integrating sustainable practices
- Greening of the electric vehicle (EV) sector
- Internet of Things (IoT)
- Industry 5.0
- Quantum computing
- Nanotechnology
- Biotechnology
- Automobile industry
- Aviation industry

15.4.1 Artificial intelligence and data analytics

Green manufacturing is about to undergo a revolution thanks to data analytics and artificial intelligence (AI), which will make industrial processes more sustainable and effective. Large data sets produced by manufacturing processes can be analyzed by these technologies [17], which can then be used to find trends and insights that can significantly increase resource and energy efficiency.

Manufacturers may optimize energy use, material utilization, and equipment performance, among other aspects of their production processes, by utilizing AI algorithms [18] and machine learning approaches [19]. With the use of AI-powered predictive analytics, manufacturers may reduce resource consumption, environmental pollution, carbon emissions, and waste by planning ahead, optimizing production schedules, and even anticipating maintenance needs.

Additionally, data analytics can offer insightful information on the environmental impact of industrial processes, allowing businesses to pinpoint problem areas and put focused plans in place to lessen their environmental effect [20]. Green manufacturers may increase their efficiency, productivity, and sustainability by utilizing AI and data analytics, which will ultimately help create a future that is more resource- and environmentally conscious.

15.4.2 Sustainable product development

A crucial component of the future of green manufacturing is sustainable product development, which focuses on the conception, design, and lifetime management of goods with low environmental effect. Improvements in environmentally friendly product creation are anticipated to have a major impact on lowering resource consumption, pollution in the environment, carbon emissions, and waste in the manufacturing process [21–23]. The following are some important areas where progress is anticipated:

Life Cycle Assessment (LCA): LCA techniques will develop further, enabling producers to evaluate how products will affect the environment at every stage of their lifecycle, from the extraction of raw materials to recycling or disposal at the end of the product's useful life [24]. More precise and thorough evaluations will be possible because of developments in LCA technologies and methodologies. This will assist manufacturers in finding areas for improvement [25] and informing their decisions throughout the product development life cycle.

Design for Environment (DFE): With the goal of minimizing environmental consequences at every step of a product's lifespan, DFE concepts will be progressively included into product design processes [13]. This entails taking into account elements like the choice of materials, the robustness of the product, energy efficiency, recyclable nature, and simplicity of disassembly [26]. Designers will be able to produce goods that are not only aesthetically pleasing and useful but also ecologically sustainable [27] and economically feasible [28] because of advancements in DFE methodology and tools.

Digital Twin Technology: By enabling manufacturers to virtually simulate and optimize product designs prior to the production of physical prototypes, digital twin technology will play a vital part in sustainable product development [29]. Manufacturers can find ways to increase productivity, save waste, and lessen environmental impact early in the design process by digitally recreating products and production processes [30,31].

15.4.3 Robotics and automation

Future green manufacturing is anticipated to be greatly impacted by developments in robots and automation, which will help achieve the objectives of lowering resource consumption, waste production, carbon emissions, and environmental pollutants [32,33]. The following are some significant methods that automation and robotics are anticipated to progress green manufacturing:

Energy Efficiency: By decreasing energy-intensive jobs, increasing equipment operation, and decreasing idle time, robotics and automation technologies can optimize energy utilization in manufacturing processes [34]. Manufacturers can achieve greater energy efficiency and lower overall energy consumption, which will reduce carbon emissions and the impact on the environment, by automating repetitive processes and streamlining production schedules.

Resource Optimization: Robotics and automation offer precise control over material utilization and industrial processes, minimizing waste and resource consumption. Automated systems can correctly measure and dispense resources, decrease material handling errors, and optimize material consumption to minimize waste formation [35]. Manufacturers can minimize their environmental impact and achieve greater resource efficiency by reducing material waste and streamlining production processes.

15.4.4 Integrating sustainable practices

Improvements in the integration of sustainable practices are anticipated to be critical in the future of green manufacturing in order to meet targets for lowering waste, carbon emissions, resource consumption, and environmental contaminants. The following are some important areas where progress is anticipated:

Sustainability of the Supply Chain: It will be more crucial than ever to incorporate sustainable practices all the way up the supply chain. This entails working with suppliers to ensure sustainable and ethical production methods, reduce transportation emissions, and source resources appropriately [22]. Improvements in supply chain traceability, transparency, and certification schemes will empower manufacturers to prioritize sustainable sourcing methods and make well-informed decisions.

Lean and Green Manufacturing: To maximize resource efficiency and reduce waste in production processes, it will be crucial to combine lean manufacturing concepts with green manufacturing techniques. This entails streamlining processes and getting rid of non-value-added tasks by putting strategies like just-in-time production, continuous improvement, and waste reduction programs into place [27]. Manufacturers will be able to operate more sustainably and efficiently because of developments in automation, digitization, and process optimization.

15.4.5 Greening of the EV sector

It is anticipated that developments in the greening of the EV industry will play a major role in the future of green manufacturing in accomplishing the objectives of lowering resource consumption, pollution, waste, and carbon emissions [36]. The following are some important areas where progress is anticipated:

Battery Technology: Improving the sustainability of EVs will need significant advancements in battery technology. This entails lowering the usage of rare and environmentally delicate minerals like cobalt and lithium and creating batteries with a greater energy density, longer lifespan, and faster charging speeds [37]. The goal of research will be to maximize battery performance while reducing environmental effects at every stage of the battery's lifecycle, from extraction of raw materials to recycling or disposal at the end of use.

Integration of Renewable Energy: Including wind and solar energy into the infrastructure for charging electric vehicles is a key component of the EV industry's environmental transformation. This entails setting up solar panels at charging stations, offsetting electricity use with renewable energy credits, and putting smart grid technology into place to maximize energy efficiency and reduce carbon emissions [38]. Developments in the production, delivery, and storage of renewable energy will facilitate the shift to a transportation system that is more environmentally friendly and carbon neutral.

Lifecycle Assessment and Recycling: In order to green the EV industry, extensive lifecycle assessments (LCAs) must be carried out in order to determine the environmental impact of EVs from the point of sale to the point of death. This entails evaluating the environmental impact of EV production, use, and recycling or end-of-life disposal [39]. The creation of more ecologically friendly and sustainable EVs will be aided by developments in closed-loop material recovery procedures, recycling technology [40], and LCA approaches.

15.4.6 Internet of Things (IoT)

Innovations in the IoT are anticipated to be crucial in helping green manufacturing achieve its objectives of cutting down on waste, carbon emissions, resource consumption, and environmental pollutants [41]. The following are some important areas where IoT improvements can support green manufacturing:

Real-Time Monitoring and Optimization: Real-time monitoring of energy consumption, resource utilization, and environmental parameters like air quality and pollutants can be made possible by IoT sensors integrated into manufacturing machinery, equipment, and processes [42]. Analyzing this data will help to find inefficiencies, streamline processes, and cut down on waste. Predictive maintenance solutions, for instance, made possible by the IoT, can foresee equipment faults, minimizing downtime and energy loss.

Energy Efficiency: By monitoring and managing energy-intensive processes, machinery, and systems, IoT technology can optimize the use of energy in manufacturing facilities. By adjusting lighting, heating, cooling, and ventilation in response to occupancy, outside temperature, and production schedules, smart energy management systems can save a substantial amount of energy and lower carbon emissions [43]. Energy-saving projects can be prioritized, and potential for efficiency improvements can be found via IoT-enabled energy monitoring and analytics platforms.

Smart Supply Chain Management: By offering real-time visibility into the movement and status of raw materials, components, and completed items, IoT-enabled supply chain management systems can increase efficiency and sustainability. IoT sensors can monitor transportation routes, track inventory levels, and optimize logistics processes to cut down on waste, lower emissions, and enhance supply chain performance overall [44].

Waste Reduction and Recycling: By monitoring waste streams, finding opportunities for waste minimization, and streamlining recycling procedures, IoT technology can support waste reduction and recycling initiatives in manufacturing operations [45]. IoT sensors can monitor the production, sorting, and disposal of garbage to guarantee that environmental laws are followed and to reduce the environmental effects. IoT-enabled waste management systems, for instance, can lower landfill trash, increase recycling rates, and automate garbage sorting.

15.4.7 Industry 5.0

A developing idea called "Industry 5.0" emphasizes the human-centric approach to manufacturing while expanding on the ideas of "Industry 4.0." By emphasizing human collaboration, innovation, and ethical considerations, Industry 5.0 advancements are projected to further strengthen sustainability efforts in the context of green manufacturing [46]. Within the context of Industry 5.0, the following are some significant areas where developments in green manufacturing can be anticipated in the future:

Customization and Personalization: Industry 5.0 places a strong emphasis on the value of personalization and customization in manufacturing to meet the demands and preferences of each individual customer as well as environmental objectives [47]. Advanced technologies like digital design tools, flexible automation, and additive manufacturing allow producers to create customized items with less waste and resource use. Customization makes it possible to optimize manufacturing procedures, material choices, and product design in order to reduce environmental impact and maximize efficiency [48].

Ethical and Social Responsibility: Industry 5.0 prioritizes human well-being, environmental sustainability, and community engagement, placing a heavy emphasis on these areas. It is expected of manufacturers to implement ethical and transparent business procedures that guarantee equitable treatment of employees, ethical product recycling or disposal, and responsible procurement of materials [49].

Resilience and Agility: Industry 5.0 encourages industrial operations to be resilient and agile in order to quickly adjust to shifting market conditions, resource availability, and environmental issues [50]. Advanced technologies, such as digital twins, simulation tools, and predictive analytics, are utilized by manufacturers to effectively respond to disturbances, improve resource allocation, and anticipate and reduce hazards. Resilient manufacturing systems are better able to endure environmental shocks and lessen their detrimental effects on communities and ecosystems.

Collaborative Ecosystems: Industry 5.0 promotes cooperative ecosystems in which suppliers, customers, manufacturers, and other stakeholders cooperate to meet shared sustainability objectives. In order to address complex sustainability concerns including pollution, resource depletion, and climate change, collaborative partnerships facilitate information sharing, resource pooling, and collective action [51].

15.4.8 Quantum computing

Developments in quantum computing could transform sustainability efforts in the future of green manufacturing by making it possible to optimize, simulate, and analyze complicated production processes more effectively. The following are some important domains where quantum computing developments can support green manufacturing:

Optimization Algorithms: Compared to traditional computers, quantum computing is more effective at solving complicated optimization issues. Utilizing this skill, manufacturing processes can be improved in terms of waste reduction, energy efficiency, and resource consumption [52]. For instance, it can assist in reducing transportation emissions through supply chain logistics optimization or consuming less energy through production schedule optimization.

Material Design and Discovery: The creation of new materials with improved properties can be aided by quantum computing. The creation of environmentally friendly materials for manufacturing processes can be accelerated by more accurately and quickly mimicking molecular structures and properties. These materials might use fewer resources and have a smaller environmental effect because they are stronger, lighter, and more sustainable [53].

Energy Optimization: By evaluating and improving energy use in real-time, quantum computing can help with energy optimization in manufacturing plants. It can provide more accurate energy demand forecasting, energy distribution system optimization, and opportunity identification for energy-saving initiatives [54]. Both operating expenses and carbon emissions may be significantly reduced as a result.

Sustainable Supply Chain Management: By streamlining transportation routes, cutting inventory, and cutting waste all the way down the supply chain network, quantum computing can make supply networks more sustainable. Quantum algorithms can offer more reliable and ecologically friendly supply chain solutions by taking into account several factors and restrictions at once [55].

Predictive Maintenance: By more efficiently analyzing massive volumes of data from sensors and equipment, quantum computing can improve predictive maintenance practices in manufacturing facilities. Manufacturers can minimize unscheduled downtime, optimize maintenance schedules, and save energy consumption linked to inefficient machinery performance by anticipating equipment faults before they happen [56].

Water Management: Especially in water-intensive industries like textile, agricultural, and semiconductor production, quantum computing can optimize water utilization in manufacturing processes [57]. It can find opportunities for water conservation, pollution avoidance, and recycling, hence lowering the environmental effect of water consumption, by more correctly simulating water distribution systems and treatment processes.

15.4.9 Nanotechnology

Reducing waste, carbon emissions, environmental contaminants, and resource consumption are the goals of green manufacturing. It is anticipated that green manufacturing would lead to breakthroughs in the following fields of nanotechnology.

The field of green manufacturing, which aims to minimize resource consumption, environmental effect, and waste generation, is expected to gain a great deal from nanotechnology developments [57]. The following are some ways that nanotechnology can support green manufacturing in the future:

Development of Nanomaterials: The design and manufacturing of sophisticated materials with specific features is made possible by nanotechnology. These materials can take the place of traditional ones, providing better results with reduced waste and resource use.

Nanoscale Catalysis: For a variety of chemical reactions, nanoparticles act as effective catalysts. Green manufacturing can minimize the formation of hazardous byproducts, increase reaction speeds, and use less energy by utilizing nanocatalysts [58].

Nanomachines for Resource Optimization: Real-time monitoring and optimization of manufacturing processes can be achieved via nanoscale devices and sensors. Green production may maximize efficiency while minimizing resource usage and emissions by modifying settings based on reliable data [59].

15.4.10 Biotechnology

Green manufacturing techniques could be greatly advanced by biotechnology by using biological processes and creatures to create sustainable solutions [60]. Biotechnology is anticipated to make the following contributions to green manufacturing in the future:

Biobased Materials: The creation of biobased materials from renewable sources including plants, algae, and microbes is made possible by biotechnology [61]. By substituting petroleum-based polymers and chemicals, these materials can lower resource use and carbon emissions.

Fermentation and Bioprocessing: Enzymes, microbes, and fermentation processes are used in bioprocessing procedures to create chemicals, fuels, and medications. Green manufacturing can accomplish more sustainable production processes with lower energy usage and waste generation by utilizing biocatalysts and microbial fermentation [62].

Biodegradable items: Biotechnology makes it easier to create and produce items and materials that decompose naturally after use, preventing trash from piling up in landfills and the ocean. A circular economy is facilitated by biodegradable textiles, polymers, and packaging materials that encourage composting and recycling [62].

Biotechnology is essential to the generation of biofuels and biogas from renewable biomass sources, including organic waste, algae, and agricultural residues. By lowering dependency on non-renewable resources and reducing carbon emissions, bioenergy technologies provide greener fuel substitutes for fossil fuels.

Bioremediation and Waste Treatment: To clean up contaminated areas and handle industrial waste streams, biotechnological techniques such as enzymatic breakdown and microbial bioremediation can be used [57]. Green manufacturing can reduce environmental impact and decrease pollution by utilizing microorganisms' metabolic capabilities.

Biological Sensors and Monitoring Systems: Biotechnology has made it possible to construct biological sensors and monitoring systems, such as biosensors and biomonitors, for real-time environmental monitoring in manufacturing plants [61]. By detecting contaminants, tracking process variables, and optimizing resource use, these biological sensing systems can improve sustainability and regulatory compliance.

15.4.11 Automobile industry

Green manufacturing in the automotive sector seeks to lessen the effects of vehicle usage and production on the environment. The following are some fields where progress is anticipated:

Lightweight Materials: Using lightweight materials like carbon fiber, aluminum, and advanced composites is a key component of green production in the automotive sector [63]. Vehicle weight reduction from these materials results in increased fuel economy and reduced carbon emissions while driving.

Hybrid Vehicles: The development of electric and hybrid vehicles is essential to the future of green manufacturing in the automotive sector [64]. Manufacturers may drastically cut their reliance on fossil fuels and greenhouse gas emissions by switching to electric powertrains from internal combustion engines.

Recycling and the Circular Economy: The implementation of recycling and the circular economy principles in the automotive sector is emphasized by green manufacturing [65]. In order to minimize waste output and lessen the need for virgin resources, automakers are looking into ways to recycle end-of-life vehicles and repurpose materials like steel, aluminum, and plastics in the construction of new vehicles.

Advanced Powertrain Technologies: One aspect of green manufacturing is the creation of advanced powertrain technologies, such as renewable biofuels and hydrogen fuel cells. Throughout the course of a vehicle's lifecycle, these alternative power sources help reduce emissions and the environmental effect of standard gasoline and diesel engines by providing cleaner alternatives [65].

15.4.12 Aviation industry

The use of green propulsion and combustion technologies is essential for cutting emissions from the aviation industry as well as other transportation-related sectors [66]. There will likely be several developments in this field:

Biofuels: As a potential replacement for conventional jet fuels, biofuels made from renewable resources like algae, leftover cooking oil, or plant biomass [67] seem promising. When utilized in airplane engines, these fuels can dramatically lower carbon emissions and other pollutants without requiring major changes to the current infrastructure.

When compared to traditional jet fuels, biofuels have the potential to drastically reduce carbon emissions [68]. They can lessen the total environmental effect of airlines and assist them in complying with ever-tougher emissions rules.

Biofuels can improve energy security for airlines and nations by diversifying fuel sources away from petroleum and lowering reliance on imported oil [69].

15.4.13 Cement industry

The cement industry's high energy consumption and reliance on fossil fuels make it a major contributor to environmental pollution and carbon emissions [70]. Green manufacturing, however, has the potential to change the cement business and make it a more ecologically friendly and sustainable one. For the cement industries of the future, the following are some possible developments in green manufacturing:

Alternative Fuels and Raw Materials: In order to lessen dependency on conventional fossil fuels and natural resources, green manufacturing can encourage the use of alternative fuels and raw materials in the production of cement. This includes replacing coal and virgin raw materials with biomass, waste-derived fuels, and recycled resources.

Enhancements in Energy Efficiency: The use of energy-efficient technologies and procedures can drastically lower the carbon footprint associated with the production of cement. This can entail utilizing renewable energy sources, including solar and wind power, to power cement plants, as well as improving heat recovery and kiln systems.

Carbon Capture and Utilization (CCU): Green manufacturing can investigate solutions for capturing and using CO_2 emissions from cement plants, either by subterranean storing the emissions or converting them into valuable goods [70]. Carbonation and mineralization are two CCU processes that can help offset carbon emissions and lessen their negative effects on the environment.

Low-Carbon Cement Production: One of the most important ways to lessen the environmental effect of cement manufacturing is to develop low-carbon or carbon-neutral cement formulations. New cementitious materials with lower embodied carbon and energy intensity, including supplemental cementitious materials (SCMs) and alternative binders, can be the focus of green production.

Waste Heat Recovery: In order to collect and make use of surplus heat produced during the cement-making process, green manufacturing might install waste heat recovery systems. By using this recovered heat for heating, power generation, or other industrial uses, energy consumption and greenhouse gas emissions can be decreased.

15.4.14 Transportation industry

The transportation industry is a major contributor to greenhouse gas (GHG) emissions worldwide, accounting for 15% of total emissions, of which 23% are attributed to CO_2 alone [71]. The substantial emissions this industry produces highlight how urgent it is to take steps to lessen its negative environmental effects. The phrase "green transportation" was created by researchers to refer to initiatives meant to lower greenhouse gas emissions from transportation-related activities.

In order to predict future trends, a system dynamics model to examine road transport in Indonesia was employed [72]. This revealed a worrisome prediction: the sector's overall fuel consumption and road emissions could rise significantly, possibly reaching levels that are 65 and 62 times greater than those observed in 2013.

The transportation industry is a major contributor to greenhouse gas (GHG) emissions worldwide, accounting for 15% of total emissions, of which 23% are attributed to CO_2 alone [71]. The substantial emissions this industry produces highlight how urgent it is to take steps to lessen its negative environmental effects. The phrase "green transportation" was created by researchers to refer to initiatives meant to lower greenhouse gas emissions from transportation-related activities.

Mobility management, intelligent transportation systems (ITS), and advancements in engine and fuel technology are emphasized as key tactics in reducing the environmental impact of transportation [71]. In order to lessen the negative effects on the environment, these developments concentrate on improving fuel economy, cutting emissions, and streamlining transportation networks.

Evangelista et al.'s exploratory case study analysis shows how important third-party logistics (TPL) is to advancing green manufacturing practices across industries. They discovered that by offering resources and experience to streamline supply chains and lessen environmental effects, TPL can help promote the adoption of green projects [73] environmental restrictions are a major factor in the adoption of green manufacturing methods by industries.

Green initiatives and environmental legislation have several advantages, as demonstrated by a case study on Dalsey, Hillblom, and Lynn (DHL) in Liverpool [74]. The study revealed that DHL's adoption of environmentally conscious efforts yielded several benefits, including higher efficiency, cost savings, productivity gains, and better product quality. This demonstrates the ways in which industries can reap financial and ecological rewards from adopting green manufacturing techniques.

15.5 CONCLUSIONS

The emergence of green manufacturing signifies a profound change in industrial paradigms, propelled by the growing consciousness of environmental issues and the requirement for sustainable growth. Green manufacturing has evolved from its late 20th-century beginnings, which were centered on waste minimization and adherence to European market laws, to a comprehensive strategy that includes both process- and product-oriented tactics. Reusable and recyclable items have been developed as a result of the incorporation of sustainability into production processes, which reduces waste, emissions, and the need for raw materials.

A thorough analysis of the literature indicates that there is a dual emphasis on theoretical knowledge and real-world application in green manufacturing. This equilibrium has made it easier to create cutting-edge methods and instruments that improve environmental performance across the manufacturing lifecycle, like green MRP, DfE, and LCA. In addition, the implementation of environmental management systems like ISO 14001 emphasizes how crucial formal frameworks are to reaching sustainability objectives.

Future developments in green manufacturing are anticipated to bring about significant progress in a number of important sectors. Production processes will become more sustainable and efficient as AI, data analytics, and the IoT are integrated. Nanotechnology, biotechnology, and quantum computing are examples of emerging technologies that will open up new possibilities for environmental stewardship and creativity. In addition, cutting carbon emissions and resource consumption would require a greening of businesses including the automotive, aerospace, and EV sectors.

To sum up, green manufacturing is a vibrant and developing area that is critical to solving today's environmental issues. Industries may meet consumer needs for eco-friendly products, improve their competitiveness, and contribute to a more sustainable future by utilizing technical breakthroughs in conjunction with sustainable practices.

REFERENCES

1. Ackerman, F. 1997. *Why Do We Recycle? Markets, Values, and Public Policy.* Washington, D.C.: Island Press.
2. Deloitte and Touche, 1992. *Business Strategy for Sustainable Development: Leadership and Accountability for the 90s.* Winnipeg: IISD.
3. Aboulnaga, I.A. 1998. Integrating quality and environmental management as competitive business strategy for 21st century. *Environmental Management and Health Journal*, 9(2):65–71.
4. Baumann, H., Boons, F., and Bragd, A. 2002. Mapping the green product development field: engineering, policy and business perspectives. *Journal of Cleaner Production*, 10(5):409–425.
5. Chen, Y.S. 2008. The driver of green innovation and green image – green core competence. *Journal of Business Ethics*, 81(3):531–543.
6. Cagno, E., Trucco, P., and Tardini, L. 2005. Cleaner production and profitability: analysis of 134 industrial pollution prevention project report. *IJO Cleaner Production.* 13(6):593–605.
7. Mohnty, R.P., and Deshmukh, S.D. 1998. Managing green productivity, some strategic directions. *Production Planning and Control*, 9(7):624–633.
8. Naderi, A. 1996. Productive design: a new design attitude. *Proceedings of APO World Conference on Green Productivity*, Manila, 178–182.
9. Jovane, F., Koren, Y., and Boer, N. 2003. Present and future of flexible automation: towards new paradigms. *CIRP Annals*, 52(2):543–547.
10. Wang, L., and Lin, L. 2007. A methodological framework for the triple bottom line accounting and management of industry enterprises. *International Journal of Production Research*, 45(5):1063–1088.
11. Burk, S., and Goughran, W., 2007. Developing a framework for sustainability management in engineering SMEs. *Robotics and Computer Integrated Manufacturing*, 23:696–703.
12. Melnyk, S.A., Sroufez, R.P., Montabon, F.L., and Hinds, T.J. 2001. Green MRP: identifying the material and environmental impacts of production schedules. *International Journal of Production Research*, 39(8):1559–1573.

13. Fiksel, J. 1996. *Design for Environment: Creating Eco-Efficient Products and Processes*. New York: McGraw-Hill.
14. Hui, I., He, L., and Dang, C. 2002. Environmental impact assessment in an uncertain environment. *International Journal of Production Research*, 40(2):375–388.
15. Krishnan, N., Raoux, S., and Dornfeld, D. 2004. Quantifying the environmental footprint of semiconductor process equipment using the environmental value systems (EnV-S) analysis. *IEEE Transactions on Semiconductor Manufacturing*, 17(4):554–561.
16. Avila, M., Gardner, J., Reich-Weiser, C., Vijayaraghavan, A., Tripathi, S., and Dornfeld, D. 2005. Burr minimization strategies and cleanability in the aerospace and automotive industry. *SAE Transactions Journal of Aerospace*, 114(1):1073–1082.
17. Bag, S., Pretorius, J.H.C., Gupta, S., and Dwivedi, Y.K. 2021. Role of institutional pressures and resources in the adoption of big data analytics powered artificial intelligence, sustainable manufacturing practices and circular economy capabilities. *Technological Forecasting and Social Change*, 163:120420.
18. Benzidia, S., Makaoui, N., and Bentahar, O. 2021. The impact of big data analytics and artificial intelligence on green supply chain process integration and hospital environmental performance. *Technological Forecasting and Social Change*, 165:120557.
19. Cioffi, R., Travaglioni, M., Piscitelli, G., Petrillo, A., and De Felice, F. 2020. Artificial intelligence and machine learning applications in smart production: progress, trends, and directions. *Sustainability*, 12:492.
20. Bag, S., and Pretorius, J.H.C. 2022. Relationships between Industry 4.0, sustainable manufacturing and circular economy: proposal of a research framework. *International Journal of Organizational Analysis*, 30(4):864–898.
21. Rusinko, C. 2007. Green manufacturing: an evaluation of environmentally sustainable manufacturing practices and their impact on competitive outcomes. *IEEE Transactions on Engineering Management*, 54(3):445–454.
22. Abualfaraa, W., Salonitis, K., Al-Ashaab, A., and Ala'raj, M. 2020. Lean-green manufacturing practices and their link with sustainability: a critical review. *Sustainability*, 12:981.
23. Sarvesh, P.S., and Rajput, S.D. 2020. Sustainable and green manufacturing – a narrative literature review. *Materials Today: Proceedings*, 26(2):2515–2520.
24. Shankar, K.M., Kannan, D., and Udhaya Kumar, P. 2017. Analyzing sustainable manufacturing practices – a case study in Indian context. *Journal of Cleaner Production*, 164:1332–1343.
25. Hami, N., Muhamad, M.R., and Ebrahim, Z. 2015. The impact of sustainable manufacturing practices and innovation performance on economic sustainability. *Procedia CIRP*, 26:190–195.
26. Karuppiah, K., Sankaranarayanan, B., Ali, S.M., Chowdhury, P., and Paul, S.K. 2020. An integrated approach to modeling the barriers in implementing green manufacturing practices in SMEs. *Journal of Cleaner Production*, 265:121737.
27. Waheed, A., Zhang, Q., Rashid, Y., Tahir, M.S., and Zafar, M.W. 2020. Impact of green manufacturing on consumer ecological behavior: stakeholder engagement through green production and innovation. *Sustainable Development*, 28:1395–1403.

28. D'Angelo, V., Cappa, F., and Peruffo, E. 2023. Green manufacturing for sustainable development: the positive effects of green activities, green investments, and non-green products on economic performance. *Business Strategy and the Environment*, 32(4):1900–1913.
29. Madu, C.N. 2022. Sustainable manufacturing: strategic issues in green manufacturing. In *Handbook of Environmentally Conscious Manufacturing*, 2nd ed., ed. Madu, C.N., (Chapter 1, 1–11). Cham: Springer.
30. Despeisse, M., Mbaye, F., Ball, P.D., and Levers, A. 2012. The emergence of sustainable manufacturing practices. *Production Planning & Control*, 23(5):354–376.
31. Fum, E., Agyabeng-Mensah, Y., Sun, Z., Frimpong, B., Kusi, L.Y., and Acquah, I.S.K. 2020. Exploring the link between green manufacturing, operational competitiveness, firm reputation and sustainable performance dimensions: a mediated approach. *Journal of Manufacturing Technology Management*, 31(7):1417–1438.
32. Chiarini, A. 2021. Industry 4.0 technologies in the manufacturing sector: are we sure they are all relevant for environmental performance? *Business Strategy and the Environment*, 30(7):3194–3207.
33. Arents, J., and Greitans, M. 2022. Smart industrial robot control trends, challenges and opportunities within manufacturing. *Applied Sciences*, 12:937.
34. Javaid, M., Haleem, A., Singh, R.P., and Suman, R. 2021. Substantial capabilities of robotics in enhancing Industry 4.0 implementation. *Cognitive Robotics*, 1:58–75.
35. Vrchota, J., Pech, M., Rolínek, L., and Bednář, J. 2020. Sustainability outcomes of green processes in relation to Industry 4.0 in manufacturing: systematic review. *Sustainability*, 12:5968.
36. Alptekin, B, Tunaboylu, B, Zaim, S, and Perlo, P. 2021. Smart manufacturing of electric vehicles. In *Digital Conversion on the Way to Industry 4.0*, eds. Durakbasa, N.M., Gençyılmaz, M.G. Lecture Notes in Mechanical Engineering (pp. 767–773). Cham: Springer.
37. Kieckhäfer, K., Wachter, K., and Spengler, T.S. 2017. Analyzing manufacturers' impact on green products' market diffusion – the case of electric vehicles. *Journal of Cleaner Production*, 162:S11–S25.
38. Dik, A., Omer, S., and Boukhanouf, R. 2022. Electric vehicles: V2G for rapid, safe, and green EV penetration. *Energies*, 15:803.
39. Ghosh, A. 2020. Possibilities and challenges for the inclusion of the electric vehicle (EV) to reduce the carbon footprint in the transport sector: a review. *Energies*, 13:2602.
40. Shrivastava, A.R., Narayana, J., Chiranjivi, V.N.S.L., and Borole, Y.D. 2021. Green energy powered charging infrastructure for hybrid EVs. *2021 9th International Conference on Cyber and IT Service Management (CITSM)*, Bengkulu, Indonesia, 1–7.
41. Naim, A., Rashid Hussain, M., Alelyani, S., and Saleh Alsaqer, M. 2022. Applications of IoT in industrial transformation and green manufacturing. In *Communication and Intelligent Systems*, eds. Sharma, H., Shrivastava, V., Kumari Bharti, K., Wang, L. Lecture Notes in Networks and Systems, 461. Singapore: Springer.

42. Sharma, N., and Panwar, D. 2020. Green IoT: advancements and sustainability with environment by 2050. *2020 8th International Conference on Reliability, Infocom Technologies and Optimization (Trends and Future Directions) (ICRITO)*. Noida, India, 1127–1132.
43. Hu, R., Shahzad, F., Abbas, A., and Liu, X. 2022. Decoupling the influence of eco-sustainability motivations in the adoption of the green industrial IoT and the impact of advanced manufacturing technologies. *Journal of Cleaner Production*, 339:130708.
44. Albreem, M. A. M., El-Saleh, A.A., Isa, M., Salah, W., Jusoh, M., and Azizan, M.M. 2017. Green Internet of Things (IoT): an overview. *2017 IEEE 4th International Conference on Smart Instrumentation, Measurement and Application (ICSIMA)*, Putrajaya, Malaysia, 1–6.
45. Naim, A., Muniasamy, A., Clementking, A., and Rajkumar, R. 2022. Relevance of green manufacturing and IoT in industrial transformation and marketing management. In *Computational Intelligence Techniques for Green Smart Cities*, eds. Lahby, M., Al-Fuqaha, A., Maleh, Y. Green Energy and Technology (pp. 395–419). Cham: Springer.
46. Yin, S., and Yu, Y. 2022. An adoption-implementation framework of digital green knowledge to improve the performance of digital green innovation practices for Industry 5.0. *Journal of Cleaner Production*, 363:132608.
47. Maddikunta, P.K.R., Pham, Q.-V., Prabadevi, B., Deepa, N., Dev, K., Gadekallu, T.R., Ruby, R., and Liyanage, M. 2022. Industry 5.0: a survey on enabling technologies and potential applications. *Journal of Industrial Information Integration*, 26:100257.
48. Fraga-Lamas, P., Lopes, S.I., and Fernández-Caramés, T.M. 2021. Green IoT and edge AI as key technological enablers for a sustainable digital transition towards a smart circular economy: an Industry 5.0 use case. *Sensors*, 21:5745.
49. Sharma, M., Sehrawat, R., Luthra, S., Daim, T., and Bakry, D. 2022. Moving towards Industry 5.0 in the pharmaceutical manufacturing sector: challenges and solutions for Germany. *IEEE Transactions on Engineering Management*, 71, 13757–13774.
50. Iqbal, M., Lee, C.K.M., and Ren, J.Z. 2022. Industry 5.0: from manufacturing industry to sustainable society. *2022 IEEE International Conference on Industrial Engineering and Engineering Management (IEEM)*, Kuala Lumpur, Malaysia, 1416–1421.
51. Narkhede, G., Pasi, B., Rajhans, N., and Kulkarni, A. 2023. Industry 5.0 and the future of sustainable manufacturing: a systematic literature review. *Business Strategy & Development*, 6(4):704–723.
52. Konstantopoulos, G., Koumoulos, E.P., and Charitidis, C.A. 2022. Digital innovation enabled nanomaterial manufacturing; machine learning strategies and green perspectives. Nanomaterials, 12:2646.
53. Andersson, M.P., Jones, M.N., Mikkelsen, K.V., You, F., and Mansouri, S.S. 2022. Quantum computing for chemical and biomolecular product design. *Current Opinion in Chemical Engineering*, 36, 100754.
54. Luckow, A., Klepsch, J., and Pichlmeier, J. 2021. Quantum computing: towards industry reference problems. *Digitale Welt*, 5:38–45.

55. Olatunji, O.O., Adedeji, P.A., and Madushele, N. 2021. Quantum computing in renewable energy exploration: status, opportunities, and challenges. In *Advances in Nonlinear Dynamics and Chaos (ANDC), Design, Analysis, and Applications of Renewable Energy Systems*, eds. Azar, A.T., Kamal, N.A., 549–572. San Diego, CA: Academic Press.
56. Ajagekar, A., and You, F. 2022. Quantum computing and quantum artificial intelligence for renewable and sustainable energy: a emerging prospect towards climate neutrality. *Renewable and Sustainable Energy Reviews*, 165, 112493.
57. Khan, S.H. 2020. Green nanotechnology for the environment and sustainable development. In *Green Materials for Wastewater Treatment. Environmental Chemistry for a Sustainable World,* eds. Naushad, M., Lichtfouse, E., 38. Cham: Springer.
58. Soni, R.A., Rizwan, M.A., and Singh, S. 2022. Opportunities and potential of green chemistry in nanotechnology. *Nanotechnology and Environmental Engineering*, 7:661–673.
59. Malik, S., Muhammad, K., and Waheed, Y. 2023. Nanotechnology: a revolution in modern industry. *Molecules,* 28:661.
60. Lin, S., and Juang, Y. (2014). Selecting green suppliers with analytic hierarchy process for biotechnology industry. *Operations and Supply Chain Management: An International Journal*, 1(2):115–129.
61. Bradu, P., Biswas, A., and Nair, C. 2023. Recent advances in green technology and Industrial Revolution 4.0 for a sustainable future. *Environmental Science and Pollution Research*, 30:124488–124519.
62. El-Desouky, N., Shoueir, K.R., El-Mehasseb, I., and El-Kemary, M. 2021. Bio-inspired green manufacturing of plasmonic silver nanoparticles/Degussa using Banana Waste Peduncles: photocatalytic, antimicrobial, and cytotoxicity evaluation. *Journal of Materials Research and Technology*, 10:671–686.
63. Wells, R.O. 2007. The automobile industry & sustainability. *Journal of Cleaner Production*, 15:989–993.
64. Nunes, B., and Bennett, D. 2009. Green technologies in the automotive industry. In Management of Green Technology, ed. Sherif, H., 192–107. Miami, FL: Industry Committee of the International Association for Management of Technology.
65. Bennett, B.N. 2010. Green operations initiatives in the automotive industry: an environmental reports analysis and benchmarking study. *Benchmarking: An International Journal*, 17(3):396–420.
66. Lu, F., Zhou, G., Liu, Y., and Zhang, C. 2022. Ensemble transfer learning for cutting energy consumption prediction of aviation parts towards green manufacturing. *Journal of Cleaner Production*, 331:129920.
67. Tang, D.Y.Y., Yew, G.Y., and Koyande, A.K. 2020. Green technology for the industrial production of biofuels and bioproducts from microalgae: a review. *Environmental Chemistry Letters*, 18:1967–1985.
68. Qiu, R. Hou, S. Chen, X. and Meng, Z. 2021. Green aviation industry sustainable development towards an integrated support system. *Business Strategy and the Environment,* 30:2441–2452.
69. Moreira, N., de Santa-Eulalia, L.A., Aït-Kadi, D., Wood–Harper, T., and Wang, Y. 2015. A conceptual framework to develop green textiles in the aeronautic completion industry: a case study in a large manufacturing company. *Journal of Cleaner Production*, 105:371–388.

70. Juan Cagiao, B.G. 2011. Calculation of the corporate carbon footprint of the cement industry. *Journal of Cleaner Production*, 3:192–202.
71. Lipman, S.A. 2007. Reducing greenhouse emissions and fuel consumption – sustainable approaches for surface transportation. *IATSS Research*, 31(1):26–35.
72. Sukarno, I., Matsumoto, M., and Susanti, L. 2016. Transportation energy consumption and emissions - a view from City of Indonesia. *Future Cities and Environment*, 2(6):1–12.
73. Pietro Evangelista, E.S. 2010. Green supply chain initiatives in the transport and logistics service industry: an exploratory case study analysis. *Proceedings of the 14th Annual Conference of the Logistics Research Network*. Leeds, Logistic Research Network, 195–203.
74. Silvia Cosimato, O.T. 2014. The influence of green innovation in logistics competitiveness and sustainability: the DHL case study. *17th Toulon Verona International Conference*, Liverpool, Liverpool John Moores University, 65–69.

Index

Kolibri 360 GmbH
Ferdinand-Jühlke-Straße 7
99095 Erfurt, Deutschland
produktsicherheit@kolibri360.de